Landscape Ecology

Dean L. Urban

Landscape Ecology

A Task-Oriented Perspective

Springer

Dean L. Urban
Nicholas School of the Environment
Duke University
Durham, NC, USA

ISBN 978-3-031-72250-9 ISBN 978-3-031-72251-6 (eBook)
https://doi.org/10.1007/978-3-031-72251-6

This Springer imprint is published by the registered company Springer Nature Switzerland AG
The registered company address is: Gewerbestrasse 11, 6330 Cham, Switzerland

Preface

Landscape ecology is a big, expansive discipline or, rather, a grand Venn diagram that highlights the intersections among a wide range of disciplines from geography to ecology to psychology. Perhaps not surprisingly, textbooks on landscape ecology either tend to be rather conceptual and superficial, or they tend to be narrowly specific in their focus. For many years I have struggled to find the right balance in teaching graduate and professional students about the ecological foundations of and essential tasks in *doing* landscape ecology.

My solution has been two books. The first, *Agents and Implications of Landscape Pattern: Working Models for Landscape Ecology* (Urban 2023), is an ecology book that emphasizes how ecological processes and patterns play out at the spatial scale of landscape management. There are two main themes in this. The first is how finer-scale processes—processes that ecologists typically embrace at the scale of field studies—are integrated into the larger spatial scales at which management happens. The second theme is how to deal with the spatial heterogeneity and couplings that we encounter when working at these larger scales.

In this companion book, I turn to the tools of the trade: the technical skills that practitioners need in order to do the work of landscape ecology and management. The problem, of course, is that landscape ecology encompasses a variety of tasks borrowed from other disciplines—or entire disciplines in themselves—and it is hard to find a text that weaves these different threads into a whole cloth. For example, landscape ecologists use the techniques of species distribution modeling, ordinations for summarizing multivariate trends in inventory and monitoring data, structured decision-making in prioritizing sites for management, and various models in forecasting landscape change. Each of these areas is a well-developed discipline, but these are only partially overlapping areas of expertise. This text is my attempt to bring these tools together in a coherent framework for landscape analysis and management.

Landscape management entails a mix of expertise in the natural and social sciences. For example, conservation practice ranges from community-level negotiations that rally stakeholders around management aims and ratify core values, to the ecology of identifying conservation targets based on their perceived conservation value, to the financial and political dealings that protect land (including the legalities of land use policies, easements, and the tax implications of donor contributions of land value as well as the opportunity cost of foregone property taxes or other land uses), to the long-term stewardship and management of protected lands. Managing working lands (i.e., for agriculture and forestry) involves a similarly wide range of activities. The cultural, economic, and policy context for landscape management varies substantially across regions, even within the USA; these factors are even more heterogeneous in other countries. While recognizing the critical importance of integrating all of these perspectives and disciplines, I have framed this book as ecological. This reflects my own experience, but it also reflects my belief that ecology is at the core of landscape management when this is practiced as a science- or evidence-based approach.

This ecology can and should be informed by stakeholder values, and the results of ecological analyses can and should be translated into terms that can be appreciated by an audience that does not have the same technical background that this book affords. To this end, I have tried to insert, where appropriate, pointers to where other disciplines or community engagement should be brought into the process. Landscape management is, by its nature, an iterative process in which we recursively set goals and objectives, attempt to realize these, and take stock of how we are doing.

Adaptive Management and Landscapes

The fundamental tasks of landscape ecology all share the confounding constraints that the study areas tend to be unmanageably large and the data are often disappointingly sparse in coverage or thin in information content. Thus, many tasks present themselves as logistical challenges, challenges that invite modeling as a framework for marshaling available information for maximum efficiency. Another consequence of these logistics is a rather high level of uncertainty in landscape-scale applications, an uncertainty compounded by the lack of control over large-scale drivers (weather and climate, disturbances, land use change).

In this book, I adopt a model-based approach to landscape ecology. This approach invites the framework of the adaptive management[1] cycle*: plan, act, monitor, react* (Fig. 1). In this, an initial model—a declared hypothesis about how the system works—constitutes the *plan*. The model might be rather simple. For example, an inventory of a species of concern (say, a rare bird) might be based on an expert opinion on what sorts of habitat should support the species. But the model could also be more explicit about the factors shaping the distribution of the target

[1]There is an enormous literature on adaptive management, including several classics (e.g., Holling 1978, Walters 1986, Lee 1993) as well as texts specialized on conservation and ecosystem management. I assume here that readers are familiar with at least the basic tenets.

species. For example, a bird habitat model that includes explanatory variables such as patch edge/area ratios and isolation is more than a habitat model; it also invokes higher-level hypotheses about biotic processes (interspecific interactions, dispersal) that might influence species distribution beyond the constraint of the availability of potential breeding habitat. Data collected in the field should test these hypotheses explicitly, which leads directly to the *monitor* and *react* stages of adaptive management.

I should emphasize here that my own conceptual models about landscapes—the *plans* that might underpin the tasks presented here—are themselves elaborated separately in the companion book (Urban 2023). While that book is not required to appreciate this one, the conceptual models in the first book are intended to provide a useful foundation for the tasks developed here.

In landscape ecology, the *act* stage of adaptive management might not be an actual management intervention over large spatial extent; landscapes are simply too large to do this scale of management routinely. Yet increasingly, management or restoration activities are replicated at fine grain over large areas, subject to regional or national policies. For example, fire management activities aimed at hazard reduction are being implemented over much of the western United States and elsewhere, and a great deal of restoration and stewardship is aimed at improving habitat quality or connectivity as an aim of conservation practice. While the scale of individual activities might not be large, the aggregate effect can be quite extensive. Thus, it will be appropriate to examine these activities as landscape-scale management experiments.

Often, landscape ecologists find themselves playing the role of detective, trying to make inferences about past management or natural events that have shaped our current landscapes. These instances are akin to experiments without proper controls, or with controls contrived after the fact. For example, hydrologists try to select matched "control" and "experimental" watersheds for retrospective analyses of the impacts of land use change on water quality. We can envision this process of attribution as contriving the *plan* and *monitor* stages of adaptive management around a preexisting *act*. Assessing the impacts of habitat fragmentation on forest bird communities is a similarly contrived, retrospective "experiment" into the impacts of land cover change over decades. These are *quasi-experiments*.

Dynamic models of landscape change do not fit easily into the conventional cycle of adaptive management. Rather than force this fit, I will note simply that dynamic models can enrich the adaptive management framework by emphasizing the longer-term temporal context of landscape change, whether natural or resulting from human activities. Models also provide the capacity to conduct "virtual management experiments" by exploring the possible consequences of alternative management scenarios. Given the logistical (and sometimes ethical) difficulties of conducting actual management experiments at the landscape scale, this ability to assess model-based *acts*—scenario analysis—is crucial to landscape ecology and management.

Ecological assessment is not a task of landscape ecology itself, but rather, a more fundamental task aimed at verifying the working model that underlies research and management. This assessment—the *react* stage—provides the follow-through that

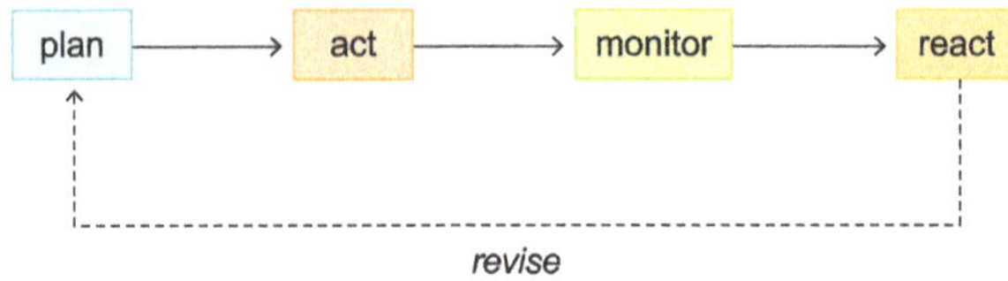

Fig. 1 The adaptive management cycle as an organizing framework for the tasks of landscape ecology. Here, the *plan* consists of an explicit model of how we believe the system works (or a hypothesis to be tested). The *act* might be large-scale management experiments, but also might be a retrospective "experiment" contrived to make inferences about the impacts of prior landscape events (disturbance, land use change). This stage also can include "virtual acts" as model-based scenarios. The *monitor* stage includes initial inventory as well as follow-up monitoring, both designed in accordance with the initial model (*plan*). The *react* stage amounts to integrated assessment, in which observations collected from the system are evaluated with respect to the model, and so which also tests the model itself

closes the adaptive management cycle, by informing us how to respond to our observations. Importantly, this assessment must include a capacity to identify significant departures from the expected, departures that would trigger a reactive decision to revisit the initial model and revise it as suggested by the evaluation.

As much as I would like to structure this book in the framework of adaptive management (and I have tried!), this has not worked very well for me. The awkward fit stems from the reality that landscape management is not a linear flow as in Fig. 1 but more of a braided stream of tasks (Fig. 2).

Workflows

I think of this in terms of *workflow paths*, and different agencies who manage landscapes are more or less invested in various paths. For example, a conservation agency whose primary mission is land protection might focus largely on monitoring the condition (and perhaps the connectivity) of protected lands, with most of the tasks related to inventory and monitoring; the assessment is done relative to a conceptual or formal model that spells out the objectives of the program. This path (the middle path in Fig. 2) might also wish to forecast observed changes in the future. By contrast, another conservation agency might be more invested in active stewardship and restoration of protected lands, implementing direct interventions to improve habitat connectivity or condition (top path in Fig. 2). By contrast, another agency might be more research-oriented and conduct experiments to test hypotheses about, for example, the relative importance of habitat geometry or isolation as influences on target species (bottom path in Fig. 2). These paths, of course, are not exclusive, and many agencies that manage land probably do some of each of these.

I offer Fig. 2 not to denigrate adaptive management, as the classic cycle easily can be embedded within any more complicated path in Fig. 2. Rather, I present the paths because they underscore the *sequencing* of actual work on landscapes, and that these

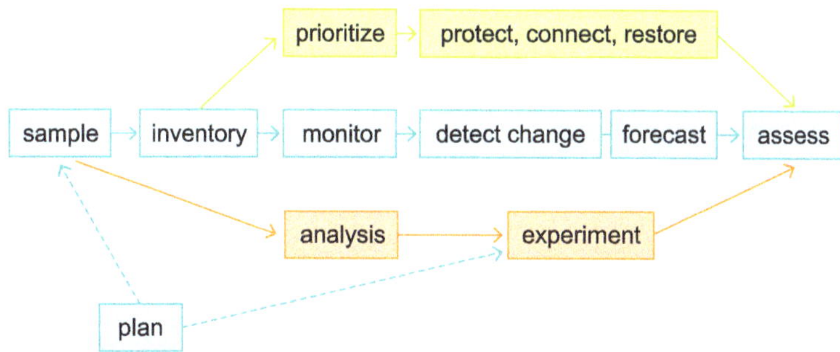

Fig. 2 Flow diagram that organizes the fundamental tasks of landscape analysis and management according to (nonexclusive!) management paths. For simplicity, recursive loops or feedbacks are not drawn, though most of these tasks would naturally be iterative. All paths flow from an initial conceptual model (the plan) that would inform inventory and monitoring, site prioritization, or hypotheses to be tested

workflows are assembled from a rather small set of fundamental tasks. These tasks include sampling, inventory and monitoring, site prioritization, and so on. The tasks logically connect, in that the outputs from one task become the inputs of the next task.

This braided-stream workflow of landscape analysis and management outlines the content of this book. In the book, I present what I think are some fundamental tasks in landscape ecology. These include sampling design, inventory and monitoring, habitat classification or species distribution modeling, site prioritization, change detection and forecasting, and ecological assessment. There are other tasks, to be sure, but I believe these few provide a solid foundation.

Workflow Within Tasks

Just as the collective tasks of working with landscapes invite a logical workflow, each of the individual tasks also has its own workflow. For example, species distribution modeling (Chap. 2) involves fitting a statistical model of some form (there are several options). But the setup before the statistics and the interpretation and evaluation after model fitting are also crucially important: indeed, the *pre-* and *post-processing* are often more work than the (statistical) *processing* itself. The same is true of the other tasks.

In this book, I have tried to emphasize the logical workflow of conducting each of the tasks highlighted in the chapters. While the details vary among tasks, there is generally a flow from preparatory work and framing, data wrangling and editing, the analysis itself, and then evaluating or interpreting the results—and then reporting all of this. For most tasks, the setup is crucial, and if the analysis is set up thoughtfully, the analysis itself is straightforward. Reporting includes a concise and precise detailing of the methods, one that will meet the litmus test of reproducibility. But reporting also entails extracting appropriate figures and tables to communicate the main results (and *which* bits are the main results is not always obvious!). Finally, the narrative interpretation of the analysis must be in terms that a nontechnical audience can understand and appreciate.

For most tasks, I summarize the workflow as a flowchart and the reporting in the form of a checklist: which steps to follow, critical decision points, and reporting requirements. These are often itemized as bulleted checklists:

☑ Report any data transformations and explain why these were selected.

Analysis as Translation

Many analytic tasks are themselves exercises in *translation*. For example, in species distribution modeling, we translate from ecological concepts to a statistical space, and then back-translate into ecological terms. Sometimes, we also translate the statistical model into geographic space, by mapping the predictions of the statistical model in a geographic information system (GIS). Similarly, in site prioritization, we begin in a conceptual space defined by stakeholder values and then translate these into empirical indicator variables. We then do the analysis in an empirical space and then back-translate to values. Again, the results are typically then mapped into a GIS to help communicate those values.

In these chapters, I have tried to emphasize the individual steps in the workflow of each task. Each task, in turn, is itself typically part of a larger workflow (Fig. 2).

Preview of Chapters

The fundamental tasks are organized in chapters that run more or less in a sequence as implied in Fig. 2, but as the figure suggests there is not a single, simple workflow that includes all the tasks. Further, some of the tasks require some more foundational groundwork. The book is structured as follows.

In Chap. 1, we begin by collecting some data. In this, we encounter the fundamental issue with landscapes: they are too big to sample efficiently by leaving coverage to chance (as with purely random samples). The alternative is to *stratify* samples somehow: in space, over a conceptual model, or a combination of these. This chapter begins by considering the basic elements of sampling design and then extends these to large and heterogeneous study areas. By carefully specifying the stratification, sampling designs can be tuned to a variety of applications ranging from a simple inventory, to targeted monitoring programs, to quasi-experimental designs. This chapter, on how to generate data, thus provides a foundation to subsequent chapters.

Chapter 2 presents the ecological and statistical basis for *species distribution modeling* (or *habitat classification*) and mapping, with a particular emphasis on applications that can be developed within the framework of a geographic information system (GIS) and evaluated at landscape scales. This is perhaps the most fundamental task in natural resource management. There are myriad statistical approaches to species distribution modeling, and in this chapter, we consider just a few alternatives while underscoring the common workflow logic of all such models. In a larger sense, this task is a useful starting point for this book because this primary task introduces most of the empirical complications that vex landscape ecology. These issues stem from the nature of ecological data: they are noisy, multivariate, redundant (the multiple variables are correlated among themselves), and almost certainly structured spatially (autocorrelated). These complications motivate the next few chapters.

Chapter 2 also introduces a narrative structure that is repeated in several subsequent chapters. In this, I illustrate the focal workflow with an extended example, from data preparation through the actual analysis to post-processing and reporting. I then mention alternative tools for the same application workflow. In some cases, these alternatives are illustrated as a second example; in one chapter, the alternatives are tightly tied to the main narrative but are included as an appendix to the chapter. In a few cases, there is a *lot* of other information available—on alternative tools, or details about tools featured in the chapter. These extra details are provided as digital supplements to these chapters.

Having collected (virtually) some landscape-scale data in Chap. 1 and applied these in Chap. 2 (with, perhaps, some frustration with the data), we then spend some time exploring these data in preparation for the analytic and management tasks to follow. In Chap. 3, we examine the nature of ecological data through exploratory data analysis (EDA), and familiarize ourselves with the kinds of data sets that ecologists typically use. This chapter also considers various editing or transformations of the primary data and how these might affect analyses. Ecologists often do not analyze the raw data themselves, and all transformations of the data influence the results of any subsequent analysis. So it will be good to pay attention to how we might modify the data. More importantly, an understanding of the data garnered through EDA will temper our interpretation of all subsequent analyses.

Ecological data are multivariate: We tend to collect data on multiple species as they occur across our sampling locations, and we measure multiple environmental variables at these same locations. Ecologists use two general approaches to dealing with multivariate data. To illustrate general trends in multivariate data, ecologists use *ordinations* of various kinds. These are the tools of the trade in gradient analysis, a long tradition in community ecology. These are the topics of Chap. 4. As a complementary approach, *classification techniques* are used to construct discrete groups from ecological data. When the data are species, this results in community types; if the data are environmental factors, the result is habitat types. Classification techniques are covered in Chap. 5. This includes the techniques popular in community ecology, but also methods such as image classification that are more common in landscape ecology. Many of the tools covered in Chaps. 4 and 5, long popular with ecologists, are seeing a new popularity under the rubric of "big data," in ecology and beyond.

Sampling for inventory and monitoring can be extended to consider more nuanced working models or hypotheses about species distribution patterns or other ecosystem processes on landscapes. These designs lead to inferential tests of the underlying model, a discussion begun in Chap. 1, revisited in Chap. 2, and considered more fully in Chap. 6. In Chap. 6, the emphasis is on statistical tests that are sensitive to issues of spatial structure, embracing this as a feature that is interesting in its own right rather than as a nuisance to be avoided (as we do in Chap. 2!). The main tools for this are a logical extension of ordination techniques introduced more generally in Chap. 4. This approach also includes tools used to partition variability across multiple spatial scales.

Chapters 3, 4, 5 and 6 are a bit of a detour from the overall workflow of the book, but an important side trip. One purpose of these chapters is to delve into the nature of ecological data—to learn some common tools for dealing with such data, and to address some analytic issues that arise repeatedly in ecological analyses. These chapters also help develop skills for discovering, summarizing, and especially communicating the main features of rich ecological data sets.

In Chap. 7, we address more explicitly the problem of correlations in multivariate data sets, by embracing this complexity rather than distilling it into simpler ordination axes or classified types. *Structural equation models* (SEMs) are one method for embracing the tangled web of correlations in ecological data sets, to depict a more nuanced model of the interactions among variables. SEM, the modern incarnation of path analysis, poses a pattern of interactions—interpreted as causal—that can include *indirect pathways* among variables (e.g., *X1* causes *X2* which then causes *Y*). SEM often is invested in *latent variables*, which are constructs (e.g., "water quality") that are measured indirectly via *indicator variables* (e.g., turbidity or nutrient loadings). This chapter picks up the concept of latent variables (*aka* factors) from factor analysis in Chap. 4, and provides a bridge, via *path models*, to structured decision-making as presented in Chap. 8.

Inventory (Chap. 1) provides a natural basis for *site prioritization*, the focus of Chap. 8. In the simplest sense, a ranking of sites according to their modeled habitat suitability (Chap. 2) provides for site prioritization. More typically, other factors come into play. This chapter begins with the task of site selection for conservation planning, reviewing the logic and computational algorithms used to select sites for systems of nature reserves. The initial focus here is the *minimum representation* problem: how to capture the most conservation targets (e.g., species or habitat types) in the fewest sites or minimum total area. The minimum representation problem is then extended to consider other factors such as species rarity, redundancy, and habitat connectivity. This discussion is then generalized to other conservation targets including watershed integrity and aesthetics (as viewsheds) and other ecosystem services. This leads to an overview of structured decision-making and multi-criteria decision analysis, the logical and analytic foundation of site prioritization in which decisions must be made in contexts where any decision might generate co-benefits or force trade-offs among competing objectives. Prioritizing ecosystem services also broaches the issue of stakeholder values and their preferences for various outcomes—which extends prioritization from ecological to social science.

Monitoring consists of repeating a sampling or inventory program (Chap. 1). This leads naturally to more formal models of landscape change, developed in Chap. 9. To begin, we spend some time on *trend detection* in monitoring data: to interpret change, we must first be sure that we are observing it. We might model landscape change by simply describing the trends in various land cover types. But because landscape area is constant, an increase in one land cover necessarily implies a decrease in something else. That is, the model rightly should address all cover types at once. One way to capture such changes is as a first-order Markov chain. This model often is too simple to be realistic, but it serves as a useful point of departure for more nuanced models that are too realistic to be simple. A sampling of models used in landscape ecology illustrates the range of applications (and models) used in the discipline. This chapter also provides a convenient platform for a more general discussion of the role of models in ecology. This discussion is very much in the spirit of the aphorism, *All models are wrong but some are useful* (Box 1976).

In Chap. 10, integrated *ecological assessment* provides the formal evaluation of the working model reflected in the conceptual model that began this sequence of tasks. The framework is multivariate, and ordination provides a set of tools designed for this type of application. In this chapter, ordinations are used as a general framework to summarize inventory, visualize monitoring data (including the effects of natural disturbance or management activities), evaluate experimental treatments (as a before/after, control/intervention design), and explore forecasts of landscape change. This approach to assessment closes the cycle of adaptive management, and also finishes the task-oriented coverage of this book.

Intended Audience

This book is essentially a course I have taught for many years to graduate and professional students. My intent here is to provide a convenient reference for similar courses, and for working professionals with a similar level of training. While some familiarity with statistics is helpful, I have tried to present the material in a way that does not presuppose that background. In my teaching, many of the example applications and lab exercises are done in the R computing environment (R Development Core Team 2021[2]), but this book is not presented with that narrow focus. I hope that the book is presented in such a way that the software-specific details of actual implementation are not distracting in their omission.

While the book has a logical structure to it and the tasks are interrelated to some degree, most of these chapters can be used as stand-alone references. Thus, teachers or practitioners can skip around and use what they choose.

[2] I will cite R and its packages often throughout this book. The cited dates are rather arbitrary as they change with each software update. That is, the citation dates could range from the 2010s to the current year without much changing the int[illegible] of results.

My aim here is to present the logic and protocol of the tasks, with an emphasis on helping the user design, implement, interpret, and present the analyses in a way that can be appreciated by a nontechnical audience. This reflects my opening comment that landscape-scale work is typically done in collaboration with social scientists and other team members, who will need to appreciate the takeaway results of the analyses but not necessarily the technical details. In my own classes, I emphasize this through writing-intensive exercises aimed at motivating the choice of tools, explaining the conceptual basis and actual analysis, and presenting the results in a compelling way. After all, technical prowess is not very useful if nobody else can understand or appreciate it.

Teaching with This Book

I have packaged this material into semester-long courses in various ways over the years. While I have recently packed the entire book into a single semester, I cannot really recommend this packaging: there is simply too much material (I skip or gloss over several topics in doing this).

Instead, I might recommend splitting this into multiple courses. Chapters 3, 4, 5 and 6 could easily comprise a course on "big data." Many of the topics in this book could easily be expanded into a single course. Especially, species distribution modeling (Chap. 2) is an enormous topic that invites a deeper dive; I have taught this as a full semester course using only Chap. 2 as a text.

I should note that I do not include GIS or remote sensing tools in this book. This is largely because my faculty colleagues teach these as separate courses and my students typically get this information in those classes. While I do refer to geospatial data and applications in this book, I do not presume any technical GIS expertise of the reader.

All of this is to say: there is a lot here, and users should feel free to pick and choose. No matter the focus of particular applications, the other chapters in this book can serve as handy references or background. While I feel comfortable with the scope of information in this book, I have also included a section in each chapter that points to additional resources. I hope readers will find this useful.

Acknowledgments

I have been teaching some of this material for 35 years or so, and in this structure for more than 20. Over those years I have taught hundreds of students, mostly graduate professional students in the Nicholas School of the Environment's Master's in Environmental Management and Master of Forestry programs at Duke University. My students have never been shy about letting me know what worked and what did not. All along, they have been willing to work hard on anything—no matter how technical or arcane—if I could convince them that the effort would be worth it and the outcome would be useful. These students are largely responsible for this course evolving from a survey of methods to a course structured by practical workflows. I thank them for that.

Similarly, my teaching assistants over the years have been especially helpful in revising the presentation of this material. This included instances where they have (rightly) pointed out when things were not working as intended and I should step back and try again. I am particularly grateful to Sarah Goslee, who wrote the first version of the R package *ecodist* (Goslee and Urban 2007) while a teaching assistant for my class; she continues to maintain that package, and I continue to rely on it.

A number of colleagues provided constructive comments on draft chapters. These include Janet Franklin (Chap. 2); Sarah Goslee (parts of Chaps. 3, 4, 5 and 6); Marie-Josée Fortin (Chap. 6); Jim Grace (Chap. 7); Betsy Albright, Chuck Peoples, and Lydia Olander (Chap. 8); Joe Sexton and Ross Meentemeyer (Chap. 9); and Bruce McCune and Lori Bennear (Chap. 10). I appreciate their willingness to wade in with constructive criticism and helpful advice. Shortcomings are my own.

As in the companion book, Regina Barnhill (B Design Studio, LLC) managed all the graphics. I appreciate her talent, good instincts, and patient support.

Durham, NC, USA Dean L. Urban

References

Box GEP (1976) Science and statistics. J Am Stat Assoc 71:791–799

Goslee SC, Urban DL (2007) The ecodist package for dissimilarity-based analyses of ecological data. J Stat Softw 22:7

Holling CS (1978) Adaptive environmental assessment and management. John Wiley & Sons, Chichester

Lee KN (1993) Compass and gyroscope: integrating science and politics for the environment. Island Press, Washington

R Core Team (2021) R: A language and environment for statistical computing. R Foundation for Statistical Computing, Vienna. https://www.R-proj-ect.org/

Urban DL (2023) Agents and implications of landscape pattern: working models for landscape ecology. Springer-Nature, Cham

Walters CJ (1986) Adaptive management of renewable resources. Macmillan, New York

Contents

Chapter 1
Sampling Designs for Landscapes

Abstract Landscapes are large and heterogeneous, and so collecting data on landscapes puts a premium on efficiency. In this chapter, we begin with the basic elements of sampling design and then apply these to inventory and monitoring. These simple designs are then extended to hypothesis-driven sampling designs that are more tightly coupled to inferential designs, especially ANOVA and partial regression. A fundamental decision in sampling landscapes is whether to avoid spatial autocorrelation or to embrace spatial structure explicitly in sampling. The key to most sampling designs is deliberate stratification over the explanatory variables of interest, including space itself. Sampling design provides a foundation for most of the tasks developed in subsequent chapters.

1.1 Introduction

Landscapes are large by most definitions and data at such scales are dearly bought. Yet, science-based management demands decisions based on empirical evidence. That means collecting data to support the analysis and application. Sampling designs for landscape-scale applications must put a premium on efficiency: We will want to retrieve as much information as possible from the sample measurements. That, in turn, means giving careful thought to *which* data we need most crucially and then designing a sampling scheme to focus on those data.

We can motivate this concern by considering how a simple sampling design might play out in a study aimed at developing a species distribution model as described in the following chapter. In the strictest case, this is a *case/control design*: We would randomly visit a number of (independent) sample locations, and at each location, we would tally the presence (case) or absence (control) of the focal species. We would also measure a suite of candidate predictor variables at each location, and then proceed to fit the model, which would distinguish the case from the control samples. But for an uncommon species, this approach likely yields far more absences than presences; indeed, for a rare species of conservation concern, we might not encounter any presences at all: statistically correct, but not very informative and certainly not very efficient. There are more effective alternatives.

D. L. Urban, *Landscape Ecology: A Task-Oriented Perspective*,
https://doi.org/10.1007/978-3-031-72251-6_1

The information content of samples will depend on the aims of the application, and so it will be useful to consider some of the primary purposes of sampling landscapes. One purpose of sampling landscapes is simply to tally the resources represented within the study area—a naive *inventory* of how much there is and where it occurs. Repeated over time, an inventory becomes a *monitoring* program; and because initial decisions about the inventory design are carried forward in time, it is important to get it right the first time. Often, a simple inventory design (i.e., to cover the study area) is elaborated to provide more nuanced information on the distribution of a resource. For example, if the target resource is a focal species, we might naturally target locations where we have some reason to believe the species might occur.

Beyond this, we might also want to know whether the places it occurs reflect habitat area or isolation effects often expected of landscapes: Is the species restricted to very large patches? Do patches of appropriate habitat isolated from other source patches tend to be unoccupied? This intuitive logical progression from a simple inventory to a targeted survey to a test of hypotheses about area or isolation effects invites questions about how to collect the samples and whether these different purposes also imply qualitatively different sampling designs for collecting data.

Posing these questions broaches the issue of *inferential* design—whether the data will support valid statistical inferences about the effects of interest. As statistical methods are extended to studies in landscape ecology, sampling design becomes increasingly important because sampling design tends to be coupled directly to inferential design.

Importantly, inferential designs for landscapes tend to adopt one of two rather exclusive approaches: to explicitly attend spatial structure in the data or to censor that spatial information in order to use conventional parametric statistics (which assume independence of the observations, i.e., that there is no spatial autocorrelation; Legendre 1993). Designs that attend spatial structure in the data—especially at multiple scales—present a particular challenge but also an opportunity in landscape-scale applications. Because this can be crucial to spatial inference from ecological data, it will be appropriate to consider issues of sample design as these support landscape-scale analyses.

In this chapter,[1] we begin with an overview of the basic elements of sampling design, such as might be invoked for a simple inventory. We then elaborate the inventory to attend to additional factors of interest, and the sampling designs evolve to become the geographic translation of an inferential model (e.g., an analysis-of-variance model represented in a geographic information system). These model-guided designs can become rather complicated and so often invite follow-up surveys and broach issues of multiphase sampling and other adaptive approaches.

Data collected from landscapes, whether from a simple inventory or a more elaborate inferential design, will then provide the basis for several subsequent tasks. These include inventory itself (collecting data is only the first part), hypothesis

[1] Some of the material in this chapter is reworked and updated from Urban (2002).

testing (i.e., evaluating the data according to an inferential model), site prioritization (ranking the best sites for conservation or, the worst, for restoration), and monitoring (trend detection) or forecasting landscape change. We will pick up these tasks in subsequent chapters.

1.2 Elements of Sampling Design

A *sample* is a set of sample *units* collected within a *sampling frame* (within the study area). A sampling design is defined by three components:

The sample unit Sample units are defined by their size (how big?) and shape (circles or squares or rectangles?), orientation with respect to environmental gradients (across or along hillslopes?), and so on. The particulars of sample units define the amount of variability *within* each sample as compared to variability *among* all samples, which can contribute to inferential power. In general, variability depends on quadrat size: smaller sample units will tend to have lower within-sample variability and higher among-sample variability (see chapter 3 in McCune and Grace 2002).

While the choice of sample units can be an important issue, for our purposes it will be assumed that this decision has already been made judiciously. Indeed, decades of trial and error have established best practices for many cases (e.g., for trees, herbaceous plants, birds, etc.), and these norms are typically obvious from even a casual review of recent applications reported in the literature.

Sample arrangement Arrangement refers to the location and spacing of samples relative to each other and across the study area. There are only a few basic approaches to sample arrangement (e.g., random, uniform), but a potentially infinite number of variations on these basic designs (see below).

Sample intensity With any design, one might also vary sample *intensity* or the number of samples located within the study area. In many cases, sample intensity and arrangement will be coupled for logistical reasons, so that decisions about arrangement imply decisions about intensity.

Repeat or aggregate samples For many targets, sampling might include repeat visits to the same site. For example, in bird surveys, it is common to census the same site multiple times to increase the likelihood that all birds are discovered. Similar concerns arise for other mobile or cryptic species. These multiple visits are then aggregated into a single observation for each point (i.e., they are not processed as time-varying); e.g., a birder might tally the maximum number of birds observed per species over several visits. When the repeat visits themselves are of interest, the design is more about monitoring than sampling (see below).

In general—but not always—sampling *efficacy* (the ability to capture information accurately) increases with sample intensity. At the same time, sampling *efficiency* decreases with sample intensity because of the cost of collecting the data. The challenge is to somehow balance the spatial arrangement and density of samples to best capture the necessary information with a reasonable number of samples. This is especially crucial for large landscapes, where data collection often is limited by available time and resources.

Sampling Design and Geospatial Data
As an aside, it might be noted that many data sets over landscapes are now collected using remote-sensing technologies. Satellite imagery might seem to make decisions about sample arrangement and intensity moot: why not simply measure *everywhere*? But such data sets can be awkwardly large and difficult to process in their entirety, and so we often sample these continuous coverages anyway, for processing efficiency. And so sampling design is still important.

Remotely sensed data also introduce the issue of information content or thematic resolution: what exactly is being measured? Compared to field measurements, remotely sensed data are often comparatively limited in their information content (consider a classified land cover map as compared to vegetation details we would measure in the field). As always, there are trade-offs between resolution and spatial extent. We will return to these issues of sample grain and resolution in the following chapter, when we consider data used to model species distributions.

1.2.1 Basic Sample Arrangements

Most sampling designs are variations on a few simple arrangements, and it will be convenient to begin by focusing on the basic building blocks. We will then extend these to more complicated arrangements for particular applications. As we proceed from simple inventories to more elaborate sampling designs intended to test hypotheses about factors such as habitat isolation or disturbance, sampling designs will reflect elements of inferential design as well: stratification, blocking and replication, and so on. We will introduce these elements later as needed.

Uniform Samples
Uniform samples are arranged in a regular pattern (transect or grid) over the study area (Fig. 1.1a). This design offers the advantage of being straightforward and efficient in the field. A second advantage is that the design can be tailored explicitly to cover the entire study area, missing no region. Sample intensity (number of samples) depends directly on the resolution of the sampling grid: a high-resolution grid implies a lot of samples (and higher cost), while if only a small number of samples is affordable, then the samples unavoidably will be farther apart.

In spatial applications, uniform samples have some disadvantages. For example, in geostatistical applications, analyses are based on the distances between samples, and a uniform design provides a finite number of between-sample distances (e.g., for a unit grid, distances of 1, $\sqrt{2}$, 2, and so on). Further, the sampling interval specifies a minimum grain in the data that the sampling can capture; if the spatial structure of the data is not known in advance, this design limits the possible structures that can be

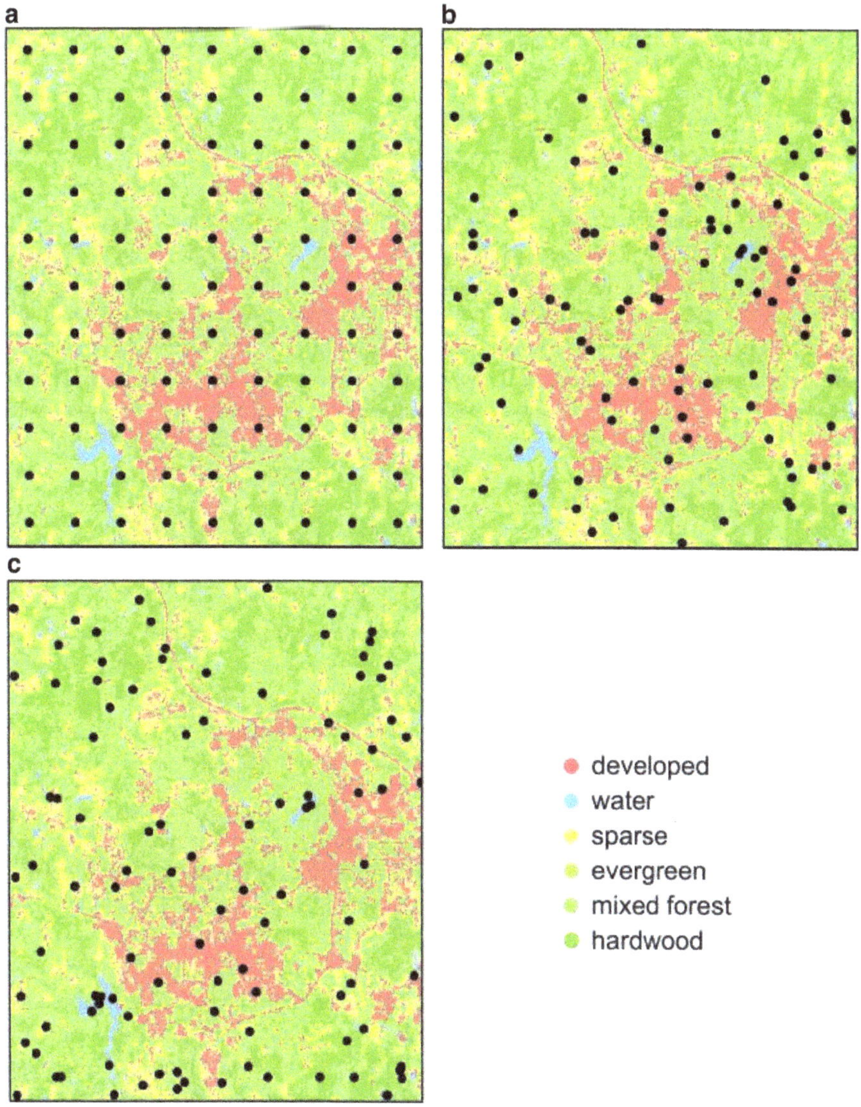

Fig. 1.1 Examples of simple sampling designs: (**a**) uniform (100 points), (**b**) random (100 points), and (**c**) stratified (96 points allocated equally to six land cover types)

observed. For periodic data structures, inappropriate sample intervals out-of-phase with the data will actually produce a false periodic structure (called *aliasing*).

In the field, uniform samples are often collected by locating samples at equal intervals along transects that are themselves equally spaced. This is an efficient method equivalent to locating samples on a grid: equivalent, because each sample can be considered the lower-left corner of a grid cell, efficient, because one requires only a long tape and a compass (or, more recently, a global positioning system, GPS, or a smartphone) to locate the sampling points in the field.

Random Samples

The most common approach to sampling, random samples is simply that: samples are located at random (*X*,*Y*) coordinates (Fig. 1.1b). For dense samples, random sampling generally covers the study area adequately (as well as uniform samples would), but for very sparse designs, there might be, purely by chance, "holes" in the study area that are missed. For spatial analyses, random samples typically provide a full range of between-point distances, and such samples are very unlikely to alias a periodic data structure. Random samples are also appealing from a statistical perspective because measurements are unbiased by before-the-fact assignments of sampling locations (i.e., the measurements are random samples of the population), thus satisfying one of the most basic of statistical assumptions for population-level inference. Note, however, that random locations do not mean that the measurements will not be autocorrelated; if they are, the samples will not be independent even if random (and see below).

In the field, random samples often are located in terms of distance and direction from easily locatable reference points. The sample points are typically chosen as (*X*, *Y*) coordinates and transferred to a map that can be carried into the field. If one takes the time in advance to chart a course through all sample points, random sampling can be made a bit more efficient (this is an orienteering exercise: e.g., point A is 37 m bearing 240° from the road intersection; from point A, go 79 m at 48° to point B; from there, 131 m at 335° to point C; and so on). Recent advances in the precision of GPS units (including smartphones) make this a much simpler exercise in the field.

Stratified Designs

A combination of a stratified uniform and random, a spatially *stratified random* design often provides a useful compromise between field efficiency and thorough coverage of the study area. A full range of between-sample distances is assured by the randomness, and the stratification assures full coverage.

In the field, a stratified random design is reasonably straightforward. One approach entails establishing the uniform baseline (a single or a series of transects) or grid and then locating samples at a random distance and direction from reference positions. For example, one might locate a sample at a random distance and direction from a series of reference nodes regularly spaced along a transect line, with the reference points georeferenced using GPS and the random points offset in the field from the reference points (and then GPS'd themselves).

Another common approach is a *nonaligned block* design, in which a sample is located randomly within each cell of a regular grid; a sparser design might sample only a subset of the cells. Nested nonaligned block designs are an efficient means of stratifying samples across a range of spatial scales (and see below).

These designs are a stratification over geographic space and so depend on an underlying grid of spatial regions (grid cells or polygons). This invites the more general case of stratification, in which samples are allocated over *any* partitioning of the study area. A common example would be to stratify samples randomly over a

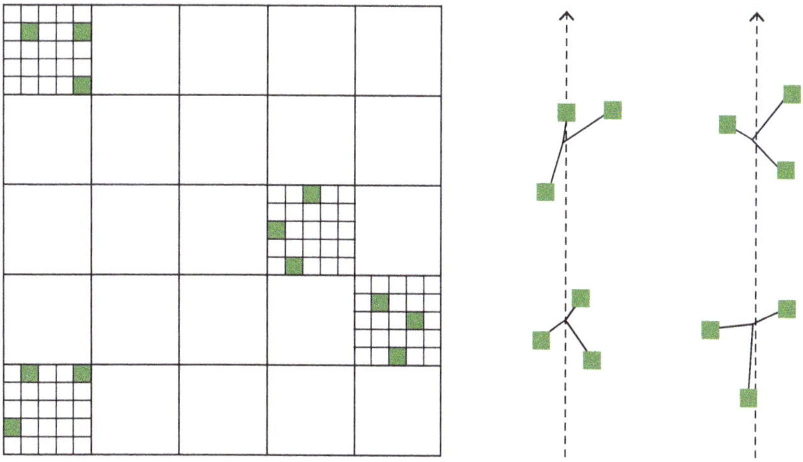

Fig. 1.2 Two approaches to a multi-scale sampling design: nonaligned blocks (left) and clusters stratified along transects (right)

map of land cover or habitat types, so that each was represented equally in the sample (Fig. 1.1c). We will rely on such logical stratifications to design sampling schemes to target specific kinds of sites to assess research questions in a subsequent section (Sect. 1.3.2).

Hybrid Designs

An almost infinite variety of hybrid designs can be customized from these basic building blocks. These are typically tailored specifically to the demands of a particular study. Common approaches include nested designs and cluster samples (Fig. 1.2).

Nested designs

This design subsamples a study area with similar designs at different scales, with the scales often differing by an order of magnitude. For example, one might use a nested nonaligned block design of 100*x*100-m grid cells, choosing grid cells randomly at some proportion (say, 25%). For this subset of the cells, one might then divide each large cell into 100 10*x*10-m cells and resample the smaller cells with their own nonaligned block design, and so on. At larger scales, this basic motif might be replicated spatially within a larger grid.

Cluster sampling This is a multi-scale design in which constellations of sample points are located at each point of a larger design. For example, the US Forest Service inventories forests nationwide with a design of fine-scale uniform constellations which are themselves located on a large-scale uniform grid. Of course, one could also use random clusters at random reference points, or uniform clusters at stratified-random points. The clustering provides measurements over a much wider range of between-sample distances than single-level sampling. In particular, clusters and nested designs provide measurements at the "close" distances often undersampled by regular or stratified samples. The challenge in any two-level sampling design is to devise a means of streamlining the process of locating samples in the field. In Fig. 1.2, the nested nonaligned blocks require that the underlying grid be delineated in the field (e.g., with stakes or flags). The transect-based equivalent requires only a long tape and a compass. Again, the increasing availability of high-precision GPS units can streamline the plot-location process considerably. For example, a complicated design can be generated in a GIS and then downloaded to a smartphone to use in the field.

It should be obvious that by choosing the spacing along a transect (or grid cell size) in addition to the number and spacing of points located at each stratification point (from points along a transect or within a grid cell) that there is an infinite variety of sample layouts available from these very basic building blocks. Some of these will be especially amenable to different kinds of landscape-scale studies, and so it will be appropriate to consider some of these applications.

1.3 Sampling Designs for Applications

Here we develop, as a heuristic exercise, a logical evolution of sampling designs, from the simplest goal of covering a study area to much more complicated designs, informed by models, that aim to provide maximum information in as few samples as possible. Along the way, we will need to digress to consider the special case of capturing information at particular spatial scales.

1.3.1 Inventory

Inventories can be quite simple, or they can be targeted at a specific resource or focal species. The latter is a straightforward extension of the former.

Simple Inventory In what we might term a naive inventory, we simply wish to know what resources are represented in or supported by the study area. How many bird species breed in the study area? How abundant are they, and are there unusual species from particular habitat types, or foraging guilds, or what? This is the sampling approach we might adopt as an initial inventory of a study area, to establish an ecological baseline.

Because the goal is complete coverage of the study area, any sampling design can provide useful data: uniform, random, or spatially stratified random samples. Here the likely trade-off will be sampling intensity (number of samples) as compared to how thoroughly the study area can be covered. Random samples will suffice unless the number of samples is so low that some areas are missed by chance; uniform samples will control coverage as long as the between-sample distances are not too large; and stratified-random samples offer a convenient compromise.

An inventory repeated over time is a monitoring program. Because subsequent measurements will be collected according to the initial design (and carry forward any mistakes!), it is important to get it right the first time.

Focused or Constrained Inventory Often we will have some idea or expectations about inventory, and it will be efficient to tailor an inventory to reflect this prior information. For example, we might be interested in a particular species—a rare plant, a charismatic wildlife species, or a species of economic concern for commodity or recreational value. In such cases, the inventory is not naive but rather tempered by some prior expectations or constraints on its distribution. We might, for example, know that the species occurs only in certain cover types (e.g., evergreen forest), or we might have a statistical model of habitat suitability for the species mapped into a geographic information system (see Chap. 2). In this case, it would make sense to bias the inventory relative to predicted habitat suitability. In this, we might bin habitat suitability into a few classes ("very good," "suitable," and "marginal" habitat and perhaps "not habitat" just to be cautious—and see below) and stratify samples over these classes to get an idea of relative species density in habitats of varying suitability.

It is worth emphasizing here that an inventory of the expected abundance of the focal species can be estimated directly from this habitat-stratified sample: The expected number of individuals of the species is the density (numbers per unit area) of the species in each level of habitat suitability, multiplied by the area of each suitability class in the entire study area—a simple weighted summation:

$$\widehat{N} = \sum_{i=1}^{k} D_i \cdot A_i \qquad (1.1)$$

where D is density and A is area for each of the k cover types or habitat classes. Here, careful stratification of the sampling makes the actual inventory quite straightforward and efficient.

It is important to note that stratifying samples randomly over levels of habitat suitability is exactly analogous to stratifying samples over geographic space. The only difference is in how we define the strata: in parameter space (habitat suitability) in one case or in geographic space in the other. This flexibility to define strata for particular applications will generate substantial inferential leverage for landscape-scale sampling designs. And the key to this will be the facility with which we can translate between parameter space (in concept) and geographic space (on the ground). Typically, we will define strata in parameter space and then use a GIS to translate these in geographic space. McGarigal and Cushman (2002) refer to this as *GIS-based filtering*. The simple inventory implied by Eq. 1.1 can be extended accordingly.

The example of stratifying an inventory over levels of presumed habitat suitability also illustrates another important consideration in sampling designs: the potential for bias. We can probe this example for some insight into the issue. We already noted that a strict case/control sampling design for a focal species might not yield very many presences. An intuitive alternative would be to target settings where we expect the focal species to occur (e.g., in particular, forest types as represented in a land cover map). This approach would certainly garner more presence points... but it would also be a self-fulfilling prophecy about what constitutes "habitat": we would only observe the species in places that we already thought looked like "habitat." To prevent this bias, we would want to also sample in locations that we think do *not* look like habitat, just to be sure. At this point, the sampling design slips away from inventory and becomes more inferential, designed to test a hypothesis.

1.3.2 Inventory Extended to Explore Hypotheses

As Austin (2002, 2007) has discussed, we have some specific expectations of species distributional patterns at the landscape scale. From metapopulation theory (Pulliam 1988; Urban 2023, Chapter 6), we expect that a good habitat that is isolated from other habitat patches might be unoccupied. Reciprocally, marginal habitat that is near occupied good habitat might also be occupied because of local dispersal subsidies. That is, while we expect to encounter more individuals of the focal species in patches of "habitat," we also have perfectly valid expectations about the lack of occurrence in habitat and occurrences in marginal or poor habitat. Likewise, we might have expectations about species response to patch size or geometry (e.g., Burgess and Sharpe 1981). It would be perfectly logical to sample in such a way that we could explore these effects along with the "habitat" effect itself.

This extension requires only that we define new sampling strata in terms of relative isolation from or adjacency to other patches of suitable habitat. For example, if we could compute (in a GIS) the distance from every patch of suitable habitat to every other patch of suitable habitat, we could then partition these distances into discrete ranges (bins) and stratify over the bins as well as over habitat suitability. From the field inventory, our expectation would be that "near" habitats would be occupied disproportionately (and perhaps even for marginal or unsuitable habitat), while suitable "habitat" far from a source might well be less frequently unoccupied. This is another stratification in the sampling design, to now consider "habitat suitability" and "distance" (or "isolation") as two main factors influencing species distribution. This logic could be extended to consider patch area or geometry (e.g., percent edge) as another factor influencing occupancy. Importantly, this stratification is no more complicated in principle than stratifying over geographic space or any other variable. It is a straightforward exercise to query a geographic information system (GIS) to find locations that meet the desired conditions for the stratification. (Note that this task, of distinguishing habitat effects from geometry or isolation, can be hugely complicated in application: Urban (2023) reviewed decades of effort on this topic, much of this led by Lenore Fahrig—see Fahrig 2003, 2017, 2018, 2019).

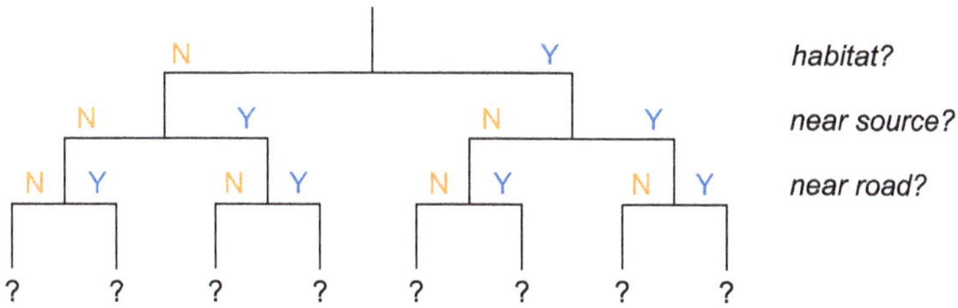

Fig. 1.3 A decision tree to isolate the effects of habitat suitability, dispersal, and disturbance (here, as proximity to roads) on species distribution (after Urban 2002). The tree has eight branches, and locations representing each case can be isolated using simple GIS queries. Note that only two end nodes are reasonably unambiguous: node 2 (numbered from left to right) should be unoccupied and node 7, occupied. Note also that only a comparison of samples representing nodes 7 versus 8 would support a clear inference about disturbance

Stratify to populate an ANOVA model Specifically addressing habitat suitability and distance (or isolation) is a straightforward extension to a stratified random design in space. This invites more nuanced stratifications and designs aimed to explore various multivariate hypotheses about species distribution. But these designs can be focused even more, depending on the aims of the application.

Consider, for example, a study in which we are interested in habitat suitability, distance effects (isolation or dispersal subsidy), and disturbance as constraints on a focal species. Further, consider the simple binary case wherein we assign each variable into two levels: suitable habitat or not, isolated or not, and disturbed or not. This is an analysis-of-variance (ANOVA) design, whereby we will want to explain the occupancy of various patches in terms of their suitability, isolation, and disturbance. There are 2^3 or eight combinations of factors in the fully crossed design (Fig. 1.3). To populate this inferential design, we would need to locate some replicate samples of each of the eight cases within our study area. This can be done via a series of queries to the GIS data by highlighting a series of intersections of the combinations of values.

The tree model helps underscore the relative value of information collected to address particular questions. In this illustration, only two of the eight end nodes (leaves) of the tree are straightforward: node 2 is unsuitable habitat, isolated, and disturbed, and we would not expect the focal species to be there; node 7 is suitable habitat, not isolated, and undisturbed, and we would expect the species to be found under those conditions. The other six cases are interesting precisely because they are ambiguous.

To partition the effects of all three factors, we need samples representing all eight of the nodes: a fully balanced ANOVA design that will allow us to estimate the relative importance of each effect (and perhaps interactions as well). Indeed, to do the ANOVA, we would need the same number of samples in each bin. By contrast, if we are primarily interested in disturbance effects, then the only cases that provide this leverage is a comparison of nodes 7 and 8: all other nodes are confounded by uncertainties due to habitat suitability or isolation.

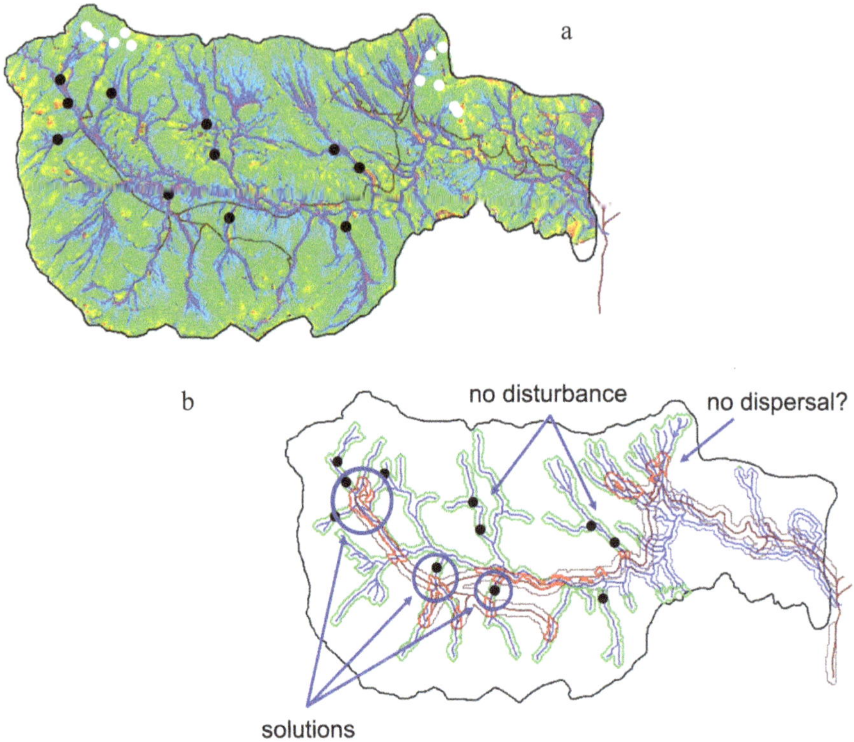

Fig. 1.4 A hypothetical sampling example, focusing on a riparian plant (black dots in maps) that might be dispersal limited (occurring in only a few watersheds) and is subject to disturbance by hikers (near trails, brown lines in the maps). (**a**) Background on the left is topographic convergence, emphasizing the riparian habitat in blue and violet. (**b**) Potential solution space emphasizes that only three locations can provide inferential leverage in this case: green buffers delineate "riparian habitat" in occupied watersheds; brown buffers, "near trails"; catchments with known occurrences (black dots) offer evidence of dispersal access. Potential sampling sites must be in green buffers, in occupied catchments, and allow for paired sites that can be located in disturbed versus undisturbed locations in reasonable proximity

A slightly more tangible illustration of this logic also allows us to add elements of inferential design to the discussion. The illustration is based on a (fictional) rare plant that occurs in riparian habitats, might be dispersal limited, and is known to be decimated by hikers who pick the flowers (Fig. 1.4, after Urban 2002). In geographic space, the sampling challenge is to find locations that can provide inferential leverage on the importance of disturbance by hikers. To do this, we must find locations that are suitable habitat and apparently not isolated (in this case, by finding small watersheds where the species is known to occur). Then, we must contrast disturbed (i.e., near trails) and undisturbed (far from trails) locations for paired sampling. The solution set underscores the reality that *most locations in the study area actually are not very informative* about the question at hand.

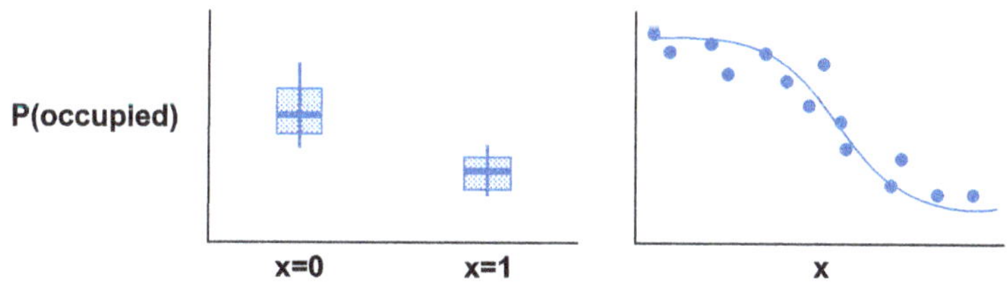

Fig. 1.5 Schematic contrast of an ANOVA design in which dependent variable *Y* is modeled in terms of a categorical predictor *X* (with 2 levels, on the left) as compared to a regression design in which *X* is a continuous predictor (on right). The regression design can suggest not only *whether* the response differs but *how* it varies with *X*

Inferential Design on Landscapes

The basic elements of inferential design include blocking, randomization, and replication. *Blocking* is an attempt to control the "all else being equal" clause that qualifies our attempts at generalization. In this case, blocking entails localizing the contrast of interest (in Fig. 1.4, disturbed or not) in the same general location so that other potentially confounding variables such as soil factors can be controlled by holding them as constant as possible (this makes some assumptions, of course, about the spatial structure of these variables—see below).

Typically, *randomization* is the assignment of treatment versus control cases randomly within blocks. The aim is to hedge bets that the blocking actually did control all unaccounted covariates and also to ensure (as much as possible) that the sample cases are independent of each other. In this case, randomization also would entail locating the actual sampling locations stochastically within the focal watersheds (but near streams and either near or far from trails).

Finally, the experimental contrast of "disturbed versus not" needs to be replicated over multiple instances of the same contrast, in multiple watersheds. *Replication* increases the precision with which the effect can be estimated. If the contrast due to disturbance is similar across all watersheds, we would have more confidence in the result and a greater ability to estimate the "disturbance effect". Reciprocally, if each watershed behaves idiosyncratically, we would see a "watershed" effect dominating the results and the "disturbance" effect would be lost. (In the statistical model, a "watershed ID" variable would serve as a *dummy variable* for any unmeasured covariates that vary among occupied watersheds.)

ANOVA Versus Regression Designs These examples have focused on an ANOVA design in which the explanatory variables are categorical factors (e.g., "near stream" or "not"). In the ANOVA case, the fully stratified design balances the ANOVA so that each of the effects can be estimated. It is perfectly easy to generalize this to a regression design in which the explanatory variables are continuous (interval scale). Analytically, the two are essentially the same: the *t*-test that assesses the differences between means for two groups is the same as the test of the regression slope on that variable used as a continuous predictor (Fig. 1.5). In the regression case, the stratification ensures that the full range of variability on each factor is sampled efficiently; by contrast, random sampling would load more samples into the middle of the distribution (assuming normality) and under-sample the tails of the distribution.

The regression design offers two advantages in applications. First, while it might be conceptually appealing to use categorical contrasts in an ANOVA design, it is often difficult to do this cleanly except in simple manipulative experiments (i.e., "treatment" or "control"). In other cases, such as "disturbed" or "not," "near" or "far," and "large" and "small," it can be difficult to find unambiguous realizations of the desired contrast: what we end up with are samples with measurements over a range of values, which we arbitrarily bin into categories. Our inability to find replicates of exactly the same value (disturbance state, distance, size) introduces noise into the analysis in the form of within-group (replicate) variability. At the same time, the arbitrary grouping into categories implies that we know in advance how to partition a continuous variable into ecologically relevant categories: if the variable of interest is "patch size," we must assume we know what defines a "large" as compared to a "small" patch—from the perspective of the dependent variable (e.g., occurrence of a focal species). It seems presumptuous to assume that we would know this in advance in many cases.

Second, an ANOVA model can tell us only *whether* the response differs across levels of the categorical factor and which levels correspond to higher or lower response values. A regression design provides the same information, as well as offering the ability to describe the *shape* of the response: is it linear, nonlinear, a threshold response, or what? Thus, regression models are often simpler to implement in the field, and they can return more nuanced information. It seems logical to use the regression design more often in ecological field studies. Importantly, one can have the best of both worlds: use a GIS to stratify samples within discrete levels of values on an explanatory variable of interest (e.g., "small," "medium," and "large" patches) and then use a regression design for the actual analysis, in which the distinctions among categorical levels are erased. This is analogous to the approach modelers use in sampling over complicated parameter distributions in various Latin or orthogonal hypercube sampling designs (e.g., Iman et al. 1981); but it is equally compelling for logistically complicated field studies at the landscape scale.

1.3.3 Hypotheses that Explicitly Embrace Space

It should be obvious that stratifying over an explanatory variable such as "isolation" is completely analogous to stratifying over locations specified in terms of (X,Y) locations: the aim, in both instances, is to cover the range of values available in the study area. In analysis, this is a regression problem. But it is confounded by the reality that the samples are drawn from spatial locations—explicitly, in geographically stratified designs and, implicitly, in cases where the stratification is over an explanatory variable that is not obviously spatial but yet is spatially structured for ecological reasons (e.g., because of local topography; Urban 2023, Chapter 4). This will invite us (or perhaps require us) to distinguish logically between explicit spatial effects and implicit effects due to spatial structure in data.

Before delving into spatial effects, we should digress briefly to be clear about what "spatial" might mean in terms of sampling. First, we might measure variables such as "distance to" some feature (e.g., water, roads). These are *implicitly* spatial variables that can only occur with reference to particular locations; we would interpret these statistically in terms of *local trends* (e.g., an increase in the likelihood of species occurrence with decreasing distance to water).

We might also measure the location itself, as latitude/longitude or other *(X,Y)* coordinates. These are *explicitly* spatial variables, and we would interpret them in terms of *geographic trends* (e.g., a relationship between temperature and latitude). In this instance, samples that are farther north are different than samples farther south.

In this discussion, we will be especially interested in spatial effects that depend on the distances *between* samples: we will want to interpret patterns such as the tendency for measurements to be more similar for samples that are closer together. That is, the interpretation will be in terms of "distance apart." By contrast to the geographic trend considered above, this pattern of *autocorrelation* might be true no matter where in the study area the samples occur; the spatial structure is localized but replicated broadly over the study area.

Spatial autocorrelation arises from environmental dependencies such as topographic influences on ecological variables that affect species, spatially contagious disturbances or stressors (fire, pests, disease), or spatial processes (especially dispersal) (Legendre 1993; Chapter 4 in Urban 2023; and see Chap. 6).

Spatial structure, as autocorrelation in the data, will confound the inferential design of many studies, in that spatial structure will violate the assumption of independence among samples that randomization seeks to provide. That is, if the data are spatially structured, randomization does *not* ensure that the samples are mutually independent (Legendre 1993). In such cases, a key assumption of parametric statistical analysis (i.e., that samples are independent) cannot be met.

There are two and only two responses to this problem. One is to embrace space explicitly and adopt analytic methods that account for spatial structure in the data—in sampling and in analysis. The alternative is to detect and quantify the spatial structure in the data and then sample in the study area in a way that ensures that the samples are indeed independent—that is, to embrace space in sampling so that space can be ignored or avoided in analysis. Either way, one *must* characterize the spatial structure in the data before proceeding.

A key element in this approach is to know the spatial resolution of the structure of the data. That is, if we want to sample over a range of values of some explanatory variable, we will need to know how that variable varies geographically: how it is scaled.

Spatial Grain and Sample Independence

The spatial resolution of variability in a variable—its *grain*—is defined by the distances over which it varies in a spatially dependent manner, the scale of its spatial autocorrelation. We would anticipate that measurements of a variable collected very close together will be similar, while measurements separated by some distance will tend to be less similar or more dissimilar. The are two general methods for describing this distance-dependency in the data. One focuses on the *similarity* of measurements as a function of distance apart, while the other approach focuses on *dissimilarity* (Urban 2023, Chapter 4, §4.3.1; and see Chap. 6).

Autocorrelation Spatial statisticians focus on autocorrelation in measured variables, the tendency of measurements to be similar when collected at locations that are close together. A common estimate takes the form:

$$I(d) = \frac{\frac{1}{W}\sum_i \sum_j w_{ij} z_i z_j}{\frac{1}{(n-1)} \sum_i z_i^2} \tag{1.2}$$

which is the estimator for Moran's I (Moran 1950; Legendre 1993). Here, the measurements are converted into z scores (deviation from the mean, divided by the standard deviation) to rescale the measurements. The w term is an *indicator variable* that takes on a value of 1 if two samples are within some specified range of distances apart (i.e., in distance class d), else it takes on a value of 0. Through this indexing, the formula provides an estimate of autocorrelation for each distance class d. The term W is the sum of the indicator weights (sample size) in each distance class, which along with the overall sample size n rescales the index to vary on the range $[-1,1]$, just like the familiar Pearson correlation coefficient.

The expected value of Moran's I is:

$$E(I) = \frac{-1}{(n-1)} \tag{1.3}$$

which, for large sample sizes, approaches 0 (as with a familiar correlation coefficient). A plot of Moran's I versus separation distance (i.e., for each discrete distance class d) yields a *correlogram.* A correlogram summarizes *direction* (positive or negative), *intensity* of the pattern (absolute value of autocorrelation), and the *scale*(s) (i.e., distance classes) at which this pattern is expressed.

A correlogram for ecological variables typically shows positive autocorrelation, meaning that samples that are close together tend to take on similar values. Negative autocorrelation, in which nearby samples have dissimilar values, is rare in natural systems. Typically, autocorrelation decreases with increasing distance apart until the index does not differ statistically from 0 (no autocorrelation, which means the samples are independent). This test can be approximated based on a large-sample normalization, or (more typically) the confidence limits around 0 can be estimated via a randomization procedure.

As a scaling technique, autocorrelation identifies pattern as distance classes within which samples tend to be similar. Legendre and Fortin (1989) provide heuristic examples of correlograms for a variety of visually distinctive patterns.

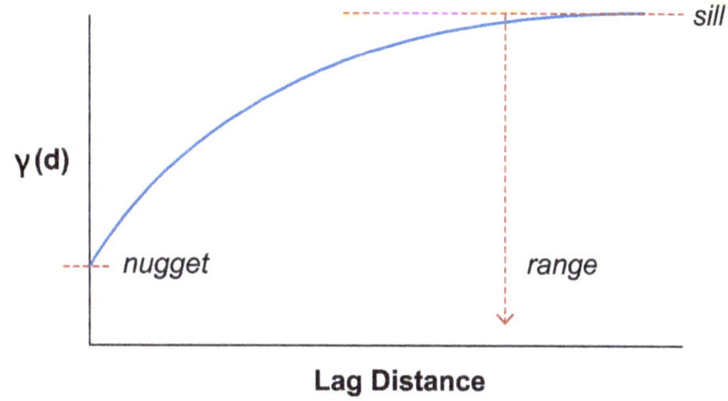

Fig. 1.6 A variogram, defined by its nugget, sill, and range. The range indicates the distance over which measurements of a variable are dependent and so dictates the sample spacing needed to capture that grain or to avoid autocorrelation and ensure independence of the measurements

Semivariance Somewhat separately from spatial statistics, the field of geostatistics (mostly at home in engineering) developed an approach to index scaling in terms of the *dis*similarity of measurements as a function of separation distance. Translating from the somewhat disparate notation of geostatistics into a format consistent with autocorrelation (Eq. 1.2), semivariance (gamma) is estimated:

$$\gamma(d) = \frac{1}{2W}\sum_i\sum_j w_{ij}\left(x_i - x_j\right)^2 \tag{1.4}$$

where the indicator variable w acts as in Moran's I to subset sample pairs by distance class and W is the number of sample pairs in distance class d. Dividing by two rescales the index so that it converges on simple variance as autocorrelation decreases to 0.0 (i.e., as among-sample variability approaches independence).

A *semivariogram* (or simply, *variogram*) plots dissimilarity as a function of separation distance (Fig. 1.6). A variogram is described in terms of three attributes. The curve tends to asymptote to a plateau value, which is its *sill* (units: semivariance). The distance at which this occurs is the *range* of the variogram, which here is the item of interest because it indicates the scale of spatial dependency. The Y-intercept of the curve is its *nugget* variance. In a perfect world, the nugget would be 0, indicating that samples measured in the same location would have identical values. For most ecological measurements, this is not the case and so the nugget suggests the natural replicate variability of the measurement. But the nugget also is affected by other factors. In particular, because the variogram plots average semivariance within a distance class against the average separation distance for pairs of samples within that class, the nugget probably does not actually intercept the Y-axis but instead occurs at some distance from 0—the average minimum distance between the samples. This means that we have no information about variability in measurements at a grain finer than this distance. This contributes to the effective grain of the data set, because the spacing of samples influences the minimum spatial resolution.

From a sampling perspective, the grain of a variable suggests the distances over which measurements should be distributed to either capture the local dependence or conversely, to avoid this. To begin with the latter case: To ensure independence of

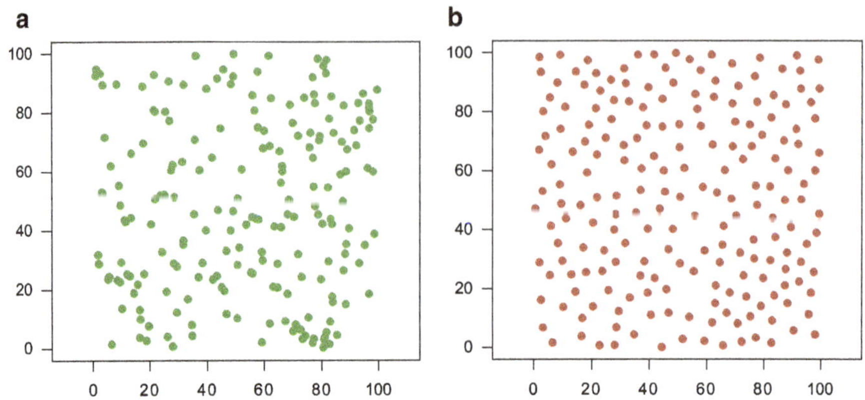

Fig. 1.7 Two sampling designs with 200 random points: (**a**) simply random and (**b**) a sequential interference design with an exclusion distance of 5

observations in a sampling design, samples should be located randomly and no closer together than the grain of that variable. This can be done using a *sequential interference* sampling design. In the simplest version of this, an initial point is chosen randomly. Then, a second point is chosen randomly and its location is compared to that of the first point; if farther than a user-specified distance (the *exclusion distance*), the point is retained; otherwise, the point is discarded and another random point is selected. Then, a third point is selected and compared to the locations of the first two points, and so on. This sequential process results in a set of random points, none of which is closer together than the exclusion distance (Fig. 1.7). This sample would then meet the assumption of independence that is required of parametric analyses.

Of course, this procedure would need to be conducted relative to the minimum grain over all variables of interest (or perhaps the most limiting). Most GIS packages include tools for generating random points with a user-specified exclusion distance in this way.

Explicitly Spatial Models

The alternative to sampling to avoid spatial autocorrelation is to sample to explicitly capture this spatial structure, perhaps over multiple scales. This means that samples must be collected at separation distances that correspond to the natural scaling of each of the variables of interest. In principle, a random or stratified random sampling design will accomplish this—if the sampling is sufficiently intense. In practice, it is often more efficient to use a multi-scaled design (e.g., Fig. 1.2) with spatial structures corresponding to the variables of interest. This approach ensures an adequate representation of "close" distances even for sparse sampling designs.

We will consider the statistical tools for spatially explicit inference later (Chap. 6). But it should be sufficient for now to emphasize that we will be unable to make spatial inferences if we collect data that are not spatially structured over the scales of interest. Note that in the special case in which we wish to document the

Fig. 1.8 Stratification over elevation (brown gradient) and microtopography, in which several topographic positions (colored dots) are sampled locally within each elevation band

effect of the spatial structure in one variable relative to that in a second variable—i.e., to estimate a spatial dependence—we will want to sample the variables at separation distances over which the measurements show dependence. That is, we want to sample over the "steep" part of their correlogram or variogram.

Partial Regression Designs in Geographic Space

In many cases, the spatial structure of the variable is explicitly of interest. In one case, the aim is to stratify over this variable while nesting this stratification within a (larger-scale) stratification over another variable. For example, elevation and local topographic convergence both affect soil moisture in montane systems, but at very different spatial scales: elevation varies over many hundreds of meters while microtopography varies over tens of meters (Urban et al. 2000). To isolate the effects of each of these variables on plant distribution, the sampling design must nest microtopography within elevation, capturing a range of topographic positions at multiple elevations. Analytically, the aim is to be able to describe the importance of microtopography as a partial regression problem: at any given elevation, microtopography exerts an additional (local) effect. To isolate this possible effect, we will require samples that capture relative higher (drier) and lower (wetter) topographic positions across a range of elevations.

This sampling challenge requires a multi-scaled sampling design. In this instance, the aim is to stratify samples over distances (scales) that capture the variability in elevation while also nesting within this pattern a subset of samples that capture the variability in topographic position. This, in turn, implies a clustered or randomized block design that explicitly varies on the two scales of elevation and microtopography (Fig. 1.8, and see Urban 2023, Chapter 4). One way to achieve this in the field is to first stratify over elevation (e.g., with sampling locations spaced to capture the range of variation in elevation) and then to nest clusters of samples at each of these locations, with the clusters structured to capture the local variability in microtopography (as depicted generically in Fig. 1.2).

Again, to emphasize the simpler case of independent samples, we would want the local cluster of points aimed at microtopography to be separated by distances at least as far as the range of autocorrelation. In that way, each microtopography sample would tend to be independent.

Spatial Structure and Virtual Pilot Studies

It might be obvious by now that spatial structure in landscape-scale data will be of interest in every case: when we wish to embrace the spatial structure with explicitly spatial statistical models, when we want to partition the effects of multiple variables structured at a variety of scales, or when we wish to avoid the complexities of autocorrelation so that we can use simpler parametric statistics that assume independence. That is, *we need to be explicit about scale whether we want to attend it deliberately or not.*

This implies that, absent this information in advance, we will need to do pilot studies to discover the natural scaling of ecological variables of interest. Pilot field studies can be as costly as the actual sampling, and so any alternatives that increase efficiency would be welcome. One approach is to use "virtual" pilot studies to explore the spatial scaling of the study system and to assess alternative sampling designs. In a virtual pilot study, geospatial biophysical proxies (Urban 2023, Chapter 1) are used to represent ecological variables of interest. For example, we might use various terrain-based indices of soil moisture, or proxies for radiation loading, and so on. It is straightforward to generate and then sample these in a geographic information system, to discover their natural scaling as well as to explore alternative sample arrangements and to capture this structure (Urban 2002). This information can then be used to inform the actual field samples (Urban et al. 2000, 2002).

1.3.4 Heuristic Sampling

One of the challenges of working empirically at the landscape scale is that the data tend to be sparse relative to the extent of the study area. This implies that inferences based on these data, while perhaps statistically significant in the conventional sense, might be a bit uncertain. One way to respond to this uncertainty is to use the model to help guide follow-up sampling, to collect new data that will best resolve uncertainties arising from the initial sampling and analysis. We might term this *heuristic sampling*, in the sense that the sampling design can "learn" from its earlier instances to inform subsequent efforts. That is, an initial sample helps build a model, and the model helps to inform a follow-up sampling design—data from which samples might efficiently improve the model.

For example, Urban et al. (2002) used a field survey to explore the environmental associations of several forest community types in the Sierra Nevada of southern California, USA. They used a classification and regression tree (Chap. 2, Supplement 2S.2) to do this analysis and generated a plausible tree that suggested that two high-elevation forest types could be distinguished largely according to slope aspect. While reasonable enough, the relatively small sample sizes suggested some uncertainty in the model. They then used the model to predict locations at (unsampled) high elevations that would contrast the model predictions at logistically convenient locations (i.e., contrasting and nearby slopes at similar elevations). New samples at

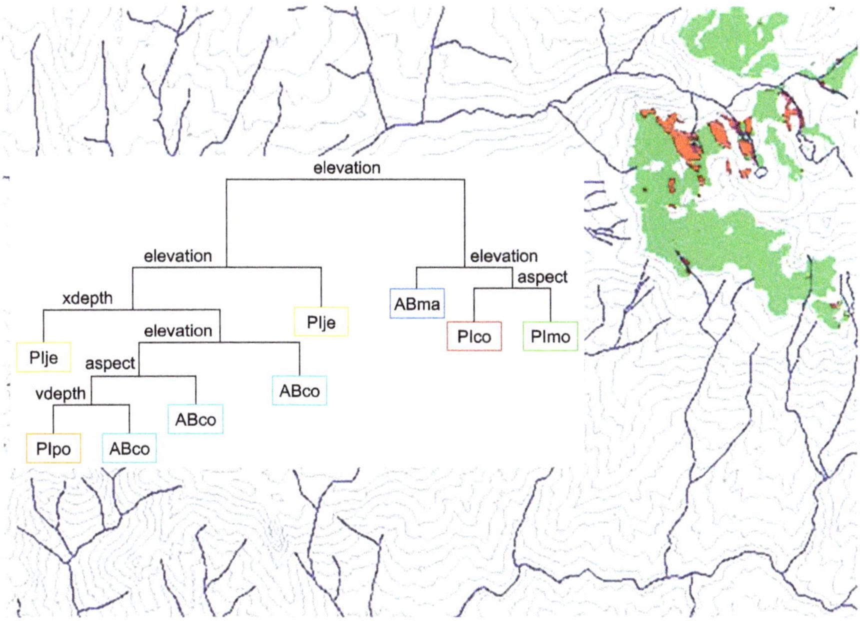

Fig. 1.9 A heuristic sampling approach, in which a classification tree (inset) is used to identify sample locations that will test the model. The red and green locations on the map correspond to the branch of the tree (right side) that contrasts western white pine (PImo, in green) and lodgepole pine (PIco, in red) forests. (Redrawn from Urban et al. (2002) with permission of Taylor and Francis, Ltd.; permission conveyed via Copyright Clearance Center, Inc.)

these locations would either corroborate the initial model or provide new information to revise and improve the model (Fig. 1.9).

The use of models to guide sampling designs can be extended to arbitrarily more complicated cases. For example, Urban et al. (2000) developed a forest simulator to synthesize the state of knowledge of how climate interacted with forest process and the fire regime in mixed conifer forests of the southern Sierra Nevada (this was part of the same project illustrated in Fig. 1.9). The simulation model provided estimates of the sensitivity of these forests to changes in temperature and precipitation. Despite its complexity, the model does not incorporate important processes such as lateral hydrologic flow, a source of uncertainty in model simulations. As a guide to establishing a monitoring program in the Park in anticipation of climate change, the investigators summarized the sensitivity of the model system to slight variation in temperature and precipitation, capturing this sensitivity in regressions based on geospatial predictors (e.g., elevation, slope, aspect). Uncertainty related to topographic position was indexed in terms of the local variability (within 100 m) of a terrain-based index of hydrologic convergence, intended to capture the ability to measure such differences efficiently in local surveys. An initial map product identified regions of the study area that might be most sensitive to changes in temperature

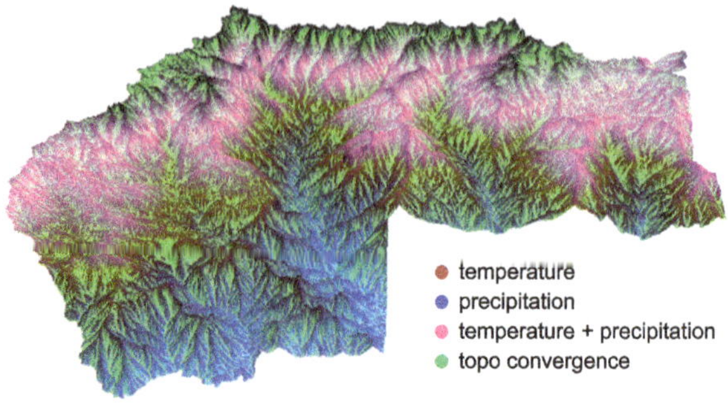

Fig. 1.10 Sampling locations targeted for monitoring in anticipation of climate change in Sequoia National Park in the southern Sierra Nevada of California (after Urban 2000). Draped on digital terrain, the intensity of red indicates sensitivity to temperature change, and intensity of blue indicates sensitivity to changes in precipitation. Intensity of green indexes local variability in topographic convergence, a source of uncertainty in the model. Colors are additive, so white (all three colors) shows locations that are climatically sensitive and with sufficient local variation in topography to capture these effects and resolve uncertainty. If constrained further for logistical reasons (road or trail access), the intersection zone amounts to <2% of the study area

and precipitation, tempered by the mapped uncertainty due to local topography (Urban 2000, Fig. 1.10).

An extension to this map added logistical considerations such as reasonable proximity to roads and trails (access to sampling sites is expensive, in terms of human resources and time). The final map suggested the most informative sites for efficient monitoring for climate change. Importantly, these selected sites represented less than 2% of the study area. This example underscores the potentially large increases in sampling efficiency—measured in terms of information return per unit sample—that can be achieved by using a model to guide the sampling design.

1.3.5 Multiphase Sampling and Other Iterative Approaches

Thus far, this discussion has focused on a single sampling episode or, in the case of heuristic designs, an initial sampling campaign followed by a second campaign informed by sensitivities or uncertainties suggested by the first. There are other approaches to *multiphase designs*. One approach, motivated by the sometimes high cost of measuring variables, uses an initial sampling during which a set of easily measured (or inexpensive) variables are collected. A second set of samples is then extracted from the initial set; the subset is typically selected so that they represent the full range of variability in the initial pool but with far fewer samples. At these samples, a larger set of (more expensive) variables is measured. These variables are then cross-walked to the initial variables via regression. In this, the approach leverages the high information content of the second set of samples to the initial set, increasing the information content of the full set.

In long-term monitoring programs, samples can degrade over time simply due to the frequency with which the sample locations are revisited. Imagine, for example, the long-term impacts on plant species composition of repeated episodes of trampling during site visits. Such impacts are possible even when great care is taken to minimize the impacts of field crews. Another problem arises in long-term studies if the initial samples were not an adequate and robust sample of the target resource or if the initial samples were adequate but the focal species disperses over time (or is dispersed by disturbance events or other vectors extrinsic to the target populations). In the latter case of dispersal, long-term monitoring would show a gradual decline in the population even if it were stable (but inhabiting new locations).

A *rotating panel* sampling design helps avoid these problems. Borrowed from work with surveys of human subjects where the issue is burnout by respondents, the aim is to "refresh" the sample by adding new respondents at each iteration or the survey. For example, in the second survey, one might reuse 80% of the initial respondents while adding 20% new respondents. In the next survey, 80% would again be resampled and another 20% new respondents added. In ecological applications, the implication is that most of the initial sampling locations would be revisited while some new samples are added at each iteration. Over time, the sampling frame (i.e., total number of locations eligible for sampling) increases in size while a constant number of samples are surveyed at each iteration (and so sampling costs remain constant). While not used commonly in ecology, this approach would seem to warrant more consideration for ecological monitoring programs.

1.4 Workflow for Sampling Design

Collecting all of this wide-ranging discussion leads us to a general workflow for sampling designs for landscapes. The workflow has two stages (Fig. 1.11). In the first stage, the study goals and objectives determine a set of ecological factors of interest; these will form the basis for sample stratification. In this, each factor will be partitioned into a few discrete levels (e.g., from low to high) and samples will be allocated over these levels. This might also include a geographic stratification if even coverage of the study area is desired. In a geographic information system, the levels of all stratification variables would then be intersected, to yield discrete combinations of all levels of all factors (although some combinations might be rare or absent in the study area). The result of this intersection is a set of *regions* that represent the stratification (Fig. 1.11, right end of the top tier).

In practice, it is convenient if the regions are coded deliberately for ease of interpretation. For example, we might stratify over elevation, a geographic proxy for radiation loading, and local topographic convergence (see Urban 2023, Chap. 1 for a discussion of such proxies). If each factor is partitioned into three levels (low/medium/high), these can be coded 1/2/3. The factors are combined into

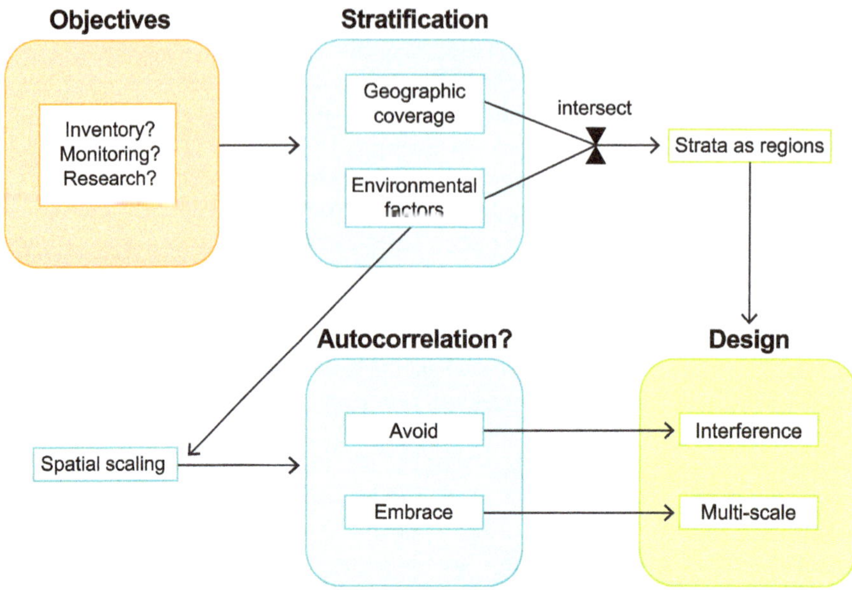

Fig. 1.11 Workflow for sampling designs for landscape applications. The process occurs in two stages. In the first stage (top tier), project goals suggest environmental variables for stratification; this stage might also stratify across geographic space to ensure full coverage of the study area. This stage results in the intersection of strata, as spatial regions in the study area. The second stage (bottom tier) addresses spatial structure—specifically, whether the aim is to avoid or embrace autocorrelation. This decision leads to either an interference or multi-scale sampling design

three-digit codes[2] by multiplying appropriately: 100*x* the elevation code +10*x* the radiation code + the convergence code. Thus, a region coded "113" is low elevation, low radiation, and high convergence (a low-elevation cove); "331" is a sunny high-elevation ridge; and so on. Each combination is readily interpretable ecologically.

The second stage of sampling design must attend to spatial autocorrelation. This means that the scaling of the environmental factors must be determined; this can be done via a digital pilot study using geospatial proxies. The key decision point is whether to avoid autocorrelation or to embrace it (Fig. 1.11, lower tier). In the former case, the sampling design would use a sequential interference model to ensure sample independence, by excluding samples within the range of autocorrelation for the environmental factor(s). In the latter case, one would use a multi-scale design to ensure that there would be samples within distances corresponding to the scale (s) of interest.

We will return to the analytic treatment of spatial data in Chap. 6.

---

[2] In teaching, my colleagues and I refer to these coded regions as *environmental zip-codes*, because they are localized but interpretable regionally.

1.4.1 *Reporting*

Presentation and reporting of a sampling design follows the workflow above. In particular, each of the decisions implied by the workflow should be justified and explained in terms of the program's goals:

☑ The goals and specific objectives of the sampling program
☑ Environmental factors that are being targeted in stratification, and why (if proxies, what they are intended to represent)
☑ Whether the sample units are stratified over geographic space, and how
☑ How the strata are combined or intersected to generate sampling regions
☑ The characteristic spatial scaling of key environmental factors (stratification variables or target resources being sampled)
☑ Whether the design seeks to avoid autocorrelation or embrace it explicitly
☑ The final design: layout, number of samples, and measurement schedule (e.g., repeat visits, etc.)

These details would naturally precede any discussion of exploratory data analysis in preparation for the application, to which we turn in Chap. 3.

1.5 Further Reading

For a more in-depth treatment of sampling designs, there are several excellent sources available that emphasize monitoring designs for natural resources (Duncan and Kalton 1987; Goldsmith 1991; Schreuder et al. 1993; Ringold et al. 1996; Lesser and Kalsbeek 1997; Gitzen et al. 2012; McDiarmid et al. 2012; Loos et al. 2015). McCune and Grace (2002, their Chapter 3) provide guidance on the implications of various sampling decisions for the analysis of community data. Nusser et al. (1998) present a useful illustration of the relationship between sampling design (i.e., data collection) and experimental design (i.e., parameter estimation and inference) for natural resource applications. Urban (2002) discusses sampling designs as applied to landscape-scale studies characterized by fine grain and large extent (some material presented here is largely reworked from that chapter); that overview, in turn, reflects a large-scale research and management program described elsewhere (Urban 2000; Urban et al. 2000, 2002). Fortin et al. (1989) provide a more general discussion of sampling designs and their implications for autocorrelated data, especially data with fine-grained spatial structure over large spatial extent. Dale and Fortin (2014) include a chapter (their chapter 1) on sampling design for landscapes. Bellehumeur and Legendre (1998) emphasize sampling designs for ecological studies concerned with spatial patterns expressed at multiple scales. Plant (2012) provides a thorough discussion of spatial sampling designs (his chapter 5).

1.6 Summary and Prospectus

Sampling designs for landscape-scale applications put a premium on efficiency: collecting maximum information while minimizing logistical costs. Applications illustrate a progression from simple inventory to targeted inventory, to studies designed to test ecological hypotheses explicitly. In this, sampling designs also show a progression from simple stratifications in geographic space, to more nuanced stratifications in a logical or parameter space, which are then mapped into geographic space through GIS-based queries. In some cases, models can be used deliberately to generate efficient and informative sampling designs.

Spatial structure (autocorrelation) in ecological data presents a fundamental challenge in sampling and inferential design. Two choices are available: to embrace space or to avoid the analytic complications of autocorrelation. In either case, the spatial structure of variables of interest must be known in advance. Virtual pilot studies can provide this information.

In this chapter, we have considered how to collect informative data at the landscape scale. Rather than moving directly into exploratory data analysis of such data— as we perhaps should!—we will instead leap directly our first fundamental task in landscape ecology: species distribution modeling. This experience will motivate a deeper appreciation of exploratory data analysis in ecological applications more generally. We turn to those general approaches in the following chapters.

In the next few chapters, we will use landscape-scale data to build species distribution models, make ecological inferences about landscape patterns and processes, prioritize sites for management, detect and forecast ecological trends over time, and, finally, build an inferential framework for integrated assessment of landscape-scale research and management programs.

References

Austin MP (2002) Spatial prediction of species distribution: an interface between ecological theory and statistical modelling. Ecol Model 157:101–118

Austin MP (2007) Species distribution models and ecological theory: a critical assessment and some possible new approaches. Ecol Model 200:1–19

Bellehumeur C, Legendre P (1998) Multiscale sources of variation in ecological variables: modeling spatial dispersion, elaborating sampling designs. Landsc Ecol 13:15–25

Burgess RL, Sharpe DM (eds) (1981) Forest Island dynamics in man-dominated landscapes. Springer-Verlag, New York

Dale MRT, Fortin M-J (2014) Spatial analysis: a guide for ecologists, 2nd edn. Cambridge Univ. Press, Cambridge

Duncan GJ, Kalton G (1987) Issues of design and analysis of surveys across time. Int Stat Rev 55: 97–117

Fahrig L (2003) Effects of habitat fragmentation on biodiversity. Annu Rev Ecol Evol Syst 34:487–515

Fahrig L (2017) Ecological responses to habitat fragmentation *per se*. Annu Rev Ecol Evol Syst 48: 1–23

Fahrig L (2018) Forty years of bias in habitat fragmentation research. In: Kareiva P, Marvier M, Silliman B (eds) Effective conservation science: data not dogma. Oxford Univ. Press, Oxford, pp 32–38

Fahrig L (2019) Habitat fragmentation: a long and tangled tale. Glob Ecol Biogeogr 28:33–41

Fortin M-J, Drapeau P, Legendre P (1989) Spatial autocorrelation and sampling design in plant ecology. Vegetatio 83:209–222

Gitzen RA, Millspaugh JJ, Cooper AB, Licht DS (2012) Design and analysis of long-term ecological monitoring studies. Cambridge Univ. Press, Cambridge, UK

Goldsmith FB (ed) (1991) Monitoring for conservation and ecology. Chapman and Hall, London

Iman RL, Helton JC, Campbell JE (1981) An approach to sensitivity analysis of computer models, part 1. Introduction, input variable selection and preliminary variable assessment. J Qual Technol 13:174–183

Legendre P (1993) Spatial autocorrelation: trouble or new paradigm? Ecology 74:1659–1673

Legendre P, Fortin M-J (1989) Spatial pattern and ecological analysis. Vegetatio 80:107–138

Lesser VM, Kalsbeek WD (1997) A comparison of periodic survey designs employing multi-stage sampling. Environ Ecol Stat 4:117–130

Loos J, Hanspach J, von Wehrden H, Beldean M, Moga CI, Fischer J (2015) Developing robust field survey protocols in landscape ecology: a case study on birds, plants, and butterflies. Biodivers Conserv 24:33–46

McCune B, Grace JB (2002) Analysis of ecological communities. MjM Software Design, Gleneden Beach, Oregon

McDiarmid RW, Foster MCS, Guyer C, Gibbons JW, Chernoff N (eds) (2012) Reptile biodiversity: standard methods for inventory and monitoring. Univ. California Press, Berkeley

McGarigal K, Cushman SA (2002) Comparative evaluation of experimental approaches to the study of habitat fragmentation effects. Ecol Appl 12:335–435

Moran PAP (1950) Notes on continuous stochastic phenomena. Biometrika 37:17–23

Nusser SM, Breidt FJ, Fuller WA (1998) Design and estimation of investigating the dynamics of natural resources. Ecol Appl 8:234–245

Plant RE (2012) Spatial data analysis in ecology and agriculture using R. CRC Press, Boca Raton, FL

Pulliam HR (1988) Sources, sinks, and population regulation. Am Nat 132:652–661

Ringold PL, Alegria J, Czaplewski RL, Mulder BS, Tolle T, Burnett K (1996) Adaptive monitoring design for ecosystem management. Ecol Appl 6:745–747

Schreuder HT, Gregoire TG, Wood GB (1993) Sampling methods for multiresource forest inventory. Wiley, New York

Urban DL (2000) Using model analysis to design monitoring programs for landscape management and impact assessment. Ecol Appl 10:1820–1832

Urban DL (2002) Tactical monitoring of landscapes. In: Liu JL, Taylor WW (eds) Integrating landscape ecology into natural resource management. Cambridge Univ. Press, Cambridge, pp 294–311

Urban DL (2023) Agents and implications of landscape pattern: working models for landscape ecology. Springer-Nature, Cham, Switzerland

Urban DL, Miller C, Stephenson NL, Halpin PN (2000) Forest pattern in Sierran landscapes: the physical template. Landsc Ecol 15:603–620

Urban DL, Goslee S, Pierce KB, Lookingbill TR (2002) Extending community ecology to landscapes. Ecoscience 9:200–212

Chapter 2
Species Distribution Modeling

Abstract Species distribution modeling is perhaps the most fundamental task in natural resource management and conservation practice. This task dates to the earliest days of ecology but has undergone a rapid evolution in the past several years, driven by new statistical techniques and the increasing availability of large-scale species distribution data. Here, we illustrate a workflow for this task, which weaves together ecological principles, data considerations, and statistical models. The statistical modeling itself follows a sequence that flows from exploratory data analysis to model fitting, model evaluation and calibration, model validation, and to applications. The process is first illustrated with a generalized linear model and then extended to alternative tools, including generalized additive models, tree-based models (random forests), and maximum entropy modeling (maxent). The results of species distribution modeling can provide the basis for an initial site prioritization (Chap. 8). But we begin with SDM because it introduces most of the empirical challenges that arise in working with landscape-scale data: ecological data are multivariate (and so redundant), noisy, and spatially structured. We address these challenges in the next few chapters.

2.1 Introduction

The aim of species distribution modeling is to generate a robust model that can describe the observed distribution of a target species in terms of environmental predictors. This task is central to evolutionary biology, wildlife ecology, community ecology, landscape ecology, and biogeography (Elith and Leathwick 2009; Franklin 2010). Applications vary across these disciplines, and the task travels under various disciplinary guises: environmental niche modeling, resource selection functions, habitat classification, and species distribution modeling. In this chapter, we will use the term species distribution modeling (SDM), but we also will spend some time

Supplementary Information The online version contains supplementary material available at https://doi.org/10.1007/978-3-031-72251-6_2.

D. L. Urban, *Landscape Ecology: A Task-Oriented Perspective*,
https://doi.org/10.1007/978-3-031-72251-6_2

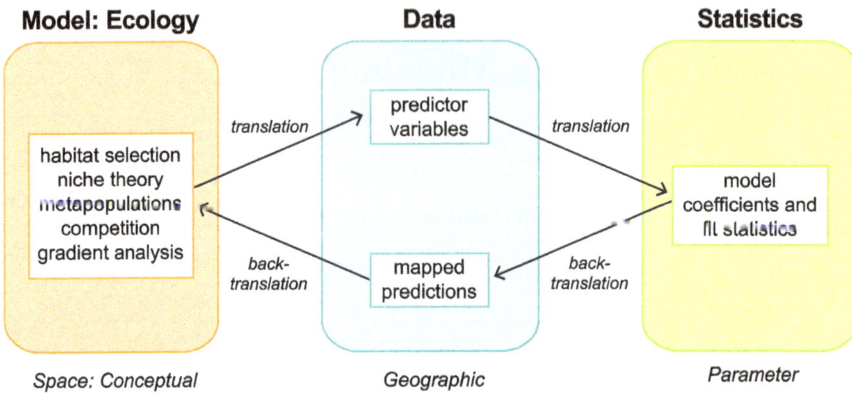

Fig. 2.1 Species distribution modeling as a confluence of an ecological, data, and statistical model. In most applications, this also entails translations between reference spaces

with the evolution of this task from local-scale applications in habitat classification to larger-scale applications in species distribution modeling.

In habitat classification, the focus is often a target species of interest to management, e.g., game species or rare species. In any case, applications typically have the explicit goal of modeling species-habitat relationships so that predictions can be made about cases beyond those observed to build the model. Implicit (and sometimes explicit) in habitat classification is the presumption that the modeled habitat association is real and robust, so that if the availability of habitats in the study area were to change due to natural processes (e.g., succession) or direct management interventions, the focal species would respond accordingly by increasing or decreasing in abundance or distribution. This application arises in wildlife management in terms of habitat suitability or quality, with quality often defined in terms of local vegetation structure or composition: a large part of wildlife management is vegetation management in an effort to improve habitat. At somewhat larger scales, the predictor variables are those that vary over landscapes (e.g., related to terrain and land cover) as well as climate variables. There has been an explosion recently in efforts to model the potential impacts of climate change on species distribution at regional scales. That these two examples are essentially the same task underscores its fundamental importance in ecology.

Austin (2002, 2007) has framed species distribution modeling in terms of three models: ecology, data, and statistics. The ecological model includes natural history or theory that accounts for our expectations about species distribution: which factors are important, the form of species-environment relationships, how we expect species to sort along environmental gradients, and so on. The data model refers to how we collect measurements to capture this ecology: sampling design, what we measure, and how we interpret these measurements. The statistical model formalizes the fit between the first two models and provides for tests of significance and inference. We will adopt this perspective as a convenient framework in which to consider species distribution models (Fig. 2.1).

This framing also invites an explicit consideration of the reference space. The ecological model is *conceptual*, phrased in terms of ideas. The data model is typically *geographic*, using geospatial predictor variables and (often) maps of the model predictions. The modeling itself occurs in *parameter* space: the fitted coefficients as well as other parameters generated in model fitting (explanatory power, goodness of fit, predictive success). Thus, the connections between models entail active *translation* between reference spaces. For example, a map of model predictions might reveal spatial patterns in prediction errors, which might be interpreted in terms of metapopulation theory.

Regardless of what statistical form the species-habitat model might take, there are several general stages in the modeling. This begins with data preparation, proceeds to fitting the model and then evaluating the fit. This typically leads to some recalibration or tuning of the model, a recursive process of refitting and re-evaluating. Ultimately, the model is accepted (tempered with a sense of its robustness) and extended into ecological applications. In this chapter, we begin with the example of one particular model, but these considerations apply to all models and we consider them generically.

It is worth reflecting that this general task has evolved in spurts for many decades. This all began when computers became widely available, so that it was possible to implement statistical versions of iconic ecological ideas such as the Hutchinsonian niche as a multidimensional construct (Hutchinson 1957). While most of the early techniques are no longer in common usage, some of the seminal papers still provide timeless insights (e.g., Green 1971; James 1971). A second wave of developments began in the 1990s, when some new statistical techniques (e.g., generalized linear models) came into vogue. We are now in the midst of another major revolution, in part facilitated by essentially unlimited computational power and a wealth of very large geospatial data sets on the distribution of biodiversity. Franklin (1995, 2023) nicely bookends this period of modeling from the perspective of biogeography. We close this chapter by benchmarking where we are now, how the tools are evolving, and where they might lead. Hopefully, this chapter will prove useful and informative no matter where the field moves from here.

Species distribution modeling is an enormous topic, and there are very many tools available. In this chapter, we focus on the logical workflow, using a single technique as an illustration. A few complementary tools are presented in an Appendix to this chapter, which tools might be substituted readily into the workflow featured here. A digital supplement (S2) includes a broader and more detailed discussion of various tools now available.

2.2 A Confluence of Models

2.2.1 The Ecological Model

A long and venerable tradition in plant community ecology invokes gradient response in explaining plant species distributions (e.g., Whittaker 1967; ter Braak and Prentice 1988; Austin and Smith 1989). In this, plant species responses to environmental gradients are overwhelmingly nonlinear and interactive (i.e., nonadditive). Importantly, competition is presumed to play a fundamental role in governing species response to gradients (e.g., Grime 1977, 1979; Smith and Huston 1989), with better competitors displacing more stress-tolerant species from their optimal habitats.

In animal ecology, a similarly venerable tradition invokes nonlinear or threshold responses in the behavioral ecology of habitat selection (e.g., James 1971). Statistical techniques used in habitat modeling vary considerably in their capacity to accommodate this ecology. Clearly, some appreciation for the essential ecology of a target species should inform the choice of tools used to model its habitat.

2.2.1.1 Plant Species Response to Environment

Austin and Smith (1989) distinguished two fundamental forms of environmental constraints on plant species distribution. *Direct* gradients are constraints with a clear ecophysiological effect on species performance (i.e., establishment, growth, or survival). Examples might include temperature and soil pH. In general, we expect species response to these gradients to be nonlinear and bell-shaped, ranging from "too little" to "too much," with "just right" somewhere in between (Fig. 2.2a).

A different form of direct gradient is a *resource* gradient, in which the resources are consumed by individuals and this consumption alters the levels of resources available to others. Examples include mineral nutrients (e.g., nitrogen), water, and light. In general, we expect species to respond to these resource levels differently than direct abiotic (non-resource) gradients: we expect the relationship to be

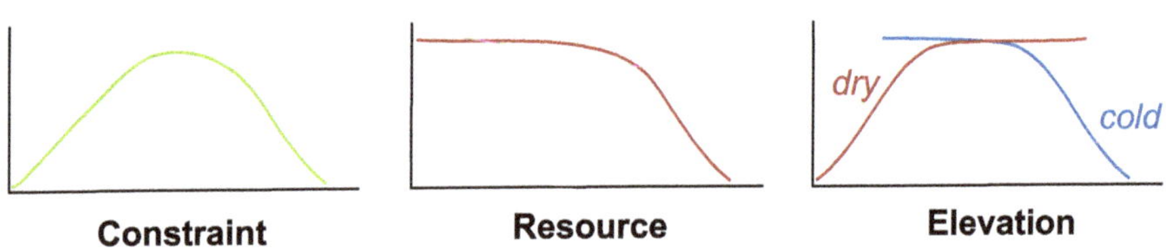

Fig. 2.2 Postulated forms of species response to (**left**) a direct abiotic gradient such as pH, (**middle**) a resource gradient, and (**right**) a complex overlay of the two (after Austin and Smith 1989). In the middle, resource is soil moisture and plant performance decreases from adequate moisture to drought conditions. On the right increasing cold temperatures at high elevations (blue line) are a direct gradient while drought conditions at warmer low elevations act as a resource gradient (red line); the result of this overlay is a curve that resembles a unimodal direct gradient as in (**left**)

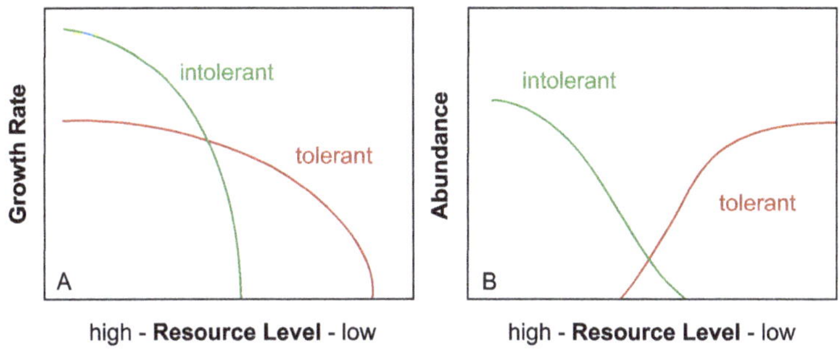

Fig. 2.3 The role of competition in governing species response to environmental gradients. (**a**) Species grown in isolation and (**b**) in competition. (Urban 2023, after Smith and Huston 1989)

monotonic, ranging between "not enough" and "plenty" (Fig. 2.2b). Depending on how the resource level is measured, this curve might be either increasing or decreasing: plant species response to available water would be an increasing curve, while response to lack of water (drought stress) would be decreasing.

Note also that resource gradients can interact with direct abiotic gradients. For example, while soil water availability acts as a resource at low to moderate levels, once the soil becomes oversaturated, it can lead to anaerobic conditions and act as a direct abiotic constraint; thus, species response to "water" would be bell-shaped: an increasing curve reflecting a consumed resource followed by a decreasing curve reflecting the impact of anoxia. It seems likely that temperature response in mid-latitudes is similarly complicated. At low temperatures, there is a direct physiological response to plant performance: the cold itself, as well as a shorter growing season. At high temperatures, the impact is likely manifested indirectly via the effect of temperature on evaporative demand; mid-latitude temperatures are not generally so hot as to have a direct impact on physiology. Thus, the combined temperature effect is an increasing curve from "too cold" overlapping with a decreasing curve toward "too dry" (Urban et al. 2000; Fig. 2.2c).

Species response to direct gradients is often modified by competition, and theory informs our expectations about how competition should modify species response. Smith and Huston (1989) used an individual-based forest simulation model to illustrate how trade-offs in life-history traits lead to shifting patterns of competitive advantage along environmental gradients (water and light). In their model, tolerance to low resource levels comes at the cost of decreased performance (lower growth rate) under high resource levels. Thus, for example, drought-tolerant species are outcompeted on mesic sites by drought-intolerant species, while drought-intolerant species are excluded from xeric sites by more tolerant species.

This means that the observed distribution (gradient position) of tolerant species is displaced from the mesic sites where that species would perform best, to more xeric sites where it performs better than a drought-intolerant species. Similarly, the drought-intolerant species has its distribution truncated on the more xeric end by competition from the more tolerant species (Fig. 2.3, and see Urban 2023, Chapter 2). This means, in the end, that the observed distribution of both species is governed by competition. This general conclusion is well-established in plant community ecology (e.g., Grime 1977, 1979; Tilman 1982).

By contrast to direct gradients, Austin and Smith (1989) recognized *indirect* gradients, in which the factors are presumed to be related to direct gradients but have no such effect themselves on species performance. Familiar examples include elevation or geographic position (e.g., latitude, longitude). In some instances, the relationship between the indirect and direct gradients is reasonably straightforward. For example, in mountainous terrain temperature decreases with increasing elevation. Similarly, continental-scale patterns in temperature and precipitation provide for clear latitudinal or longitudinal gradients (e.g., as nearly orthogonal gradients in the Central Plains of North America because of the north/south-trending Rocky Mountains and consequent rain-shadow effect). Likewise, the moderated maritime climates of coastal zones often can be modeled in terms of simple distance from the coast (e.g., Urban et al. 1993). Urban (2023, Chapter 1) reviews various abiotic variables that can serve as proxies for more physiologically important factors.

It is important to recognize that ecological theory is moot as to the form of species response to indirect gradients, except to the extent that the correlated direct gradients are known. As noted above, elevation can serve as a useful indirect gradient in the mountains, but this unimodal response proxies for an overlapping direct (cold) temperature effect and a resource gradient in available water. Likewise, the role of competition is generally unknown for indirect gradients for the same reason. In application, indirect gradients might provide useful predictions about species distribution. But under conditions of rapid environmental change, especially climate change, such indirect proxies might provide unreliable predictions to the extent that the relationship between the proxy and the actual direct gradient(s) is unknown (Urban et al. 2002; Lookingbill and Urban 2005). For example, the "cold" and "dry" components of elevation might be decoupled to an unknown extent under climate change.

2.2.1.2 Habitat Selection by Animals

The nature of species-environment relationships is reasonably well considered for plants. For animals, the distinction of direct versus indirect variables is not quite as clear. A great deal of research with animals—generally vertebrates, and especially birds—has focused on what sorts of variables are best correlated with species response. In most instances, the critical variables are related to foraging or breeding behavior. For example, Klopfer (1965) used laboratory experiments to explore behavioral cues in habitat selection by birds. In a seminal statistical study of bird habitat relationships, James (1971) tried to capture these relationships explicitly in her concept of the *niche gestalt*: this gestalt is the minimal set of structural features that characterize the breeding territories of a bird species. Similarly, studies of small mammals have emphasized variables related to nesting or foraging cover (e.g., Dueser and Shugart 1979); and studies of reptiles have used variables related to foraging substrates (e.g., Schoener 1968), thermoregulatory microhabitat (e.g., Adolph 1990), and so on.

One theme in this research has explored whether *proximate* or *ultimate* variables are better predictors of species response. An ultimate factor would be one that has a direct, physiological influence on the animal (e.g., actual prey abundance). A proximate factor would be some variable that is more easily measured but highly correlated with the ultimate factor. For birds, this distinction is often phrased in terms of *structural cues* in habitat selection. A structural cue is a physical component of habitat that informs the animal about the levels of some other resource. Smith and Shugart (1987) conducted a field study that asked whether ovenbirds (*Seiurus aurocapilla*) relied more on forest structure than actual food abundance in selecting breeding territories; they found that structural variables were better predictors than actual measures of food abundance (invertebrates in the forest floor litter). This makes sense given the timing of habitat selection: male birds establish territories early in the spring before the critical food resources would be obvious; structural cues are also likely to be less variable over time. More generally, animal microhabitat variables might be sorted into direct and indirect variables to parallel the framework for plants, if we recognize ultimate factors as direct and proximate cues as indirect variables. It seems that many microhabitat variables might be somewhat indirect indicator variables for ultimate factors that are often not easily measured.

By contrast to plant ecology—where competition is played out in real time—the role of competition in structuring animal communities is sometimes more of an evolutionary construct. MacArthur's classic (1958) study of warblers emphasized that their different niches kept the species isolated, effectively avoiding active competition. Connell (1980) referred to this as the "ghost of competition past," again emphasizing that competition over evolutionary timescales has acted to minimize competition on ecological timescales. In times of rapid ecological change or with the introduction of new (especially exotic) species into ecosystems, this presumption is much harder to support. Interactions of other sorts, however, are now being incorporated into some models of animal species distribution. For example, Giannini et al. (2013) presented a case in which they modeled the distribution of bumble bees (*Bombus*) in terms of brood parasites and the host plant that the bees pollinate.

2.2.1.3 Biotic Interactions and Species Distribution Models

Wisz et al. (2013) reviewed the evidence that biotic interactions affect species distributions across a range of scales, including interactions within- and across-trophic levels. They suggest some avenues by which these interactions might be accounted for more directly in species distribution models. Inferences about biotic interactions can be confounded by correlated (and sometimes unobserved) processes or constraints. Dormann et al. (2018) offer guidance on how to interpret the possible influence of biotic interactions on species distributions and models of these.

One emerging approach is *joint species distribution models* (Ovaskainen et al. 2010; Pollock et al. 2014; Clark et al. 2014, 2017; Warton et al. 2015; Tikhonov et al. 2017; Wilkinson et al. 2019). JSDMs model species collectively, to account for patterns of pairwise covariation among species. Such covariation might occur due to joint environmental responses (positive or negative, measured or not), or to actual interactions among species (typically unobserved). The models are constructed by estimating the environmental effects and also fitting the covariances among species on the residuals of the environmental influences. We attend these models in a supplement (S2.5) to this chapter.

2.2.1.4 Scaling Considerations

Much of the classic literature on species distributions was founded in niche theory, and a primary concern was inferring the role of competition in governing species distributions. By contrast, Pulliam (2000) notes that many recent applications in landscape ecology and conservation are couched in metapopulation theory (Urban 2023, Chapter 6). In this, source habitat patches (i.e., rich patches) are expected to subsidize poor-quality sink patches via dispersal (Pulliam 1988). This switch in theory has two important implications for habitat modeling:

1. Metapopulation theory predicts that the realized distribution of a species might be larger than its potential distribution, if individuals populate marginal or unsuitable habitat (sinks) because of dispersal subsidies from productive (source) habitat nearby. This is in contradiction to the Hutchinsonian model, which argues that competition should render the realized niche narrower than the fundamental niche. (This might be subtle to visualize: the metapopulation case requires the translation from a distribution in geographic space into a distribution in an appropriate parameter space, while the Hutchinsonian case is constructed only in parameter space. We delve into translations between spaces below.)
2. On landscapes, where dispersal is explicit of interest, dispersal subsidy and isolation both appear as model failures (misclassifications)—hence, model failures might well be as interesting ecologically as cases where the model "works." In particular, occupancy of "not habitat" might suggest dispersal subsidies, while unoccupied "habitat" patches might indicate isolation effects. Both cases are ecologically compelling.

The shift in underlying theory from niche theory to metapopulations also witnesses another key issue in habitat modeling: a change in scale of analysis. Niche theory, focused on competition, rightfully is addressed at scales at which species can interact. Thus, we began with studies of microhabitat partitioning within a habitat patch measured on the order of hectares (e.g., a forest stand). With the shift to metapopulation theory, the study area shifts accordingly to landscapes, where the process of interest is interactions among habitat patches—which habitat patches might support within-patch studies of microhabitat pattern. Recently, as conservation efforts have expanded to continental or global scales, habitat models are being invoked at much larger scales and typically at correspondingly coarser spatial resolution. This change in scale is often keyed by a change in labeling: as spatial extent increases, habitat classification becomes species distribution modeling, and the ecological context shifts from community ecology to biogeography. Species-environment relationships modeled at multiple scales—microhabitat, landscape, and regional or subcontinental—thus invoke different underlying theory and different ecological processes that govern species response (Johnson 1980; Morris 1987; Peterson 2006). The scale of analysis also has important implications for the data model invoked in the application. But while the ecological and data models might vary with scale, the statistical model is often essentially same, mechanically, at any scale.

2.2.2 *Data Models*

The data model refers to the selection of actual measurements used to fit a habitat classification model. This is informed by the underlying ecology. In some instances, the way that the data link to ecology is rather straightforward. For example, in MacArthur's classic bird studies in the 1960s, there was a direct relationship between foliage height diversity and bird species diversity (MacArthur et al. 1961; see also Wiens 1974; Cody 1981; Robinson and Holmes 1982). This is because different bird species nest and forage at different locations (heights) within a forest, and so the variety of nesting/foraging substrates readily explains bird species diversity.

In field studies of habitat relationships couched in niche theory, the choice of habitat measurements can be guided by a naturalist's appreciation of the life history of the target species. Classic studies of birds and small mammals provide good examples. In many instances, however, the link from ecology to data is less straightforward. This is especially true as we attempt to scale up from local "habitat" to more regional "species distributions."

2.2.2.1 Scaling Data from Microhabitat to Regional Applications

Today species distributions are modeled at increasingly larger spatial scales: landscapes, regions, entire continents, and even global patterns. These changes in scale have important implications for the data model underlying habitat classification.

The Landscape Scale At the landscape scale, species distribution models often rely on geospatial data held in a geographic information system (GIS). In this, two kinds of explanatory variables might be used. One kind is larger-scale versions of the same variables described above; for example, in plant community ecology, landscape-scale studies still need to address the key factors of temperature, soil moisture, and other edaphic factors. The difference is, at the landscape scale, these factors are indexed with geospatial variables derived from digital elevation models (DEMs) and rather limited information on soils (Moore et al. 1991a; Urban et al. 2002; Lookingbill and Urban 2005; Urban 2023, Chapter 1; and see Wilson et al. 2013). For animal studies, geospatial variables related to vegetation structure and composition are rather difficult, as these are often derived from land cover data with rather low information content (e.g., the National Land Cover Dataset, Homer et al. 2004). Newer remote sensing platforms such as lidar (Lefsky et al. 2002; Bergen et al. 2009) provide some hope for indexing vegetation structure more effectively; similarly, hyperspectral imagery with appropriate spatial resolution (e.g., Martin et al. 1998; He et al. 2015) holds some promise for indexing vegetation composition in terms appropriate to landscape-scale habitat models. Data-fusion methods integrating structural and compositional information (e.g., He et al. 2015; Hakkenberg et al. 2018) seem especially promising and are evolving rapidly.

One issue that commonly arises in geospatial approaches to habitat modeling is the treatment of "point" species occurrence data relative to the spatial grain of geospatial data. For example, a species occurrence would typically be recorded as an area-less point, but then would be overlain onto geospatial data extracted from a GIS. For example, the occurrence of a bird species with a 1-ha territory might be matched to land cover data with 30-m cell sizes. This invites some error because of spatial registration (the bird point might intersect the wrong land cover pixel because of locational error). Ecologically, it would make sense to rescale the land cover data to the scale at which the bird actually interacts with land cover—at 1 ha. To do this, one might simply compute the relative land cover (e.g., percent forest) within a 1-ha aggregate around the bird point. This has three effects, all positive: (1) it scales the bird data to the land cover data; (2) it converts a categorical variable (land cover type) to an interval-scale variable with more information content (percent forest instead of forest/not); and (3) it reduces the possible influence of imprecision of spatial registration.

A second set of variables emerges as new at the landscape scale: variables related to dispersal (e.g., various proximity or isolation indices) or the larger spatial context of a site (e.g., patch size, or the amount of developed land cover within some neighborhood around a focal site). These two kinds of variables naturally invite the question of the relative importance of local as compared to landscape-scale variables in explaining species distribution. Such two-scaled studies have been done for small mammals (Martin and McComb 2002; Michel et al. 2007), birds (Cushman and McGarigal 2002; Grand and Cushman 2003; Lawler and Edwards 2006), and reptiles (Moore and Gillingham 2006). Scott et al. (2002) include a number of contributions concerned with the scale of habitat relationships. We return to the issue of distinguishing spatial effects (e.g., dispersal) relative to environmental effects in Chap. 6.

Regional and Larger Scales Conservation groups and the climate-change research community have extended habitat models to ever larger scales. Indeed, these two camps are now merging as conservationists begin to weigh the implications of climate change on biodiversity. At these large scales, the spatial resolution of the analysis tends to be rather coarse (e.g., based on 1-km^2 imagery and DEMs, and often much coarser). Appropriately, the variables invoked in these models tend to emphasize regional variation in climate, such as in so-called "bioclimatic envelope" models (e.g., Busby 1991). These models are at home in the realm of biogeography, just as microhabitat studies belong to community ecology and landscape ecology provides the disciplinary context for studies at that scale.

Interestingly, in studies of plant distribution, the same factors are invoked at all scales—temperature, moisture, fertility—but the actual variables used to index these factors vary considerably across scales (Stephenson 1998). In general, at increasing scales the explanatory variables tend to become more indirect estimates of the target factors. For example, a fine-scale study might measure soil moisture directly; a landscape-scale study would use terrain-based indices of exposure (evaporative demand) and drainage indices; a regional study might use mean growing-season precipitation or a drought index; and a global study might rely on mean annual precipitation. In this, it often is logistical constraints that govern the choice of variables, even though the same underlying ecological model informs these decisions.

By contrast to landscape-scale studies that explicitly compete local versus landscape-scale variables in model evaluation, regional and larger-scale applications do not seem to have embraced a multi-scale modeling approach. This is despite the reality that the larger-scale patterns must have landscape- and finer-scale factors nested within them. In part, this might be logistical: empirical estimates of finer-scale variables are not often available at continental and global extent. This omission confers some uncertainty in model interpretation, as we cannot know if the correlations with coarse-scale factors are direct, or if the species is responding to finer-scale variables embedded within the measured factors.

Forecasting species response to climate change offers a good illustration of this dilemma. One might readily correlate the distribution of a species with regional climate variables. For example, the distribution of the Mexican spotted owl (*Strix occidentalis lucida*) in the southwestern United States might be modeled in terms of temperature. This model would side-step the question of whether owls respond to temperature directly, or if they actually respond to vegetation (itself responding to temperature), or if the owls are responding to prey species (themselves responding to vegetation or to temperature). Partial regression or multilevel modeling might help answer these questions, but only if all of the candidate variables can be measured in a multi-scale sampling design. From this perspective, rapid climate change offers itself as a grand—if poorly controlled—experiment, as it is likely to decouple climate and vegetation (and perhaps competitors or predator/prey relations). In particular, forest response to climate change might be lagged by decades or even centuries (e.g., Urban et al. 1993), while owls (and their prey) would be free to respond to temperature much more quickly. This decoupling among factors might give us the opportunity to "let the owls show us" what they are tracking as "habitat."

This hypothetical example underscores a fundamental issue in species distribution models: however elegant statistically, they are still correlational, and only experimental manipulations of the explanatory factors can reveal which variables are actually governing species response.

2.2.2.2 Exploratory Data Analysis and Variable Selection

In every case, variable selection should be informed by exploratory data analysis (Chap. 3). In particular, species-environment correlations (Chap. 3, Sect. 3.3.3) or pairs plots (Fig. 3.8) can indicate the strength of the relationships as well as whether the relationships will meet the assumptions (e.g., normality, linearity) of many regression-based species distribution models.

Exploratory data analysis can also be quite helpful when selecting among multiple predictor variables that are correlated themselves (see below).

2.2.2.3 Variable Selection Under Redundancy or Collinearity

The reality is that models are often developed from a set of variables that are indirectly related to factors that we believe shape species distributions. For example, we might consider a variety of terrain-based indices of soil moisture. Climate variables are especially vexing from this standpoint, as from standard datasets, we might consider monthly minima, maxima, and means for temperature and precipitation as well as any number of synthetic indices of temperature or moisture—dozens of variables. This means that the sets of variables we use as predictors tend to be correlated and redundant.

Correlations among predictors give rise to two complications in species distribution models. The first is statistical: given strong correlations among the predictors, the model can be degraded statistically so that the estimates are unstable or biased. The second complication is ecological, in the sense that with redundant predictors, we cannot easily gain a clear understanding of which factors define "habitat."

One solution to this is to manage the degree of redundancy among predictors, so that no pairs of variables are strongly correlated. There is no exact threshold for this, but Dormann et al. (2013) suggested an absolute value of 0.70 as a reasonable threshold correlation. As practical guidance, if a pair of predictors is more strongly correlated than $|0.7|$, we would discard one of the pair. In practice, we might choose to retain the variable with the strongest univariate correlation with the response variable (species) or the one for which it is easier to posit an ecological reason for the correlation.

A special case of redundant predictors arises in geospatial applications in which candidate predictors are generated in a GIS based on distances or spatial windows of varying scales. For example, we might posit that a pine-dwelling bird might be associated with the proportion of evergreen forest cover within the neighborhood of a sample point; that neighborhood might be computed for windows of increasing size around the point (e.g., windows of 100, 500, 1000 m radius). These measures would be correlated (because they are nested), so we would retain only the version that is most strongly correlated with the bird's occurrence.

Another solution to correlations among variables is to summarize the pattern of correlations using principal components analysis (PCA). PCA returns a set of synthetic axes, which are mutually independent and thus can serve as nonredundant predictors. We will return to this analysis in Chap. 4.

We return to this issue of correlated predictors when we consider how to interpret the relative importance of predictor variables in the fitted model (Sect. 2.4.2). (This is not as straightforward as we would like.)

2.2.3 *Inferential Design and Statistical Models*

Statistical models of habitat affinities or species-environment relationships have two components: an inferential design and the estimation (fitting) of the model. From our perspective, the estimation component corresponds to the procedural details of the model, and this depends on the model (we will consider a few). The inferential component refers loosely to the assumptions we make about how the data are collected and how these will be interpreted statistically. A key issue in this is how we intend to interpret locations where we observe the species of interest, relative to locations where we did not observe it—either because we looked but did not find it there or because we did not even look (i.e., we only have "presence" data).

2.2.3.1 Inferential Design and Contrasts to "Habitat"

Selecting predictor variables is clearly crucial to the success of habitat models. But how we will use these variables to distinguish "habitat" is also crucial. This falls into the realm of inferential design, and this is a large part of how the data model is linked to the statistical model. Depending on how the actual samples are collected, there are a few experimental or inferential designs in species-habitat analysis (Manly et al. 2002).

To begin, consider that in a perfect (statistical!) world, we would collect data for a habitat model by selecting random (independent) locations in the field; at each location, we would tally the presence or absence (or perhaps, abundance) of the focal species, along with a set of candidate explanatory variables. This is a classic *case/control design*. This approach would meet the assumptions of most statistical analyses including regression. The problem, of course, is that most species are uncommon, so the frequency of "present" (case) sites would be very low relative to "absent" (control) sites. Indeed, for rare species of interest in conservation biology, we might not encounter *any* presences at all if we sampled randomly and sparsely. Clearly, a more efficient design is needed. Jeliazkov et al. (2022) offer some perspective and guidance on sampling and modeling distributions for rare species (which, again, is *most* species). Three alternative designs are common in species distribution models.

"Habitat" Versus "Nonhabitat" Samples The observations in this design consist of two sets of samples. Locations where the species was observed (presences) are deemed "habitat." A set of contrasting samples is measured at points that were censused, and the species was not observed (absences); these samples are deemed "nonhabitat." This is a case/control design as described above, although in practice the sample locations might be stratified somehow rather than simply random (recall Chap. 1).

Clearly, there could be several reasons why a species might not be observed in a particular location at the time of sampling: it might be present but not observed (cryptic or quiet); it might be in residence but not at that precise location at that the time censuses were conducted; the location might be suitable habitat and yet unoccupied for biogeographic or other reasons; and so on. We return to these possibilities later. For now, it will be sufficient to note that these two groups provide maximum contrast for a model of species-habitat relationships.

"Habitat" Versus "Available Habitat" In this, as in the case above, "habitat" samples are locations where the focal species has been observed (presences). The second group consists of a (typically) random sample of the study area, providing a sample of the habitat types *available* to the species (i.e., where the species *might* have been observed). These sites are often termed *pseudo-absences* because they are used in model fitting in the same way as observed absences. In this approach, the second group will be interpreted as sites representing the range of options provided by the study area, i.e., the various settings to which the species has reasonable access. In this, the null hypothesis is that the species uses habitats in the same proportion or frequency that they are encountered; it exhibits no habitat preferences.

Either the *habitat/nonhabitat* or the *habitat/available habitat* model can provide a useful distinction of what constitutes "habitat" for the focal species. One key difference will be in how model errors (misclassifications) are interpreted, which we will attend later. Both approaches are *discriminatory* in that they seek to distinguish "habitat" samples from another group of samples. The "habitat/available habitat" approach is very common currently, as data on many species of conservation concern are collated as presence points only.

"Presence-Only" Models A third design is based only on the characteristics of the "habitat" samples (presences), making no assumptions about sites where the species was not observed (or was not censused). This approach is *generative* in that the model attempts to describe the "habitat" samples without reference to another group of samples. A simple example might be an "envelope" model by which we define habitat based on the minimum and maximum values observed, over the presence, for key explanatory variables. For the owl example above, that might be expressed as a *climate envelope*: the owl occurs on sites between $T_{\min}$ and $T_{\max}$ degrees in mean annual temperature and between $P_{\min}$ and $P_{\max}$ in total annual precipitation.

Note that it is hard to interpret those ranges without reference to the range in temperature and precipitation over the larger study area! In such applications, it is common to present the presence-only solution within the context of a (typically) random sample of the study area—a sample of available habitats. But in this case, the pseudo-absences are purely context; they do *not* influence the estimation of the model.

It might be noted here that many discriminatory models that rely on pseudo-absences are sometimes presented as "presence-only" models. They are not, though it is true that they only have actual observations on presences. Models that use pseudo-absences have solutions that depend on which pseudo-absences are used (different sets of pseudo-absences generate different fitted models); presence-only models depend only on the presences.

Presence-only models are appealing because they neatly side-step the ambiguities associated with "absences" in a conventional design. These models do, however, sacrifice some statistical leverage by not using the contrast available from a second group.

2.2.3.2 Inferential Issues

It might be useful to consider these alternative approaches in terms of a simple statistical test, the *t*-test. In this, consider a habitat model that consists of a single explanatory variable. In the presence/absence case, the model is a two-sample *t*-test, and the question is whether the two subsamples (presences, absences) have different means given their variances. The presence/pseudo-absence case is equivalent to a one-sample *t*-test, where the question is whether the presences differ (have a different mean, given their variances) from a random draw from the pooled sample. In the presence-only model, there is no inference to be made—the model is purely descriptive (i.e., it simply computes the sample mean and/or variance).

Because of the increasing availability of species occurrence data provided as "presence points," models that contrast presences with pseudo-absences are increasingly popular. This invites careful attention to what the pseudo-absences represent. It is worth emphasizing here that the presence/pseudo-absence model is actually *two* models: one that contrasts the presences with the pseudo-absences and a second (implicit) model that asserts that the random pseudo-absences are an appropriate representation of the habitats available within the study area. If the latter is not true, then the model itself might be misleading even if it is highly significant statistically. Barbet-Massin et al. (2012) provide some guidance on the selection of pseudo-absences, but this is still an active area of methodological research (and see Elith 2019).

Selection of pseudo-absences can provide a substantial filter on the ecological implications and interpretation of a model. For example, return to a habitat model for the Mexican spotted owl in the southwestern United States. The bird occurs in mixed-conifer forests, which themselves occur at mid- to higher elevations in the mountains (we will ignore realistic complications to this over-simplification for now). Now consider three approaches to choosing pseudo-absences:

1. If we were to sample the landscape randomly, we might include in the pseudo-absences a large number of samples in desert or semiarid grassland, as well as (perhaps) some cities and open water. A habitat model constructed from these data would show us that owls live in *forests*, and it would do so with high accuracy because the habitat contrast is so flagrant.
2. If we were to sample pseudo-absences by masking out the land covers that are clearly not habitat (desert, agriculture, semiarid grasslands), we would generate a more nuanced model that might tell us what *kinds* of forests or woody vegetation types owls prefer (e.g., mixed conifers more than pines or pinyon-juniper).
3. If we were to sample pseudo-absences even more narrowly within forests, we might learn *what kinds of mixed-conifer forests* the owls prefer. In this series of three models, the statistical significance and accuracy would likely decrease at each step while the ecological nuance would increase: a trade-off. Clearly, we should pay attention to how we sample pseudo-absences!

2.3 Fitting the Statistical Model

There is a large and growing collection of statistical tools for modeling species-environment relationships, and we need not review all of these here (but see Johnson and Gillingham 2005; Elith et al. 2006; Tsoar et al. 2007; Elith and Leathwick 2009; Fletcher and Fortin 2018; Norberg et al. 2019). Instead, the general approach for model fitting and evaluation is outlined using a relatively simple model, a logistic regression (a form of generalized linear model, GLM). We begin with the logistic regression because it is well-established in practice and reasonably intuitive. We will develop the workflow using the GLM—from model estimate, to model interpretation, to model tuning and evaluation. Much of what we consider here with GLMs applies similarly to any other model.

2.3.1 Modeling Workflow

The workflow for developing a statistical species distribution model unfolds in stages that begin with the adoption of a data model, iterates through a process of model fitting and evaluation, and then extends into model applications that return to the data and ecological reference spaces (Fig. 2.4).

While some details of this workflow depend on the statistical model, the general approach does not. Modeling comprises an initial stage of iterative model-fitting, evaluation, and calibration. This is typically focused on a single data set, and the evaluation is relative to the data used to develop the model. This is not an independent test of the model, and we will use the term model *verification* to refer to these

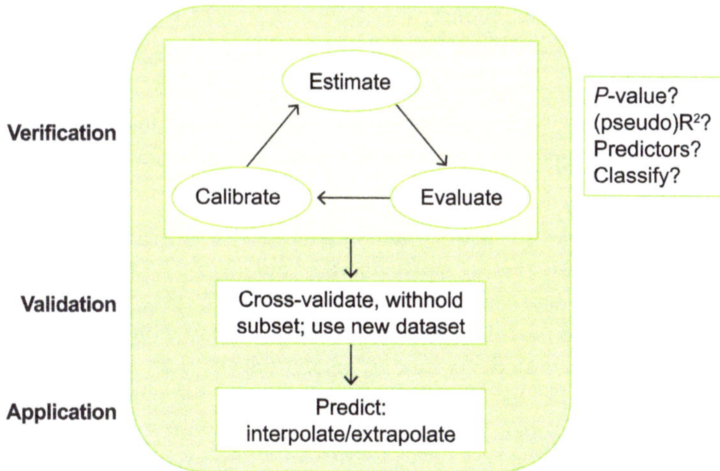

Fig. 2.4 Workflow for fitting and evaluating a species distribution model. The workflow enters from the data model and in applications exits to the data and ecological models (Fig. 2.1)

dependent tests. The model might also be tested using data *not* used to fit the model. This is an independent test of the model, a stronger test, and we will use the term *validation* to refer to such independent tests (see Chap. 9 for more on model testing). Finally, once a model is verified (and perhaps calibrated) and validated, it can be used in practical applications.

Verification Initial model fitting begins with estimating the model and then evaluating the fit. Often, there can be considerable latitude in this evaluation and ample opportunity to refit the model after making adjustments based on the initial evaluation. This fitting, evaluation, and refitting iterates until a satisfactory model is selected. In evaluating a statistical model, we will attend a few aspects of the fit:

1. Does the technique provide a test of significance for the model? That is, are the observed differences *real* with reasonable statistical confidence? While not all models provide a *P*-value, many do and this is useful (but not always crucial) to model evaluation.
2. Does the technique provide an *ecological interpretation* of "habitat"? That is, which of the predictor variables are most important in distinguishing "habitat"? Do these relationships make sense ecologically?
3. How *well* does the model work? How much of the variation does it explain? In a simple regression, this would be captured in the regression R^2 statistic, and there are similar indices for many models.

Beyond its R^2 (or equivalent), does the model provide a means to *predict* the likelihood that a new, unobserved sample is "habitat"? That is, can the model be used to classify new samples? In model verification, we will often want to know how many of the cases used to *train* the model were also classified correctly. For landscape-scale or regional studies, especially, we might wish to map these predictions over the entire study area (and see below).

Validation Model calibration with the data used to build the model (i.e., training data) is a necessary but not sufficient test of the model. A stronger test is to evaluate the model against independent observations, a model validation. There are several ways to do this.

1. When data are sparse, one way to validate a model is to use subsets of the training data to test the model. This is *cross-validation*, commonly *k*-fold cross-validation. In this, the training data are partitioned randomly into *k* subsets (folds). In tenfold cross-validation, the model is fit 10 times; each fit is based on 9 of the 10 subsets of the data, and each model is used to predict (classify) the observations in the withheld subset. After ten models are fitted, each of the observations has been classified using a model that was independent of the test data, providing an aggregate validation of the model but using the same training data set. (In many instances, the ten models themselves can be averaged into a single model, an ensemble model; see below.)
2. Another approach to validation is to withhold a proportion of the training data for testing. For example, we might set aside 25% of the training data and use it later to validate the model. This is expensive, of course, in terms of data and so for smaller data sets practitioners often resort to cross-validation to use the limited data more efficiently.
3. The highest standard for model validation is to test the fitted model using a completely independent data set—from elsewhere in the study area, a different study area, or a different time. In practice, the farther removed from the training data, the stronger the validation.

Validation increases our confidence in a model and helps us temper our interpretation of applications of the model beyond the domain of its initial development and fitting. We return to this theme, more generally in Chap. 9.

Applications Once verified and validated, a model can be used in applications that extend it in space or time. Making predictions with the model by mapping these into a GIS is one common application. If this is a mapping into the original study area, filling in unsampled locations, this is an *interpolation* of the model. Extending the model into a new region or study area is an *extrapolation* in space. Projecting the model into the future, a model *forecast*, requires an explicitly stated scenario about the assumptions supporting the application.

In the following sections, we work through the flow of fitting, evaluating, and calibrating a GLM. In appendices to this chapter, we consider a sampling of alternatives to the logistic regression illustrated here. In this, we focus on three complementary approaches that reflect the diversity of approaches currently popular. These include generalized linear models (GLMs) and their extension to generalized additive models (GAMs; Appendix A.1.1); classification and regression tree (CART) models, and especially extensions to these including random forests (A.2.2), and maximum entropy modeling as available in the software package *maxent* (A.3.2). We also consider the recent advent of *joint* species distribution models, which estimate habitat relationships for multiple co-occurring species simultaneously (Supplement S2.5). Again, most of the issues illustrated here using the logistic GLM can also be applied to these other models; they are substitutable or complementary workflows.

2.3.2 The Generalized Linear Model

Generalized linear models (GLMs) are extensions of the familiar linear model (i.e., linear regression) in two ways. First, GLMs admit a transformation from linearity, specifically via a *link* function that converts the linear prediction to a nonlinear form. Second, GLMs admit alternative distributions for the response variable. Recall that, in a linear model, the response variable (and hence the model error) is presumed normal (Gaussian). GLMs include forms with Poisson distributions suitable for count data (such as provided by field censuses) or data tallied as frequencies by category and binomial distributions for a binary response (such as in the case of habitat classification). The common distributional *families* of GLMs are associated with common link functions as well: the Poisson model with a log link, and the binomial model with a logistic link, and so on. Of course, one familiar version of GLM is the Gaussian family with an identity link: the linear model itself.

Logistic regression is a form of generalized linear model in which the response variable is binary. In the cases we will consider, this categorical response is group membership. The groups might be "presences" versus either known "absences" or "pseudo-absences." While these are ecologically different, statistically they are exactly the same model.

For several years, logistic regression was the model of choice for habitat classification (Guisan and Zimmerman 2000; Guisan et al. 2002). The logistic model is attractive in this instance for several reasons: it is rather accepting of ecological data and can use mix categorical and continuous predictor variables. The model predicts the likelihood of membership in the target group, that is, the probability, on [0,1], that a sample is "habitat." This probability is often interpreted as an index of habitat suitability.

To predict the likelihood of membership in group k, the logistic regression takes the form:

$$P(k) = \frac{e^u}{1 + e^u} \tag{2.1}$$

where u is itself a linear model on the predictor variables:

$$u = b_0 + b_1 x_1 + b_2 x_2 + b_3 x_3 + \cdots + b_p x_p + \varepsilon \tag{2.2}$$

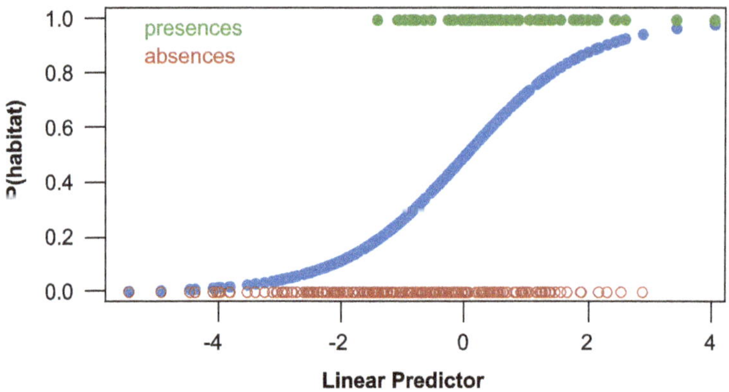

Fig. 2.5 An illustration of a GLM as a logistic regression. The data are presences (green) and absences (red), and the (blue) link function predicts the likelihood that a sample is "habitat"

for p predictor variables, where the b's are coefficients to be estimated. Algebraically, this can be written:

$$\ln\left(\frac{P(k)}{1-P(k)}\right)=u \tag{2.3}$$

where the left-hand side—the *logit*—is the log of the *odds* or *log-likelihood* ratio, that is, the probability that a sample is in group k (i.e., "habitat") compared to the probability that it is not.

In the equations above, the binomial distribution is not evident but it accounts for the reality that the data are 0's and 1's. The link function (Eq. 2.1) takes the linear model (u), which varies from minus to plus infinity, and maps it onto the domain of the data, as a sigmoidal curve ranging between 0 and 1 (Fig. 2.5).

GLMs are estimated using maximum likelihood methods, the goal of which is to find the best possible estimate of the parameters of the model (the most likely solution). In the case of the linear model, this estimator is well-known and is provided by the least-squares solution, which itself has an analytic solution. In more complicated models, the maximum likelihood estimate is not known in advance, and so the solution is estimated using an iterative approximation (iterative, weighted least squares). (Thus, the linear version of a GLM is an alternative way to estimate a model that actually has an exact analytic solution.) The form of the logistic equation lends itself to estimates of model variance in a generalized form, termed *deviance*. Deviance is equal to twice the log-likelihood ratio. This term is distributed approximately as X^2, providing the basis for tests of significance of the model.

2.3.2.1 Model Evaluation: Preview

In the following sections, we turn to various aspects of model evaluation. Again, this illustration uses the logistic regression. But these considerations can apply any model that returns an estimate of habitat suitability (i.e., on [0,1]) and an estimate of the contribution of each predictor variable. Details that depend on the idiosyncrasies of particular models are described in the appendices or supplement to this chapter.

2.4 Model Evaluation

Thus far, we have considered the ecological, data, and statistical models used in habitat classification and species distribution modeling. Now we turn to evaluating the statistical model we have generated. This evaluation will focus on several aspects of model performance. Not all of these aspects will be important for every application, but these are probably the most common considerations.

2.4.1 Model Significance and Explanatory Power

Not all habitat models provide a strict test of overall significance. In general, models that contrast two groups ("presence" versus "absence", or "presence" versus "pseudo-absence") do provide a ready test: the test essentially is that the two groups have different means on the predictor variables (this test itself being contingent on the groups' variances). The specific test depends on the details of the model. With a logistic GLM, the linear model is tested as a regression (i.e., *t*-tests for the contributions of each predictor variable in the linear model), while the overall test via the link function is tested using X^2 (converting the log-likelihood ratio into deviance).

The test of strict significance (i.e., $P < 0.05$ or whatever arbitrary level) is sometimes not very satisfying: many models might be significant without being particularly informative or useful. Partly because of this, there is a growing interest in model selection based on comparisons among alternative models, all of which might be statistically significant. For example, in assessments framed in terms of Akaike's Information Criterion (AIC) and its variants, the explanatory power of alternative models is evaluated by weighing the additional explanatory power of more complicated models against the cost of estimating parameters for the added terms (Burnham and Anderson 2002; Burnham et al. 2011). In this approach, model selection often amounts to choosing among several models, all of which are probably statistically significant, and the result is the most parsimonious model.

Note that models that are truly presence-only cannot provide a simple test of overall significance. There are also a few discriminatory models that do not provide a test of significance. For such models, the significance test is simply skipped and evaluation turns to questions of whether the model is interpretable and how well it works.

The relative explanatory power of a regression model is typically reported as it R^2 value, the proportion of the variance it captures. For a GLM and other models, the equivalent index is a pseudo-R^2 value, which is the proportion of deviance the model explains.

2.4.2 Interpreting the Importance of Predictor Variables

While significance and predictive accuracy are interesting from a statistical perspective, ecologists often are interested in *interpreting* a classification model. That is, which variables are most important in distinguishing "habitat" from other samples?

In a regression, we interpret the coefficients to predict the incremental change in the dependent variable given a unit change in the predictor variable. In the case of the logistic GLM, the interpretation of coefficients is complicated by the fact that the dependent variable is the log odds ratio (log likelihood). To interpret the coefficients in terms of their effect on the dependent variable, we exponentiate the coefficient to see the expected change in the odds given a unit change in the predictor.

In regression-based models, we often would like to infer the relative importance of the predictors by comparing their coefficients. This interpretation can be confounded by the measurement units of the variables. For example, the raw coefficients for elevation (ranging over hundreds or even thousands of meters) cannot easily be compared directly to the coefficients for pH (typically ranging between 5 and 7). A common way to make these comparisons more straightforward is to standardize the predictors to z-scores before fitting the model. The trade-off in this is that standardized coefficients cannot be compared across data sets (other studies), because the standardization is particular to a data set. Thus, raw coefficients can be compared across data sets (studies) but not within the same study, while standardized coefficients can be compared within a study but not across studies. In cases where we might want to compare coefficients both within and across studies, we would need to compute the model with both raw and standardized variables.

Regression-based models are also confounded by the fact that model coefficients (hence, the apparent importance of variables) depend on the *order* in which they are entered into the model. For sets of predictors that are themselves correlated, this is a vexing problem for which there is no simple and elegant solution (Fig. 2.6).

Stepwise solutions to model construction can help fit parsimonious models. In a *forward* stepwise solution, the first variable entered into the model is the single best predictor; the next variable is the one that most improves the model. This second variable is typically not strongly correlated with the first, and so this procedure minimizes the complications illustrated in Fig. 2.6. *Backward* selection follows the same logic but removes redundant variables. A *full* stepwise procedure begins with forward selection but can proceed either forward or backward once there are several

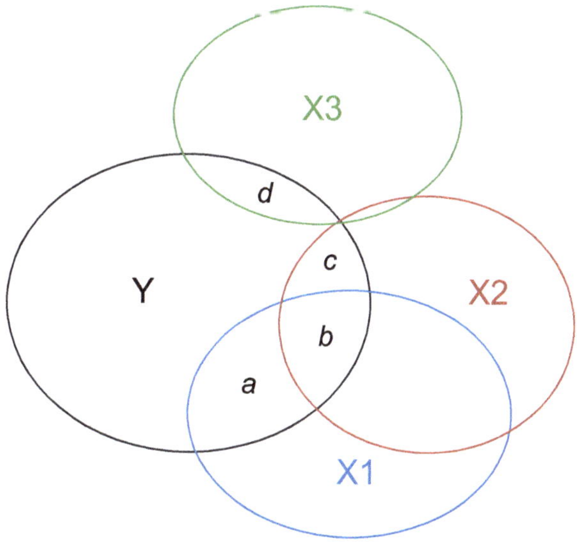

Fig. 2.6 Venn diagram illustrating partial explanatory power of predictors X1, X2, and X3 in a regression of Y. Size of ellipse indicates total variance in a variable, and overlap shows covariance between pairs of variables. If X1 is entered into a regression before X2, X1's explanatory power is $(a + b)$ and X2's is c. If the order is reversed, X2's is $(b + c)$ and X1's is a. If X1 is entered first, X3 is a better second term because its novel contribution d is larger than X2's nonredundant c

variables in the model; the solution converges on the most parsimonious model. As appealing as stepwise solutions might seem, they can be difficult to interpret because the final set of variables depends very much on the correlations among predictors; in some cases, strong correlations among variables can make the solution unstable.

In other models, there is no simple way to gauge the importance of individual variables, and so special methods have been devised for this purpose. For example, in the program *maxent* (Phillips et al. 2006; Appendix A.3), a jack-knifing procedure is provided in which each variable is evaluated in turn during model construction. The result is a summary in which the importance of each variable is indexed in two ways: (1) the explanatory power of the model when *only* that variable is included in the model and (2) the explanatory power when every variable *except* that one is included. This summary indicates the importance of each variable while also reflecting the degree to which the variable is redundant due to correlations with other variables. A variable that has a strong relationship with the species will be important by itself. A variable with a strong relationship to the species but which is also highly correlated with other strong predictors will have high importance by itself but low importance on withholding, as its effect will be compensated by other (redundant) predictors. A variable with a strong relationship to the species and which is not correlated with other predictors will show high importance by itself and high importance on withholding. In this, the paired models (by itself/withheld) reveal what we would like to know about variable importance. In principle, there is no reason why this jack-knifing approach could not be applied to any species distribution model. With random forests, a similar approach has been developed to estimate variable importance (Appendix A.2.2).

2.4.3 Predictive Modeling and Classification Success

Whatever the statistical model used to classify habitat, the result can be reduced to a comparison of the actual values from the data and the values predicted by the model. For a CART model (Appendix A.2.1), this table emerges directly from the binary classification. For models that predict likelihood or other continuous scalars (GLMs, GAMs, random forests, maxent), the prediction must be thresholded to a binary prediction. In this, the continuous prediction is set to 0 (not habitat) or 1 (habitat) relative to a user-selected threshold value (Liu et al. 2005; Jiménez-Valverde and Lobo 2007, and see below). The *confusion matrix* summarizes the model's classification success in terms of matches and mismatches between model and data (Table 2.1).

Here, the diagonal elements (a and d) are classification successes, as model predictions match the data. The off-diagonals (b and c) are misclassifications. The model successes are unambiguous. But at this point, the choice of inferential design influences the interpretation of model misclassifications. Importantly, the ecological interpretation of misclassifications often can be immensely informative about the model and also, more generally, about species-habitat relationships for the focal species.

In a model that contrasts presences and absences, the off-diagonals are unambiguous and can be interpreted at face value. Case c corresponds to the species being observed on a sample that has been classified as "not habitat." While clearly a model error, this case also is perfectly consistent with source/sink models of metapopulations (sensu Pulliam 1988, 2000) and with metapopulation-based expectations of species occurring in nonhabitat or marginal habitat patches adjacent to actual habitat. Case b corresponds to instances of the species *not* occurring on samples that are predicted to be habitat. There are several (nonexclusive) reasons why this might be the case. It is possible that the species *was* there but it was not observed during sampling (e.g., because the species is cryptic). If the species is quite rare, suitable habitat might be unoccupied simply *because* the species is rare. Alternatively, there might be biotic interactions such as competition or predation that exclude the species from some otherwise suitable habitats. Finally, and consistent with metapopulation theory, suitable habitat might be unoccupied because it is isolated and unreachable. In short, while correct model predictions are satisfying and reassuring, we often learn much more about the ecology of the system by examining model misclassifications. In these cases, we might analyze the misclassifications in terms of other candidate explanations: proximity to other (occupied) habitat, patch size or geometry, and so on (and see below).

Table 2.1 Confusion matrix for a classification model

	Data	
Model	*Habitat*	*Not*
Habitat	a	b
Not	c	d

In a model that contrasts presences and pseudo-absences, the interpretation of misclassifications is not quite as straightforward. In particular, misclassifications of pseudo-absences are to be expected, at a rate that corresponds to the proportion of the study area that qualifies as habitat, termed *prevalence* (strictly, the proportion of the study area occupied by the species). That is, if 20% of the landscape is "habitat," then a random sample of the study area should itself be 20% habitat; a perfectly accurate model would quite properly result in a 20% misclassification rate for pseudo-absences samples predicted to be "habitat" (case c).

In either model, the actual locations where these types of misclassifications occur in the study area can be immensely informative, and so mapping these predictions into the study area is a useful method for interpreting the model (and see below).

2.4.3.1 Measures of Model Accuracy

The confusion matrix provides an intuitive approach to quantifying model performance in terms of classification accuracy. For example, note that the diagonal elements a and d tally the number of cases where the model prediction matches the data. So, one easy index of model accuracy is simply the proportion of cases where that is true:

$$A = \frac{(a + d)}{(a + b + c + d)} \tag{2.4}$$

This is numerically correct, but not very satisfying for many applications. Consider, for example, the case of a rare species that occurs in just 5% of the study area (in a random sample). We can create a model that works with 95% accuracy by simply predicting that the species never occurs anywhere: nearly perfect accuracy from a model that predicts nothing!

An alternative approach indexes accuracy in terms of model successes relative to what one might expect by chance (i.e., by guessing). A common method is the Kappa statistic (Fielding and Bell 1997). Other indices of model accuracy can be derived from the confusion matrix; Fielding and Bell (1997) review a dozen or so and make several recommendations for applications. Pearson (2007) offers further suggestions on significance testing.

2.4.3.2 Receiver Operating Characteristics (ROC) Curves

In a presence/absence model, the off-diagonal elements of the confusion matrix can have different ecological interpretations. Clearly, in an optimal habitat classification we would like to minimize both types of error. Yet in ecological applications, these two errors might have qualitatively different costs or risks. For example, in the case of a rare and threatened species, we might be interested in classifying "habitat" for a variety of reasons. At one extreme, we might wish to establish a reserve for the species. In this case, we would naturally want to identify the very best habitat available. To this end, we would want to ensure that the model is not overly tainted with "nonhabitat" mistakenly included with the best (true) habitat. Alternatively, we might be interested in monitoring or perhaps restricting management activities in habitats that might possibly represent potential habitat. In this case, including some nonhabitat with the true habitat is of rather less concern, because we would want to err conservatively. This suggests that we might want a method for "tuning" a habitat classification model toward the aims of a particular application.

Receiver operating characteristics (ROC) curves provide for this tuning. ROC curves derive from signal theory, and the analogy to tuning a receiver such as a radio is both intuitive and correct. What we require of a useful receiver is the ability to capture the signal (frequency, or broadcast station) of interest while also rejecting the signal (actually, "noise") of stations at nearby frequencies.

ROC curves are based on the confusion matrix. To begin, consider some definitions based on the elements of the table (notation follows Pearce and Ferrier 2000):

1. *Sensitivity* (*Se*) = number of positives ("habitat") samples predicted correctly, divided by the total number of actual positives; "true positives" = $a/(a + c)$.
2. *Specificity* (*Sp*) = number of negative ("nonhabitat") samples predicted correctly, divided by the total number of actual negatives; "true negatives" = $d/(b + d)$.
3. *False positive fraction* = number of false positives (absent from predicted "habitat"), divided by the total number of negative samples, = $b/(b + d)$.
4. *False negative fraction* = number of false negatives (present on predicted "nonhabitat"), divided by the total number of positive samples, = $c/(a + c)$.

One way to look at this classification is to consider the result of applying a threshold probability to collapse a continuous model prediction from a logistic regression into a binary classification (Fig. 2.7). In this, each set of samples (habitat and nonhabitat) occupies a range of conditions along the probability scale, probably with some overlap. Choosing a threshold at any point along this axis dictates the balance of misclassifications.

For example, in the extreme case we could threshold the prediction probability at 0.0, in which case *every* sample would be classified as "habitat." This would be silly, of course—but at the same time it would eliminate false negatives. At the other, equally silly, extreme, we could threshold at $P = 1.0$, in which case *nothing* would be classified as "habitat" and we would eliminate false positives. Somewhere in between is a probability threshold that can "tune" classification to a balance of true and false predictions for any given application.

A graph of this balance of misclassifications as a function of varying the threshold probability is a receiver operating characteristics (ROC) curve (Fig. 2.8). In this, the false positive rate (equal to 1 minus the model's specificity, *Sp*) is plotted on the abscissa while the true positive rate (sensitivity, *Se*) is on the ordinate. Note that because the row and column totals of the confusion matrix are fixed, we can capture all of its elements with only two terms. The choice of true positives and false positives, arranged as in Fig. 2.8, is an arbitrary convention but yields a graph that

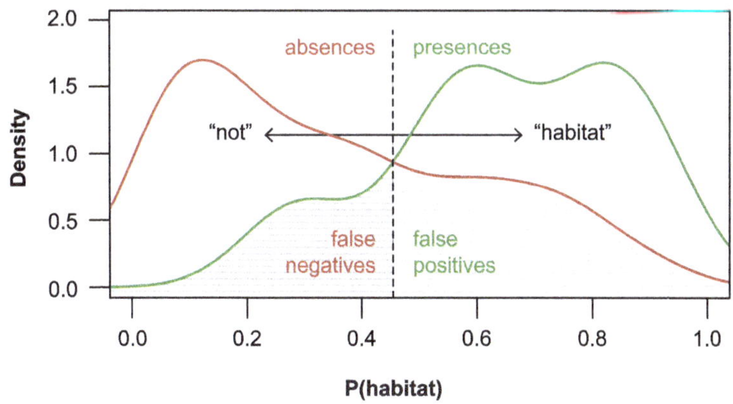

Fig. 2.7 Frequency distributions of probabilities predicted by a habitat model. In green is the "habitat" sample (presences), while the "nonhabitat" (absences) sample distribution is in red. The threshold (vertical bar) dictates the probability above which a sample is classified as "habitat". Green cases to the left of the cut-off are false negatives, while red cases to the right of the cut-off are false positives. Sliding this threshold one way or the other controls the rate of false positives and false negatives

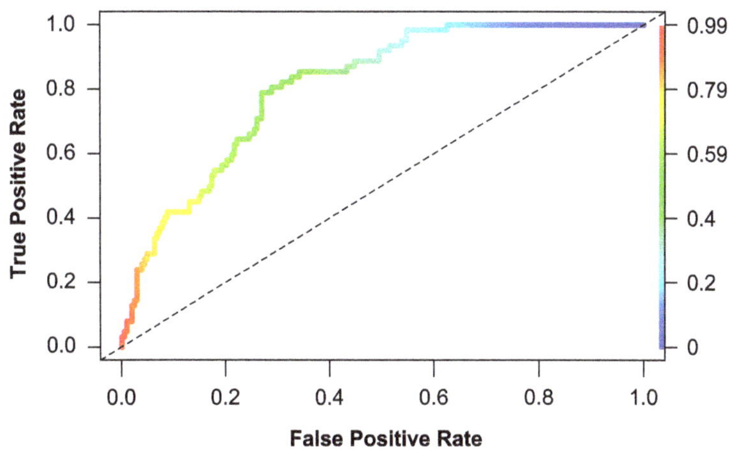

Fig. 2.8 An example of an ROC curve, as error rate for true positives (ordinate) relative to false positives (abscissa). At the origin, the model classifies everything as "nonhabitat" and so misclassifies no actual nonhabitat. At the upper right extreme, the model classifies everything as "habitat" and so misclassifies none of the presences but all of the absences. The diagonal is equivalent to guessing. (ROC curve drawn using the ROCR package in R, Sing et al. 2005)

varies monotonically with the decision threshold. Also note that while the curve is often drawn as a smooth line, this actually represents a summary of a large number of confusion matrices, one for each decision threshold. Thus, if we computed sensitivity and specificity for decision threshold ranging from 0.1 to 0.9 in increments of 0.1, there would be only nine points connected for the curve (11, if 0.00 and 1.0 are included). In practice, curves might be drawn from thresholds evaluated from 0.01 to 0.99 (i.e., at intervals of 0.01), providing more resolution. The diagonal line in Fig. 2.8 represents the expectation under random assignment (guessing).

Given this framework, an index of classification success can be derived from the ROC curve itself. The area under the curve (AUC) summarizes the overall accuracy of the model and is independent of sample sizes or prior probabilities of each group. Normalized by the total area of the frame (the bounding box), AUC ranges from 0.5 (= guessing; a truly inept model could yield an AUC less than 0.5) to 1.0 (perfect discrimination). It is intuitive to note that the best compromise, the optimal tuning of the model, is at the threshold probability that is at the tangent to the ROC curve that is farthest from the diagonal reference line (the "guessing" line in Fig. 2.8). For a presence/absence model this tuning maximizes the total proportion of correct predictions.

In practice, the optimal tuning of a model using ROC methods depends on two criteria. First, the relative abundance of "habitat" as compared to "nonhabitat" will influence the optimal threshold. Clearly, if "habitat" is extremely rare then the tendency should be to classify a novel sample as "nonhabitat" unless the counter-evidence is strong. Thus, prior probabilities are important to this decision. Second, optimal classification might depend on the relative cost or penalty of each type of misclassification. For example, in the case of a rare and threatened species, the cost of missing an actual habitat sample might outweigh the cost of misclassifying a "nonhabitat" sample. Note that if the prior probabilities (sample sizes) and relative costs of misclassification are equal for the two groups, the optimal tuning is at a threshold of $P = 0.50$. Note also that this probably will *not* be the correct optimal tuning for an actual model, as the priors and relative costs will almost never be equal.

While ROC curves are now routinely used to *optimize* classification models (e.g., Vayssieres et al. 2000; Pontius and Schneider 2001), they can also be used as an objective means of tuning a model to *desirable* but not optimal classification rules such as the application of classifying potential habitat for rare species in conservation practice. For example, we might tune a model to provide a target true positive rate (e.g., 95%), or to bias the tuning to an application-specific balance of true positives or true negatives. And while ROC methods are commonly applied to logistic regression models, the approach can be extended readily to models of other forms, as long as the model prediction is scaled as a probability on [0,1].

2.4.4 Habitat Mapping

With the increasing accessibility of geospatial data, species distribution models can be extended to classify potential habitat over very large spatial extent. In this, the model is encoded into the GIS as a database in which the independent variables are predictors held as geospatial raster coverages (e.g., terrain-based indices, soil type, land cover, and so on). In the case of logistic regression, the model takes the form of a multiple regression equation, transformed by the logit link, with the predicted value indicating the probability of group membership ("habitat"). The result is a continuous surface illustrating the probability of each location (cell) being a "habitat." Other statistical models also provide methods for mapping the predictions.

This continuous probability can subsequently be thresholded to yield a discrete (habitat/nonhabitat) classification, at an ROC-optimized or any threshold selected by the user. This after-the-fact thresholding provides an easy method for visualizing the areal extent and spatial pattern of classified habitat at the selected threshold.

In the case of CART models (Appendix A.2.1), the mapping procedure is especially straightforward because it amounts to a series of "IF … THEN …" statements in the GIS (Moore et al. 1991b). CART models also offer the interpretative advantage that individual branches of the tree can be mapped separately. There might be several alternative pathways that lead to equivalently good "habitat" for the target species, with each pathway specifying a different combination of predictor variables. Substitutable resources would be one instance leading to alternative pathways; complementary biophysical settings are another example. By mapping these alternative pathways (branches) separately, the locations of these alternatives can be seen directly. For example, Taverna et al. (2005) used CART models to model and map conditions supporting relic hardwood stands in the North Carolina Piedmont. Hardwoods persist on sites that were too poor to support crops during the region's agricultural past. These sites included bottomlands too wet to plow, uplands too steep or dry for farming, and soils with high plasticity; these sites were spatially disjunct and readily interpretable in a mapping of the separate CART branches.

Regardless of the form of the model, mapping the predictions can provide a powerful visualization of model errors. A simple way to do this is to map the model prediction (probability of being habitat, or a binary prediction) and then overlay the actual samples with the points color-coded appropriately. For example, in a binary map of predicted habitat, false negatives appear as "habitat" points outside of the mapped habitat surface; mapped "habitat" locations devoid of actual presences (false positives) are likewise interesting (Fig. 2.9). Both of these misclassifications are intriguing from the perspective of metapopulation theory: they are precisely the model "errors" expected of the theory.

Mapped model predictions can inform conservation actions. For rare species, much of the area predicted to be habitat would be unoccupied and so false positives would be common. Mapping these can suggest locations to be targeted in follow-up surveys, especially in cases where parts of the study area might not have been surveyed for logistical reasons. Mapping unoccupied "habitat" can also suggest sites that might be suitable for reintroductions of species of special concern. Bartel and Sexton (2009) illustrated this application in a study of St. Francis's satyr (*Neonympha mitchellii francisci*), an endangered butterfly.

Mapped misclassifications also can suggest potential explanations for the model errors. Are they spatially clustered? Is there a regional trend or systematic bias in the model errors? Is there another environmental variable that might explain the misclassifications? This graphical exploration of the model predictions can be an efficient route to model refinements.

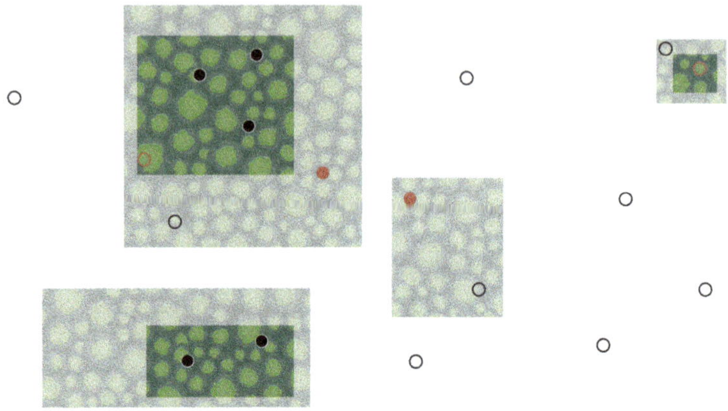

Fig. 2.9 Schematic of possible outcomes from mapping the predictions of an SDM and overlaying actual presence/absence data. Light shading, forest; dark shading, predicted "habitat"; filled circles, presences; open circles, absences; black, correct predictions; red, misclassifications. In particular: filled red circles are false negatives (occupied "nonhabitat") and open red circles are false positives (unoccupied "habitat"), both of which can be useful in interpreting the model

2.5 Benchmarking (and Further Reading)

Species distribution modeling is in the midst of an escalation that is seemingly enabled by the growing availability of geospatial data sets at various scales, along with increasing sophistication of specialized software and essentially unlimited computing power. Drew et al. (2011) provide a sampling of perspectives on landscape-scale applications. Fletcher and Fortin (2018) provide thorough coverage of the ecology and statistics (in R) of SDM. Other publications are providing detailed guidance on technical details such as sampling strategies for generating pseudo-absences (Barbet-Massin et al. 2012; Elith 2019; Elith and Franklin 2024; Soley-Guardia et al. 2024), as well as more philosophical guidance on selecting models from the perspective of matching tools to applications or the intended inferences to be drawn from the application (e.g., Elith and Graham 2009; Merow et al. 2014; Guillera-Arroita et al. 2015). Even a cursory search for papers published on SDMs reveals a daunting number of publications, and the pace is only increasing. This suggests that it is nearly impossible to keep up with developments in this area. Still, there is some consensus on where we are and how to work in this space.

2.5.1 Models and More Models: How to Choose?

There have been many comparative reviews of SDMs (e.g., Guisan and Thuiller 2005; Elith et al. 2006; Bahn and McGill 2013; Nieto-Lugilde et al. 2018; Valavi et al. 2022). Norberg et al. (2019) reviewed 33 variants of 15 different SDMs (including 12 variations on GLMs!). Their conclusions are consistent with previous reviews on several key points. Importantly, while some models consistently perform well in many applications, there is no single "best" model because model performance depends on the species being modeled, sample sizes, which environmental predictors are used, and often, geography. They suggest using a few complementary models, and then using validation tests (i.e., tests against independent data) to choose the model that performs best for that application.

The growing availability of accessible software makes it easier than ever to explore alternative models. In the R environment, for example, there are multiple packages that can fit several alternative models to the same data set while facilitating the processing (e.g., *biomod* and *biomod2*: Thuiller et al. 2021; *dismo*: Hijmans et al. 2021; *SDMtoobox*: Brown et al. 2017; *ENMTools*: Warren et al. 2021; *flexSDM*: Velazco et al. 2022; and see Fletcher and Fortin 2018).

Some caution might be due in choosing a "best" model. In applications aimed at extending the model to other locations (extrapolations) or into the future (e.g., forecasts under climate change), the model that performs best under in model fitting might not perform best in extensions. For example, while many SDMs perform better when spatial information is included in the model (Keitt et al. 2002; Norberg et al. 2019), these tend to not perform so well when extrapolated to other locations—presumably because the spatial structure is different in the new location. Similarly, models fit to current climate might not perform as well under future climate scenarios if the correlation structure among climate variables changes.

2.5.2 Ensemble Models

Because there are so many models available and their performance might vary substantially in a given application, practitioners increasingly resort to combining the predictions of various models to generate *ensemble* models (e.g., Araújo and New 2007; Forester et al. 2013). This generally improves model behavior, because ensembles reduce the idiosyncratic biases of any single model and provide a more robust solution. In forecasting, ensembled models are appealing because they can indicate not only the expected trends or patterns as predicted by the models but also the degree of consensus (or conversely, uncertainty) among the models.

Models can be ensembled in various ways. At the simplest extreme, collapsing the prediction of each model to binary ("habitat" or not) and then stacking the results yields an aggregate prediction that can range of 0 (all models agree that a sample is not habitat) to k (when all k models agree that the sample is "habitat"). For continuous predictions such as those on [0,1] (as is provided by GLMs, GAMs, random forests, maxent, and some others), averaging the prediction provides a more robust estimate of habitat suitability, as well as an estimate of the variance or range of values across models (i.e., a measure of model-based uncertainty).

In general, a current consensus seems to be to use multiple complementary models and to ensemble the results. But this recommendation is not as simple as we might wish. Intuitively, we would expect that in averaging several models, some of the individual models might perform slightly better than the average while others perform slightly worse (this is, of course, what averaging does!). This means that in any given application, there is often a single model that out-performs the ensemble (e.g., Crimmins et al. 2013; Zhu and Peterson 2017; Hao et al. 2020). In applications within the empirical domain of model calibration (i.e., the same study area, same time period), the simple solution is to choose the best single (validated) model for that application.

Valavi et al. (2022), in their comparative review, found that the details of *how* a model (e.g., the weighting of background points) was fitted could be as important as the choice of model (e.g., GAM versus random forest). They also found that ensembles performed better when the different models were calibrated individually before ensembling.

In applications to novel conditions (e.g., a new study area, projections into the future), the challenge is that we have no basis for knowing whether a single model might outperform the ensemble. Here, the conservative approach is to use an ensembled model—not because it will be most accurate but because we cannot know in advance which individual model might be most accurate. This kind of application also invites (demands) a deliberate approach to model validation, in which the model(s) are subjected to a variety of validation tests, ideally involving data sets from other regions and (if possible) other time intervals. This approach underscores the cumulative nature of model testing, by which we gain confidence in a model based on a growing body of validations using data that are increasingly separated from the training data. We return to this general theme in Chap. 9.

2.5.3 Reporting

Some of the details that must be reported will depend on the particulars of the application and the statistical model (see Appendices, below). But a number of details will be important across most applications, and these can be itemized as a checklist.

The checklist includes items related to the data, pre-processing, model fitting, and post-processing.

Data preparation and exploratory data analysis:

☑ What the data are (description of samples, observations); true presences, true absences, pseudo-absences?
☑ Sample sizes (presences, absences, or pseudo-absences)
☑ Description of predictor variables and their ecological interpretation
☑ Variable screening for correlations or redundancy
☑ Editing (missing values, outliers) or data transformations

For a GLM:

☑ Model family (distribution) and link function
☑ Relative weighting of observations?
☑ Partitioning of data for model testing (withholding test fraction, cross-validation)
☑ Model significance (*P*-value)
☑ Explanatory power (pseudo-R^2)
☑ Significance of predictor variables (linear and linked models)
☑ Estimates of variable importance (which variables matter, how this was done)
☑ Thresholding and how it was chosen; if using ROC, report tuning results (AUC or ROC curve)
☑ Classification success on training data (confusion matrix)
☑ Classification success on independent data or cross-validation

Other statistical models will have other details that matter, but the items above typically will have counterparts for alternative models.

2.5.3.1 A Call for Standard Reporting Protocols

Zurrell et al. (2020) called for a standard reporting protocol for SDM applications. In this, they suggest that practitioners address a set of questions in five general categories related to the model application: an Overview, Data issues, Model estimation, model Assessment, and Predictions. These issues are consistent with the workflow presented here (Fig. 2.4) and with other guidance (e.g., Pearson 2007; Franklin 2010; Guisan et al. 2017). Beyond this, the ODMAP framework provides a standard rubric akin to metadata reporting protocols. Their web-based guide (https://odmap.wsl.ch/) can facilitate this detailed reporting.

The ODMAP framework, if adopted generally, would solve three problems. First, it provides a checklist for practitioners developing applications and so ensures that decisions are made deliberately as the application is developed. Second, it provides thorough documentation for end-users or clients who might want to use the results. This transparency about how the model was estimated would also help with methodological reproducibility across studies. Finally, a standard reporting protocol would greatly facilitate meta-analyses and synthesis across applications.

To be clear, it might be apparent that many of the details in the protocol are already best practices, and these include many of the considerations elaborated in this chapter. What Zurrell et al. (2020) propose is not a radical change in how we work but in how we *report* our work. It is easy to endorse their recommendations (Fitzpatrick et al. 2021).

2.6 Summary and Prospectus

Species distribution modeling can be framed as the intersection of an ecological model, a data model, and a statistical model that connects ecology and data. The ecology of the focal species motivates the selection of environmental predictors and can also inform the initial choice of a statistical model consistent with the ecology. The data model includes the selection of measured predictors as well as decisions about scale, data transformations, and editing to reduce redundancy among predictors. There are very many statistical models available, including generalized linear or additive models (GLMs or GAMs), tree-based models (especially random forests), and the maximum entropy model (maxent). Joint SDMs, which model several species at once, are emerging as a promising approach.

Current best practice is to explore several complementary modeling techniques for any particular application. From these alternative models, a single model might be selected based on its performance in validation tests. Increasingly, the alternative models are combined into an ensemble to provide a robust, average model. The overall modeling process entails a series of very many decisions about ecology, data, model fitting and evaluation, and predictions and interpretation. A standard reporting protocol for SDM applications has been proposed. This is long overdue and would also be valuable in that it provides a useful framework and checklist for developing SDM applications.

Species distribution models promise to remain an active and sometimes controversial arena as new techniques are developed and competed. The field is moving so fast that authoritative papers on the topic can become dated almost as soon as they are published. But an appreciation for the ecological foundations and logical basis for applications—even if based on now-outdated techniques—can still provide useful guidance for emerging approaches and techniques. We will still evaluate new developments in terms of the match between ecology, data, and statistics; and we will still expect that statistical models can be evaluated in terms of their significance, ability to estimate the relative importance of predictors, and accuracy in prediction. The field will evolve but the foundations will remain the same.

Appendix A: Alternative Statistical Models

There are a *lot* of species distribution models out there, and more coming all the time. Rather than review all of these (if that's even possible), here we delve into a few approaches that are relatively popular and robust methods for species distribution modeling. These include extensions to the GLM, especially generalized additive models (GAMs); tree-based models, especially random forests; and maximum entropy modeling as provided by program *maxent*. Here we focus on these approaches while also pointing to some additional alternatives. This material is excerpted and condensed from digital supplement S2.

A.1 Extensions to the GLM: GAMs

The logistic form of the GLM is very common in modeling species distributions, as well as similar applications such as development probability in models of land use change (i.e., habitat models for humans, see Chap. 9). This GLM also serves as a point of departure for a variety of extensions. Several variations are described in Supplement S2.1.

One popular extension used in species distribution models is the generalized additive model. Here we briefly describe the model as an extension of a GLM, then emphasize the simplicity with which this can be substituted for a GLM in the workflow outlined in Sect. 2.3 and following (above).

A.1.1 Generalized Additive Models

A *generalized additive model* (GAM) allows for substantially more flexibility in the model fit (Yee and Mitchell 1991). In a GAM, the constant regression coefficients of the linear predictor of the GLM (i.e., the *b*'s in Eq. 2.2) are replaced with smooth (usually nonlinear) functions *s* of the predictors:

$$u = b_0 + s_1(x_1) + s_2(x_2) + s_3(x_3) + \cdots + s_p(x_p) + \varepsilon. \tag{A.1}$$

Typically, these functions are nonparametric smoothing functions such as scatter-plot smoothers or splines (Fig. 2.10). While quite flexible in generating predictors, a GAM does not produce an equation that can be "plugged in" to predict other cases (though data can be fed through the fitted model to predict new cases). GAMs are now quite common in habitat classification and species distribution modeling because of their flexibility (Guisan and Zimmerman 2000, Guisan et al. 2002). In particular, the nonlinear responses that can be modeled with a GAM can offer a distinct advantage over the linear GLM.

Like a GLM, a GAM provides an estimate of the likelihood of group membership and can be interpreted and post-processed in the same way that a GLM is evaluated.

That is, model evaluation is as described in Sect. 2.4 (above). These details are not repeated here. In the larger workflow of SDM, substituting a GAM for a GLM is a simple swap of the statistical model; the workflow is essentially the same.

A.2 Tree-Based Models

Tree-based models include classification and regression trees (CART models) as well as various extensions of these. CART analysis is a flexible, nonparametric modeling approach. The approach is actually two techniques, depending on the response variable. A *regression* tree predicts an interval-scale dependent variable,

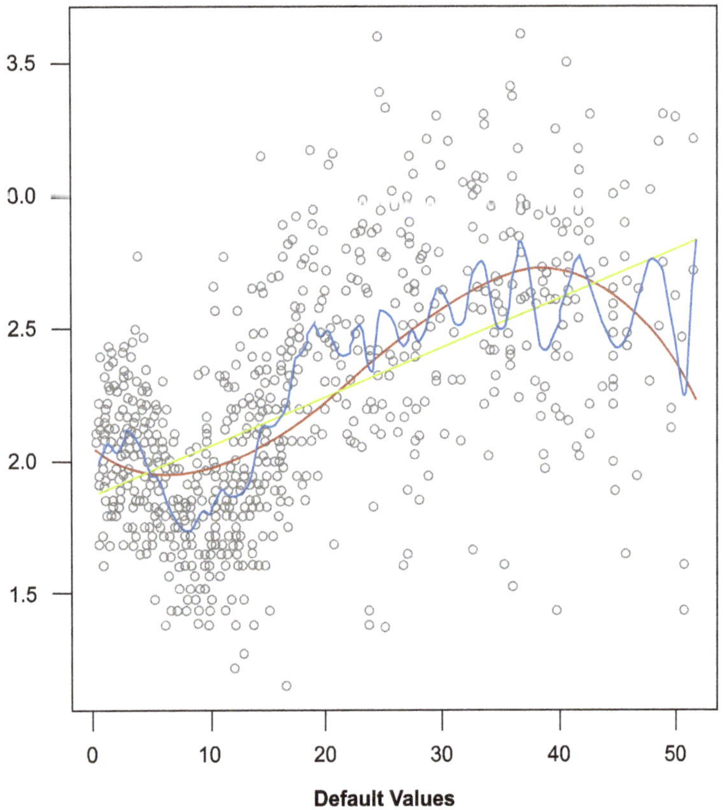

Fig. 2.10 Examples of smoothing splines of varying stiffness, from very stiff (linear, green) to moderately stiff (red) to loose (blue). (Redrawn from Perperoglou et al. 2019; permission licensed via Creative Commons)

while a *classification* tree predicts a categorical response. Thus, we might model species abundance using a regression tree but use a classification tree to model species presence/absence. Here we begin with a comparatively simple classification tree and then proceed to a more complicated but (often) more effective extension to this model, a random forest model. Tree-based models are described more broadly in Supplement S2.2. Our focus here is on the workflow of the modeling process, in parallel with the example illustrated with the GLM (above).

A.2.1 Classification Trees

The distributions of species or ecological communities often are governed by complicated contingencies involving interactions among variables, substitutable or complementary resources, or sensitivity to local context. In narrative descriptions of these cases, we tend to use qualifiers such as "an," "but," "or", "except," and so on. For example, in the Sierra Nevada of California, USA, white fir (*Abies concolor*) occurs not only at mid-elevations *but also* on higher-elevation sites with warmer (southern) exposures *or* at lower-elevation sites with deeper soils *or* cooler (northern) exposures. These contingencies tend to be difficult to capture in linear additive models such as the logistic regression we considered previously. CART models, by contrast, are well suited to exactly these contingencies, because of the way the models are constructed.

Tree models are described in detail by Breiman et al. (1984) and Venables and Ripley (1999; their Chapter 10); they are developed for ecological applications by Iverson and Prasad (1998), De'Ath and Fabricius (2000), Vayssieres et al. (2000), De'Ath (2002), and Urban (2002; see also Urban et al. 2002). Tree-based models are discussed in more detail in Supplement S2.2.

Algorithm

A CART model is a *recursive partitioning* of the data set, in which each partition aims to split the data into two groups that are as "pure" as possible in terms of the dependent variable. After each split, the resulting subgroups are each split again, and this recursion continues until the groups cannot be isolated more "purely." The algorithm requires user-defined groups. These groups might be several community types or the binary case typical of habitat classification (e.g., "habitat" versus "not"). The predictor variables can be any mix of interval-scale continuous, ordinal (rank), or categorical factors. CART models are nonparametric, in that each variable is transformed to rank scores before further analysis. One measure of group "purity" is deviance, which is a log-likelihood measure based on expected group memberships (other measures of purity are available).

The algorithm for recursive partitioning is intuitive and rather simple:

1. The procedure considers the group impurities implied by splitting the samples into two groups, using as the threshold (splitting) value the difference between each two levels of each predictor variable. That is, the algorithm considers *every possible split* of the data. The data are then partitioned on the best "splitter" (best level of the best variable) to generate two groups: the first branch of the tree. One side of the branch is predicted to be "habitat," the other "not habitat."
2. This process repeats, for each subgroup, until a stopping rule is met. The stopping rule could be that the final groups are completely pure (not common), that they cannot be made any purer via additional splits on the available predictors, or that the groups are too small to warrant further division.

Interpretation

The result of the partitioning is a tree-like structure (dendrogram) that traces the conditions that lead to membership in each group. In this *decision tree*, different

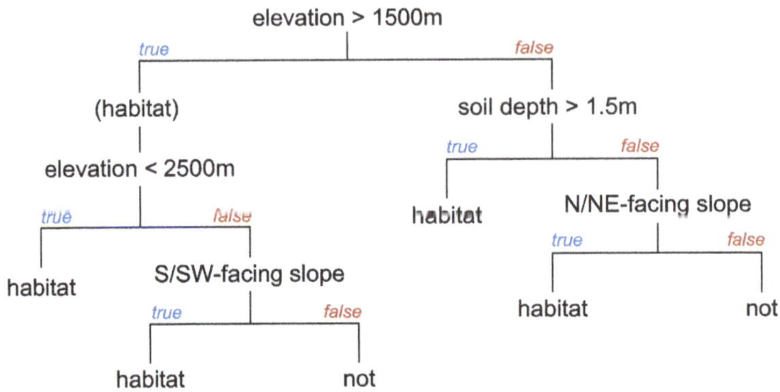

Fig. 2.11 A classification tree to illustrate the contingent or compensatory conditions leading to "habitat" for white fir in the southern Sierra Nevada. Each branch traces an interaction of environmental conditions that result in suitable habitat (This heuristic is not fitted to actual data; see text)

variables can be invoked in different branches, variables can be reused or appear in multiple branches of the tree, and a given group can appear in different branches (as end nodes or leaves). The decision tree *explains* group differences simply by tracing the conditions that lead to membership in each group. For the narrative example of white fir (above), the decision tree traces the conditions explicitly (Fig. 2.11). In the tree, the following interpretations are available:

- Interactions appear as long branches (*interaction pathways*) that itemize the combinations of conditions required to describe the group in terms of the predictors (e.g., "this *and* that *and* that . . ."). Importantly, these interactions are local to a given branch of the tree. By contrast, in a regression the interactions (indeed, all effects) are global—they apply equally to the entire data set.
- Nonlinear relationships are modeled by invoking the same variable in sequence. For example, to account for the mid-elevation distribution of a particular tree species, we might find a CART branch that splits this as "Elevation > 1500" followed by a subsequent split "Elevation < 2500." Similarly, a bimodal or more complicated distribution would simply reuse the same predictor to successively refine the description of the distribution.
- Substitutable relationships ("*either* this *or* that") appear in different branches of the tree, invoking different combinations of predictor variables. For example, alternative settings for white fir (outlined above) might appear in one branch in terms of soil depth and topographic convergence and in another branch in terms of elevation and slope aspect. Because the conditions specified within a branch are independent of other branches, there is no logical inconsistency in declaring these alternative explanations within the same model.
- Context-dependent contingencies are captured as the full set of conditions along a branch leading to a group's membership. That is, the prediction for a terminal leaf is dependent of the context of everything higher up in that branch—including interactions, compensatory relationships, supplementary conditions, or whatever.

Again, these interactions are local to that branch: they do not apply to any other branches of the tree.

Thus, the tree model explains group differences simply by reading the conditions along each branch. In general, variables that appear "high in the tree" are more important in differentiating groups, while variables that appear far down in the tree (near the terminal leaves) provide more nuanced fine-tuning of the model. Beyond this, it can be rather difficult to ascribe a partial explanatory power to each variable. One way to do this would be to build two models, with and without a selected variable, and compare the models in terms of their explanatory power. This would be roughly equivalent to the comparison of full and restricted models using GLMs but evaluated in terms of classification success for the CART model (and see below).

Prediction and Cross-Validation

Group membership for a sample is predicted by simply "dropping it through the tree." The prediction is binary: "habitat" or not. In verification, the model is often used to predict group membership for the data used to construct the model—a procedure termed *back-classification*. Back-classification of the data generates a confusion matrix, as described previously for a thresholded GLM (Table 2.1).

In a classification tree, there is additional information available for each terminal node: the number of correct and incorrect cases. These tallies can provide a means to estimate the probability that a case in that node is "habitat" or not, simply as the proportion of cases so predicted.

Because the tree model essentially describes the data as faithfully as possible, CART models tend to be rather over-fitted to the data. That is, they capture idiosyncrasies in the data that improve the model relative to those data but will degrade the model when it is applied to other, independent data. There are several ways to minimize this issue, including various forms of "pruning" by which idiosyncratic (and probably over-fitted) branches are trimmed from the tree. Another solution is especially pertinent here because it leads to a further extension that we consider below.

A common way to minimize over-fitting is to use a *cross-validation* procedure. A conventional approach is *k*-fold cross-validation. In this, the data are divided into *k* (often 10, and typically random) exclusive subsets. A CART model is then constructed using all but one subset, and this model is used to predict the withheld subset. For tenfold cross-validation, this means 10 models based on 90% of the data each time; and in this, each sample is predicted by a model that was constructed independently from that observation. This, in turn, generates ten confusion matrices and the misclassification rates can be averaged over the ten trials. In each trial, the observations being predicted are independent of the observations used to fit the model.

As with any statistical model, the overall accuracy of the model will increase as the model gets more complicated (in this case, more branches). At the same time, the misclassification rate under cross-validation will increase for larger trees, reflecting the over-fitting of the model. The goal of tree fitting is to achieve a balance between model accuracy and model robustness. As with many procedures, there might not be an exact solution to this compromise, but there are conventional "rule of thumb" approaches provided in software packages.

Ensembling trees A tenfold cross validation yields ten individual tree models. To use these, the trees are averaged or *ensembled.* Doing so uses a combination of two methods. For categorical predictors, the ensemble tree uses a "popularity vote" to choose the splitting variable and the level(s) used for splits. Thus, the split that emerges most often in the multiple model runs is the split used in the ensemble model. For continuous predictors, the split level is averaged over multiple runs. The end result is a single model.

A.2.2. Random Forests

A powerful way to generate an accurate and robust tree model involves generating a large number of trees—a forest—and aggregating the results. A popular approach, random forests, generates a very large number of trees (perhaps thousands), with each tree developed from a subset of the observations (Breiman 2001). The individual trees are not pruned, but on ensembling, the idiosyncrasies of individual trees are averaged away.

Beyond this subsetting of observations, a random tree also chooses a subset of predictor variables available for splitting. Because not all of the variables are available for each case, random forests provide a means to assess the relative explanatory power of each variable, by identifying variables whose omission dramatically degrades model performance. This provides for a straightforward ranking of variable importance over all the trees in the forest.

Elith et al. (2006) found random forests and other extensions to CART models to be quite effective as predictors in their comparison of species distribution models. These models are very popular tools in SDM applications.

CART models in general are sensitive to mismatched sample sizes for presences and absences (or more typically, pseudo-absences), and they can also be confounded by overlap between presences (true "habitat") and pseudo-absences. As random forests become more popular with practitioners, they are continually refined and updated. Valavi et al. (2021) provide the current state-of-the-art along with some suggested best practices.

A.2.3. Workflow for Tree-Based Models

One powerful way to use CART models is to begin with a simple (single) CART model, essentially as a means of exploratory data analysis (EDA). If this tree indicates clear alternative paths to "habitat" (i.e., branches that invoke different and contrasting combinations of predictors), this would suggest that these contingencies are important and, in turn, provide guidance for further analysis. First, this would argue against using other models that are fit globally and which might not capture these local contingencies very well. Second, this EDA would lead naturally to the next step of fitting a boosted tree or a random forest for the actual application. Here we focus on random forests (boosted trees are described in Supplement S2.2).

The most popular implementation of random forests is probably package *random forest* in R (Liaw and Wiener 2002). This software includes many options for fitting and reporting, and many of these options are set to default values that are reasonable starting point. See Valavi et al. (2021) for further guidance on fitting random forests using presence/pseudo-absence data typical of SDM applications.

Beyond the suggestion of using CART for EDA and a random forest for the application, the workflow proceeds as with a GLM (above): data preparation, model fitting, and model evaluation.

Model Evaluation

In terms of model evaluation, tree-based models do not yield a *P*-value nor an estimate of how much of the variability in the data they explain (i.e., a pseudo-R^2). This is because of the nonparametric algorithm by which classification trees are constructed.

Note that the dendrogram used to interpret and present a CART model—very helpful in EDA—is not typically available for a random forest based on perhaps 1000 or more trees.

The systematic exclusion of some of the predictor variables in each of the trees provides an estimate, by omission, of the relative explanatory power of each variable, somewhat analogous to the variable jack-knifing approach provided in *maxent* (below) and described more generally in Sect. 2.4.2 (above). The relative importance of each predictor variable can be summarized in either tabular or graphical form.

Note that a random forest provides a prediction on [0,1], in which "habitat suitability" is tallied as the proportion of total cases (trees) in which each sample was predicted to be "habitat." This output can be further evaluated using post-processing steps outlined for the GLM in Sect. 2.4 (above). These predictions are the result of a cross-validation process, so each case is predicted by a model for which that case was not used.

A.3 Maximum Entropy Models

A relatively recent entry into the SDM toolkit is *maxent*, a program for habitat modeling using maximum entropy methods (Phillips et al. 2006, Phillips and Dudik 2008, Phillips et al. 2009, Elith et al. 2011). In Supplement S2.3, we consider maximum entropy modeling in three passes: (1) the concept of *maximum entropy* and its logical appeal, (2) this logic applied to habitat modeling and the numerical algorithm of maxent, and (3) the software package *maxent* (Phillips et al. 2006; Phillips and Dudik 2008). This model can be confusing because the three perspectives are difficult to separate: maxent is only available in *maxent*, and the software includes many features that are not particular to that model itself.

A.3.1 Model Development: Maxent

In the maxent model as implemented in the software package *maxent*, the typical approach is to base the model on a set of presences relative to a large set of pseudo-absences, termed *background* samples. The software generates the background points from a user-provided stack of raster maps of the predictor variables. (Alternative approaches are described below.)

The statistical task in maximum entropy habitat modeling can be developed in a relatively straightforward way. It is easy to think of maxent as similar to a GAM (A.1.1 and Eq. A.1), with the difference being in the function that describes the fitted model. In a GAM, the function is a smooth nonlinear function based on local smoothing. In *maxent*, the function is fitted piecewise from a collection of available *features*. The features are shapes: linear, quadradic curves, stepwise thresholds, hinges, and so on (Fig. 2.12).

The model is fitted iteratively. At each iteration, a single feature of a single predictor variable is added to the model, with that addition selected to maximize the overall improvement of the model fit. What is being fitted is the *difference* between the distribution of the presences, over all of the predictors, relative to the distribution of the background samples. The solution is reached using a sophisticated machine-learning algorithm. In this sense, the algorithm is not very different from a CART model (above): the fit proceeds via an exhaustive exploration of all incremental improvements to the model.

The translation of the maximum entropy principle to this task is simple: the aim is to estimate the "loosest" distribution that satisfies the constraint that the expected value of the fitted distribution is the same as the observed mean over the presences, for each of the predictors. The "loosest" distribution, in principle, is a uniform distribution, but the loosest distribution in this implementation is the distribution of the background of the study area—i.e., that the presences are not different from the background.

The result of the model fitting is a set of fitted distributions for the predictors, which collectively estimate the likelihood that a given sample is "habitat." From this point forward, interpreting the model would follow the steps outlined (Sect. 2.4 and following, above) for any SDM. But some of these steps are provided in particular forms by the *maxent* software.

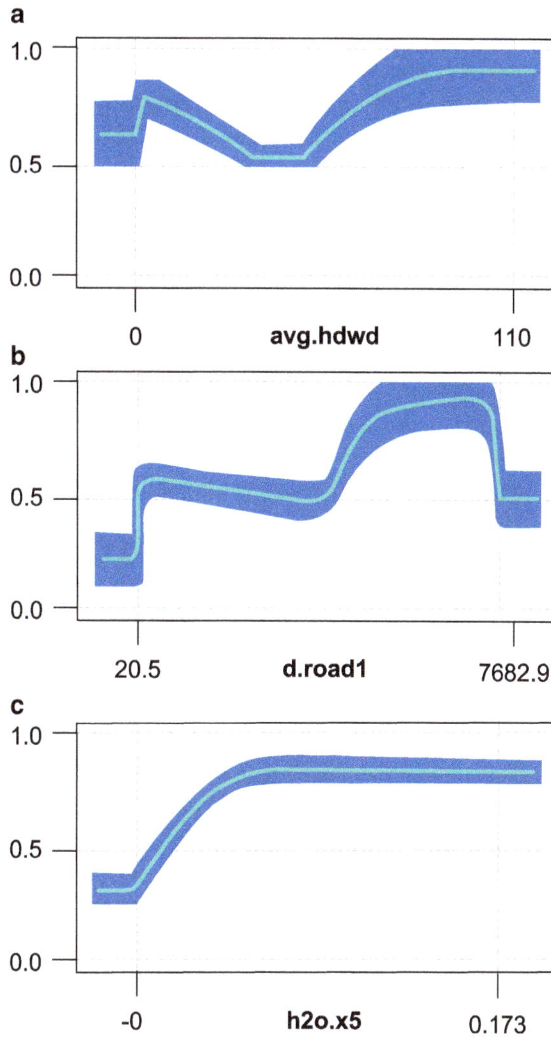

Fig. 2.12 Examples of features used in *maxent* to fit a model for the pine warbler (*Setophaga pinus*, unpublished data from Minor and Urban 2010). For each variable, the mean response is the cyan line while the blue envelope shows the variation over tenfold cross-validation. Variable avg.hdwd is average size of hardwood trees; d.road1, distance to primary road(s); h2o.x5, amount of water in a five-cell radius window (30-m cells). All variables are shown as relative response; redrawn from output from program *maxent*. (Phillips et al. 2006)

A.3.2 Workflow: Program *Maxent*

The software package *maxent* implements the model described above. But it also provides a wealth of other diagnostic and interpretative aids—none of which is particularly restricted to maximum entropy models of species distribution. These extensions are described in more detail in Supplement S2.3 (and in the software documentation). Some of these affect model inputs, some influence the fitting, and some are output options—but all are specified before the model is run. Options include:

- An option to provide both the presences and background samples (pseudo-absences) as text files ("samples with data"), instead of using map data. This option also allows the user to be very deliberate about which areas might qualify as "background" (though *maxent* also provides substantial flexibility in this by permitting masks on the study area). This option makes it easier to compare *maxent* results to those of other models or to use actual absences as "background" samples.
- An option for subsetting separate "training" and "test" subsamples of the data, by specifying a proportion of the observed cases to withhold for model testing. This proportion is sampled randomly for the observed cases.
- An option to bootstrap the model by performing a *k*-fold cross-validation of the input samples. For example, a tenfold cross-validation would create ten random partitions of the data, build ten models, test each of the ten withheld partitions in turn, and return average model performance. In particular, this provides error estimates (confidence limits) on model output.
- Given the way the model is estimated, it would be easy to overfit the model, and so there is a method for adjusting the fit to be arbitrarily "tighter" or "looser": a tighter fit better matches the data but at greater risk of overfitting; a looser fit is more conservative and so misses more of the input data but is accordingly less overfitted and more robust in extrapolation.
- An option to specify the estimated prevalence of the target species. Prevalence is the proportion of the study area (here, background) that would qualify as "habitat." This is typically unknown and *maxent* is calibrated to work best under this condition; but if an empirical estimate is available, it can be specified. This affects the likelihood that any sample should be predicted to be "habitat."
- The relative explanatory power of each predictor is summarized in terms of its cumulative contribution to the model fit. Program *maxent* also provides a further summary, by jack-knifing the predictors in the model. In this, for each predictor variable, two additional models are estimated: a model with *only* that predictor and a second model with every variable *except* that predictor. A variable that is important to the focal species will have a large effect by itself. On omission, this variable will degrade the model only if there are no other predictors in the data set that are sufficiently correlated that they can substitute for the omitted variable. This combination of focal models can be quite informative.
- A tuning of the model, based on receiver operating characteristics (ROC) curves. In this, the aim is to maximize the rate of predicting true positives (known occurrences) while minimizing the total proportion of the study area (background samples) predicted to be habitat. Note that because *maxent* lacks information on true negatives ("not habitat"), it cannot construct an actual ROC curve; it substitutes a plausible alternative instead based on the large set of pseudo-absences. Similarly, this curve is used to estimate area under the ROC curve (AUC) as an overall index of model performance. The program does not allow a user-specified tuning but instead offers a few alternatives. (The user also could post-process the output to tune the model more deliberately, as described in Sect. 2.4.3 above.)

The tuned and thresholded predictions can then be used to construct a confusion matrix (Table 2.1, above).

- The software offers a variety of output options (described in the user's guide). For applications using geospatial input variables, *maxent*'s mapped outputs are quite useful and informative (and voluminous!). It should be emphasized again that many of these options could be applied to any habitat model—these are features of the software, not of maximum entropy models. Likewise, these other options in the software (training vs test subsamples, *k*-fold cross-validation, jack-knifed estimates of variable importance, calibration using ROC methods) are generic in principle and could be applied to any SDM.

A.3.3 Best Practices

Maxent is an increasingly popular species distribution model. While initially available only in the (free) program *maxent*, it is now available in many other formats, including various packages in R; the code itself is now open-source (Phillips et al. 2017) and freely available. There is also a growing user's community that can provide technical assistance and general guidance. The software, tutorials, user's group access, and other support are available via its website hosted by the American Museum of Natural History (https://biodiversityinformatics.amnh.org/open_source/maxent).

A.4 Conclusions

There are many other models available today and more are arriving almost every day. In this Appendix, we have considered three popular models, to provide a sample of what is available. These models also underscore that many of these alternatives can be rather simply substituted: one the model is fitted, the post-processing and evaluation is remarkably similar. That is, the workflow is rather robust to the choice of any particular model.

The proliferation of modeling approaches is a good thing, in that it provides many options to the practitioner. The cost of this is complexity and uncertainty about which model(s) to choose. This underscores the importance of matching models to the ecology, to every extent possible (see Sect. 2.5.1) and perhaps ensembling predictions from multiple models (Sect. 2.5.2). And, of course, attempting to keep abreast of rapidly developing alternatives and best practices!

References

Adolph SC (1990) Influence of behavioral thermoregulation on microhabitat use by two *Sceloporus* lizards. Ecology 71:315–327

Araújo MB, New M (2007) Ensemble forecasting of species distributions. Trends Ecol Evol 22:42–47

Austin MP (2002) Spatial prediction of species distribution: an interface between ecological theory and statistical modelling. Ecol Model 157:101–118

Austin MP (2007) Species distribution models and ecological theory: a critical assessment and some possible new approaches. Ecol Model 200:1–19

Austin MP, Smith TM (1989) A new model for the continuum concept. Vegetatio 83:35–47

Bahn V, McGill BJ (2013) Testing the predictive performance of distribution models. Oikos 122: 321–331

Barbet-Massin M, Jiguet F, Albert CH, Thuiller W (2012) Selecting pseudo-absences for species distribution models: how, where, and how many? Methods Ecol and Evol 3:327–338

Bartel RA, Sexton JO (2009) Monitoring habitat dynamics for rate and endangered species using satellite images and niche-based models. Ecography 32:888–896

Bergen KM, Goetz SJ, Dubayah RO, Henebry GM, Hunsaker CT, Imhoff ML, Nelson RF, Parker GG, Radeloff VC (2009) Remote sensing of vegetation 3-D structure for biodiversity and habitat: review and implications for lidar and radar spaceborne missions. J Geophys Res 114: G00E06. https://doi.org/10.1029/2008JG000883

Breiman L (2001) Random forests. Mach Learn 45:5–32

Breiman L, Friedman JH, Olshen RA, Stone CJ (1984) Classification and regression trees. Wadsworth and Brooks/Cole, Monterey

Brown JL, Bennett JR, French CM (2017) SDMtoolbox 2.0: the next generation python-based GIS toolkit for landscape genetic, biogeographic and species distribution model analysis. PeerJ 5: e4095. https://doi.org/10.7717/peerj.4095

Burnham KP, Anderson DR (2002) Model selection and multimodel inference: a practical information-theoretic approach, 2nd edn. Springer, New York

Burnham KP, Anderson DR, Huyvaert KP (2011) AIC model selection and multimodel inference in behavioral ecology: some background, observations, and comparisons. Behav Ecol Sociobiol 65:23–35

Busby JR (1991) BIOCLIM—a bioclimate analysis and prediction system. In: Margules CR, Austin MP (eds) Nature conservation: cost effective biological surveys and data analysis. CSIRO, Melbourne, pp 64–68

Clark JS, Gelfand AE, Woodall CW, Zhu K (2014) More than the sum of the parts: forest climate response from joint distribution models. Ecol Appl 24:990–999

Cody ML (1981) Habitat selection in birds: the roles of vegetation structure, competitors, and productivity. Bioscience 31:107–113

Connell JH (1980) Diversity and the co-evolution of competitors, or the ghost of competition past. Oikos 35:131–138

Crimmins SM, Dobrowski SZ, Mynsberge AR (2013) Evaluating ensemble forecasts of plant species distributions under climate change. Ecol Model 266:126–130

Cushman SA, McGarigal K (2002) Hierarchical, multiscale decomposition of species-environment relationships. Landsc Ecol 17:637–646

De'Ath G (2002) multivariate regression trees: a new technique for modeling species-environment relationships. Ecol 83:1105–1117

De'Ath G, Fabricius KE (2000) Classification and regression trees: a powerful yet simple technique for ecological data analysis. Ecology 81:3178–3192

Dormann CF, Elith J, Bacher S, Buchmann C, Carl G, Carré G, García Marquéz JR, Gruber B, Lafourcade B, Leitao PJ, Munkekmuller T, McClean C, Osborne PE, Reineking B, Schroder B, Skidmore AK, Zurrell D, Lautenbach S (2013) Collinearity: a review of methods to deal with it and a simulation study evaluating their performance. Ecography 36:27–46

Dormann CF, Bobrowski M, Dehling DM, Harris DJ, Hartig F, Lischke H, Moretti MD, Pagel J, Poiinkert S, Schluening M, Schmidt SI, Sheppard DS, Steinbbauer MJ, Zeuss D, Kraan C (2018) Biotic interactions in species distribution modelling: 10 questions to guide interpretation and avoid false conclusions. Glob Ecol Biogeogr 27:1004–1016

Drew CA, Wiersma Y, Huettmann F (eds) (2011) Predictive species and habitat modeling in landscape ecology: concepts and applications. Springer, New York

Dueser RD, Shugart HH (1979) Niche pattern in a forest-floor small mammal fauna. Ecology 60: 108–118

Elith J (2019) Species distribution modeling. In: Oxford bibliographies. https://doi.org/10.1093/OBO/9780199830060-0226

Elith J, Franklin J (2024) Species distribution modeling. In: Scheiner SM (ed) Encyclopedia of biodiversity, vol 5, 3rd edn. Elsevier, pp 558–572

Elith J, Graham CH (2009) Do they? How do they? Why do they differ? On finding reasons for differing performances of species distribution models. Ecography 32:66–77

Elith J, Leathwick JR (2009) Species distribution models: ecological explanation and prediction across space and time. Ann Rev Ecol Evol Syst 40:677–697

Elith J, Graham CH, Anderson RP, Dudik M, Ferrier S, Guisan A, Hijmans RJ, Huettmann F, Leathwick JR, Lehmann A, Li J, Lohmann LG, Loiselle BA, Manion G, Moritz G, Nakamura M, Nakazawa Y, Overton JM, Peterson AT, Phillips SJ, Richardson K, Scachetti-Pereira R, Schapire RE, Soberon J, Willians MSW, Zimmermann NE (2006) Novel methods improve prediction of species distributions from occurrence data. Ecography 29:129–151

Elith J, Phillips SJ, Hastie T, Dukik M, Chee YE, Yates CJ (2011) A statistical explanation of MaxEnt for ecologists. Divers Distrib 17:43–57

Fielding AH, Bell JF (1997) A review of methods for the assessment of prediction errors in conservation presence/absence models. Environ Cons 24:38–49

Fitzpatrick MC, Lachmuth S, Haydt NT (2021) The ODMAP protocol: a new tool for standardized reporting that could revolutionize species distribution modeling. Ecography 44:1067–1070

Fletcher R, Fortin M-J (2018) Spatial ecology and conservation modeling: applications with R. Springer Nature Switzerland AG

Forester BR, DeChaine EG, Bunn AG (2013) Integrating ensemble species distribution modeling and statistical phylogeography to inform projections of climate change impacts on species distributions. Divers Distrib 19:1480–1495

Franklin J (1995) Predictive vegetation mapping: geographic modeling of biospatial patterns in relation to environmental gradients. Prog Phys Geogr 19:474–499

Franklin J (2010) Mapping species distributions: spatial inferences and prediction. Cambridge University Press, Cambridge

Franklin J (2023) Species distribution modelling supports the study of past, present and future biogeographies. J Biogeogr 50:1533–1545

Giannini TC, Chapman DS, Saraiva AM, Alves-dos-Santos I, Biesmeijer JC (2013) Improving species distribution models using biotic interactions: a case study of parasites, pollinators, and plants. Ecography 36:649–656

Grand J, Cushman SA (2003) Multiple-scale analysis of bird-habitat relationships: breeding birds in a pitch pine-scrub oak (*Pinus rigida-Quercus ilicifolia*) community. Biol Conserv 12:307–317

Green RH (1971) A multivariate statistical approach to the Hutchinsonian niche: bivalve mollusks of Central Canada. Ecology 52:543–556

Grime P (1977) Evidence for the existence of three primary strategies in plants and its implications for ecological and evolutionary theory. Am Nat 111:1169–1194

Grime JP (1979) Plant strategies and vegetation processes. Wiley, New York

Guillera-Arroita G, Lahoz-Monfort JJ, Elith J, Gordon A, Kujala H, Lentini PE, McCarthy MA, Tingley R, Wintle BA (2015) Is my species distribution model fit for purpose? Matching data and models to applications. Global Ecol Biogeogr 24:276–292

Guisan A, Thuiller W (2005) Predicting species distribution: offering more than simple habitat models. Ecol Lett 8:993–1009

Guisan A, Zimmerman NE (2000) Predictive habitat distribution models in ecology. Ecol Model 135:146–186

Guisan A, Edwards TC, Hastie T (2002) Generalized linear and generalized additive models in studies of species distributions: setting the stage. Ecol Model 157:89–100

Guisan A, Thuiller W, Zimmerman NE (2017) Habitat suitability and distribution modeling with applications in R. Cambridge University Press, Cambridge

Hakkenberg CR, Peet RK, Urban DL, Song C (2018) Modeling plant composition in a forest landscape using LiDAR and hyperspectral remote sensing. Ecol Appl 28:177–190

Hao T, Elith J, Lahoz-Monfort JJ, Guillera-Arroita G (2020) Testing whether ensemble modelling is advantageous for maximising predictive performance of species distribution models. Ecography 43:549–558

He KS, Bradley DA, Cord AF, Rocchini D, Tuanmu M-N, Schmidtlein S, Turner W, Wegman M, Pettorelli N (2015) Will remote sensing shape the next generation of species distribution models? Remote Sensing Ecol Conserv 1:4–18

Hijmans RJ, Phillips SJ, Leathwick JR, Elith J (2021) dismo: Species distribution modeling. R package version 1.3.5. https://CRAN.R-project.org/pack-age=dismo

Homer C, Huang C, Yang L, Wylie B, Coan M (2004) Development of a 2001 national land-cover database for the United States. Photogram Eng Remote Sensing 70:829–840

Hutchinson GE (1957) Concluding remarks. Cold Spring Harb Symp Quant Biol 22:415–427

Iverson LR, Prasad AM (1998) Predicting abundance of 80 tree species following climate change in the eastern United States. Ecol Monogr 68:465–485

James FC (1971) Ordinations of habitat relationships among breeding birds. Wilson Bull 83:215–236

Jeliazkov A, Gavish Y, Marsh CJ, Geschke J, Brummitt N, Rocchini D, Haase P, Kunin WE, Henle K (2022) Sampling and modelling rare species: conceptual guidelines for the neglected majority. Glob Chang Biol 28:3754–3777

Jiménez-Valverde A, Lobo JM (2007) Threshold criteria for conversion of probability of species presence to either-or presence-absence. Acta Oecol 31:361–369

Johnson DH (1980) The comparison of usage and availability measurements for evaluating resource preference. Ecology 61:65–71

Johnson CJ, Gillingham MP (2005) An evaluation of mapped species distribution models used for conservation planning. Environ Conserv 32:117–128

Keitt TH, Bjornstad ON, Dixon PM, Citron-Pousty S (2002) Accounting for spatial pattern when modeling organism-environment interactions. Ecography 25:616–625

Klopfer P (1965) Behavioral aspects of habitat selection: a preliminary report on stereotypy in foliage preferences of birds. Wilson Bull 77:376–381

Lawler JJ, Edwards TC Jr (2006) A variance-decomposition approach to investigating multiscale habitat associations. Condor 108:47–58

Lefsky MA, Cohen WB, Parker GG, Harding DJ (2002) Lidar remote sensing for ecosystem studies. Bioscience 52:19–30

Liaw A, Wiener M (2002) Classification and regression by random forest. R News 2:18–22

Liu C, Berry PM, Dawson TP, Pearson RG (2005) Selecting thresholds of occurrence in the prediction of species distributions. Ecography 28:385–393

Lookingbill T, Urban D (2005) Gradient analysis, the next generation: towards more plant-relevant explanatory variables. Can J For Res 35:1744–1753

MacArthur RH (1958) Population ecology of some warblers of northeastern conifer forests. Ecology 39:599–619

MacArthur RH, MacArthur JW, Preer J (1961) On bird species diversity. II. Prediction of bird census from habitat measurements. Am Nat 96:167–174

Manly BFJ, McDonald LL, Thomas DL, McDonald TL, Erickson WP (2002) Resource selection by animals: statistical design and analysis for field studies. Kluwer, New York

Martin KJ, McComb WC (2002) Small mammal habitat associations at patch and landscape scales in Oregon. For Sci 48:255–264

Martin ME, Newman SD, Aber JD, Congalton RG (1998) Determining forest species composition using high spectral resolution remote sensing data. Remote Sens Environ 65:249–254

Merow C, Smith MJ, Edwards TC Jr, Guisan A, McMahon SM, Normand S, Thuiller W, Wuest RO, Zimmerman NE, Elith J (2014) What do we gain from simplicity versus complexity in species distribution models? Ecography 37:1267–1281

Michel N, Burel F, Legendre P, Butet A (2007) Role of habitat and landscape in structuring small mammal assemblages in hedgerow networks of contrasted farming landscapes in Brittany, France. Landscape Ecol 22:1241–1253

Minor ES, Urban, DL (2010) Forest bird communities across a gradient of urban development. Urban Ecosyst 13:51–71

Moore JA, Gillingham JC (2006) Spatial ecology and multiscale habitat selection by a threatened rattlesnake: the eastern massasauga (*Sistrurus catenatus catenatus*). Copeia 2006:741–751

Moore ID, Gryson RB, Ladson AR (1991a) Digital terrain modelling: a review of hydrological, geomorphological, and biological applications. Hydrol Process 5:3–30

Moore DM, Lee BG, Davey SM (1991b) A new method for predicting vegetation distributions using decision tree analysis in a geographic information system. Environ Manag 15:59–71

Morris DW (1987) Ecological scale and habitat use. Ecology 68:362–369

Clark JS, Nemergut D, Seyednasrollah B, Turner PJ, Zhang S (2017) Generalized joint attribute modeling for biodiversity analysis: median-zero, multivariate, and multifarious data. Ecol Mongr 87:34–56

Nieto-Lugilde D, Maguire KC, Blois JL, Williams JW, Fitzpatrick MC (2018) Multiresponse algorithms for community-level modelling: review of theory, applications, and comparison to species distribution models. Methods Ecol Evol 9:834–848

Norberg A, Abrego N, Blanchet FG, Adler FR, Anderson BJ, Anttila J, Araujo MB, Dallas T, Dunson D, Elith J, Foster SD, Fox R, Franklin J, Godsoe W, Guisan A, O'Hara B, Hill NA, Holt RD, Hui FKC, Husby M, Kalas JA, Lehikoinen A, Luoto M, Mod HK, Newell G, Renner I, Roslin T, Soininen J, Thuiller W, Vanhatalo J, Warton D, White M, Zimmermann NE, Gravel D, Ovaskainen O (2019) A comprehensive evaluation of predictive performance of 33 species distribution models at species and community levels. Ecol Monogr 89:e01370. https://doi.org/10.1002/ecm.1370

Ovaskainen O, Hottola J, Siitonen J (2010) Modeling species co-occurrence by multivariate logistic regression generates new hypotheses on fungal interactions. Ecology 91:2514–2521

Pearce J, Ferrier S (2000) Evaluating the predictive performance of habitat models developed using logistic regression. Ecol Model 133:225–245

Pearson RG (2007) Species distribution modeling for conservation educators and practitioners: synthesis. American Museum of Natural History. Available at http:// ncep.amnh.org

Perperoglou A, Sauerbrei W, Abrahamowicz M, Schmid M (2019) A review of spline function procedures in R. BMC Med Res Methodol 19:46. https://doi.org/10.1186/s12874-019-0666-3

Peterson AT (2006) Use and requirements of ecological niche models and related distributional models. Biodiv Informatics 3:59–72

Phillips SJ, Dudik M (2008) Modeling of species distributions with Maxent: new extensions and a comprehensive evaluation. Ecography 31:161–175

Phillips SJ, Anderson RP, Shapiro RE (2006) Maximum entropy modeling of species geographic distribution. Ecol Model 190:231–259

Phillips SJ, Dudík M, Elith J, Graham CH, Lehmann A, Leathwick J, Ferrier S (2009) Sample selection bias and presence-only distribution models: implications for background and pseudo-absence data. Ecol Appl 19:181–197

Phillips SJ, Anderson RP, Dudik M, Schapire RE, Blair ME (2017) Opening the black box: an open-source release of Maxent. Oikos 40:887–893

Pollock LJ, Tingley R, Morris WK, Golding N, O'Hara RB, Parris KM, Vesk PA, McCarthy MA (2014) Understanding co-occurrence by modelling species simultaneously with a joint species distribution model (JSDM). Methods Ecol Evol 5:397–406

Pontius RG, Schneider LC (2001) Land-cover change model validation by an ROC method for the Ipswich watershed, Massachusetts, USA. Agric Ecosyst Environ 85:239–248

Pulliam HR (1988) Sources, sinks, and population regulation. Am Nat 132:652–661

Pulliam HR (2000) On the relationship between niche and distribution. Ecol Lett 3:349–361

Robinson SK, Holmes RT (1982) Foraging behavior of forest birds: the relationships among search tactics, diet, and habitat structure. Ecology 63:1918–1931

Schoener TW (1968) The *Anolis* lizards of Bimini: resource partitioning in a complex fauna. Ecology 49:704–726

Scott JM, Heglund PJ, Morrison ML, Haufler JB, Raphael MG, Wall WG, Samson FB (eds) (2002) Predicting species occurrences: issues of accuracy and scale. Island Press, Washington

Sing T, Sander O, Beerenwinkel N, Lengauer T (2005) ROCR: visualizing classifier performance in R. Bioinformatics 21:3940–3941

Smith TM, Huston ML (1989) A theory of the spatial and temporal dynamics of plant communities. Vegetatio 83:49–69

Smith TM, Shugart HH (1987) Territory size variation in the ovenbird: the role of habitat structure. Ecology 68:695–704

Soley-Guardia M, Alvarado-Serrano DF, Anderson RP (2024) Top ten hazards to avoid when modeling species distributions: a didactic guide to assumptions, problems, and recommendations. Ecography 2024:e06852

Stephenson NL (1998) Actual evapotranspiration and deficit: biologically meaningful correlates of vegetation distribution across spatial scales. J Biogeogr 25:855–870

Taverna K, Urban DL, McDonald RI (2005) Modeling landscape vegetation pattern in response to historic land use: a hypothesis-driven approach for the North Carolina Piedmont. Landsc Ecol 20:689–702

ter Braak CJF, Prentice IC (1988) A theory of gradient analysis. Adv Ecol Res 18:271–313

Thuiller W, Georges D, Gueguen M, Engler R, Breiner F (2021) biomod2: ensemble platform for species distribution modeling. R package version 3.5.1. https://CRAN.R-project.org/package=biomod2

Tikhonov G, Abrego N, Dunson DB, Ovaskainen O (2017) Using joint species distribution models for evaluating how species-to-species associations depend on the environmental context. Methods Ecol Evol 8:443–452

Tilman D (1982) Resource competition and community structure. Princeton University Press, Princeton

Tsoar A, Ollouche O, Steinitz O, Rotem D, Kadmon R (2007) A comparative evaluation of presence-only methods for modeling species distribution. Divers Distrib 13:397–405

Urban DL (2002) Classification and regression trees. In: McCune B, Grace JB (eds) Analysis of ecological communities. MjM Software Design, Gleneden Beach, pp 222–232

Urban DL (2023) Agents and implications of landscape pattern: working models for landscape ecology. Springer Nature, Cham

Urban DL, Harmon ME, Halpern CB (1993) Potential response of Pacific Northwestern forests to climatic change: effects of stand age and initial composition. Clim Chang 23:247–266

Urban DL, Miller C, Stephenson NL, Halpin PN (2000) Forest pattern in Sierran landscapes: the physical template. Landsc Ecol 15:603–620

Urban D, Goslee S, Pierce K, Lookingbill T (2002) Extending community ecology to landscapes. Ecoscience 9:200–212

Valavi R, Elith J, Lahoz-Monfort JJ, Guillera-Arroita FG (2021) Modelling species presence-only data with random forests. Ecography 44:1731–1742

Valavi R, Guillera-Arroita G, Lahoz-Monfort JJ, Elith J (2022) Predictive performance of presence-only species distribution models: a benchmark study with reproducible code. Ecol Monogr 92(1):e01486

Vayssieres MP, Plant RE, Allen-Diaz BA (2000) Classification trees: an alternative nonparametric approach for predicting species distributions. J Veg Sci 11:679–694

Velazco SJE, Rose MB, de Andrade AFA, Minoli I, I. and J. Franklin. (2022) flexsdm: an R package for supporting a comprehensive and flexible species distribution modelling workflow. Methods Ecol Evol 13:1661–1669

Venables WN, Ripley BD (1999) Modern applied statistics with S-plus, 3rd edn. Springer, New York

Warren DL, Matzke NJ, Cardillo M, Baumgartner JB, Beaumont LJ, Turelli M, Glor RE, Huron NA, Simoes M, Iglesias TL, Piquet JC, Dinnage R (2021) ENMTools 1.0: an R package for comparative ecological biogeography. Ecography 44:504–511

Warton DI, Blanchet FG, O'Hara RB, Ovaskainen O, Taskinen S, Walker SC, Hui FKC (2015) So many variables: joint modeling in community ecology. Trends Ecol Evol 30:766–779

Whittaker RH (1967) Gradient analysis of vegetation. Biol Rev 42:207–264

Wiens JA (1974) Habitat heterogeneity and avian community structure in North American grasslands. Am Midl Nat 91:195–213

Wilkinson DP, Golding N, Guillera-Arroita G, Tingley R, McCarthy MA (2019) A comparison of joint species distribution models for presence–absence data. Methods Ecol Evol 10:198–211

Wilson JW, Sexton JO, Jobe RT, Haddad NM (2013) The relative contribution of terrain, land cover, and vegetation structure indices to species distribution models. Biol Conserv 164:170–176

Wisz MS, Pottier J, Kissling WD, Pellissier L, Lenoir J, Damgaard CF, Dormann CF, Forchhammer MC, Grytnes J-A, Guisan A, Heikkinen RK, Høye TT, Kühn I, Luoto M, Maiorano L, Nilsson M-C, Normand S, Öckinger E, Schmidt NM, Termansen M, Timmermann A, Wardle DA, Aastrup P, Svenning J-C (2013) The role of biotic interactions in shaping distributions and realised assemblages of species: implications for species distribution modelling. Biol Rev 88: 15–30

Yee TW, Mitchell DD (1991) Generalized additive models in plant ecology. J Veg Sci 2:587–602

Zhu GP, Peterson AT (2017) Do consensus models outperform individual models? Transferability evaluations of diverse modeling approaches for an invasive moth. Biol Invasions 19:2519–2532

Zurrell D, Franklin J, Konig C, Bouchet PJ, Dorman CF, Elith J, Fandos G, Feng X, Guillera-Arroita G, Guisan A, Lahoz-Monfort JJ, Leitao PJ, Park DS, Peterson AT, Rapacciuolo G, Schmatz DR, Schroder B, Serra-Diaz JM, Thuiler W, Yates KL, Zimmermann NE, Merow C (2020) A standard protocol for reporting species distribution models. Ecography 43:1261–1277

Chapter 3
Landscape-Scale Ecological Data

Abstract Data collected on landscapes tend to be multivariate: the variables covary and so may be (at best) redundant or (worse) confounding. Ecological data also are typically spatially structured at various scales. And they are noisy. Ecologists have devised various methods for dealing with these types of data. Here we begin with an overview of the kinds of data sets that ecologists often use, focusing on species abundances and environmental variables measured at the same locations. From these primary data matrices, we generate secondary matrices that are often used in actual analyses; correlation or covariance matrices are familiar examples, along with sample *x* sample distance or dissimilarity matrices. The chapter also illustrates a variety of exploratory data analyses: relationships among species, among environmental variables, and between species and environmental variables. The results of exploratory data analyses will, in turn, inform all subsequent analyses. We explore three general approaches to ecological analyses in the next three chapters.

3.1 Introduction

We began the tasks of landscape analysis and management by looking at species distribution modeling (SDM). This was because this task is fundamental to our larger agenda, but it also motivates much of this book by raising several issues that are common to many ecological analyses. In this chapter, we begin to lay a foundation for the analysis of landscape-scale data, by revisiting some issues discovered in species distribution modeling and setting the stage for the next few chapters.

Ecological data tend to be high-dimensional or multivariate: when we sample, we measure *many* things (e.g., many species, many environmental variables, etc.). These measurements tend to be correlated among themselves, which can complicate inferences we would like to draw from them (recall inferences about variable importance in regression-based SDMs in Chap. 2, Sect. 2.4.2). And most measurements will be spatially structured (autocorrelated)—a complication we simply side-stepped in SDMs (but see Sect. 2S.1.3 in the Supplement to Chap. 2). We will get to these issues in subsequent chapters, but we begin with an overview of typical ecological data sets and their characteristics.

D. L. Urban, *Landscape Ecology: A Task-Oriented Perspective*,
https://doi.org/10.1007/978-3-031-72251-6_3

This chapter is not so much a task itself but rather a set of considerations and exploratory analyses that rightly should precede *any* analytic tasks conducted at the landscape scale. We begin by considering some generalities about ecological data and introduce useful approaches to exploratory data analysis and data screening, data transformations, and data preprocessing in preparation for further analyses covered in subsequent chapters.

3.2 Ecological Data Matrices

Landscape ecologists often work with data sets of a characteristic structure. Typically, these data comprise measurements of a set of variables as observed at a number of locations. Often, these data consist of measures of species abundances (or presence/absence) for a number of sample locations (points or quadrats). For our purposes, these data will be referred to as a *species data matrix*. There might also be an additional data set consisting of other environmental variables measured at the same locations. For example, in addition to species abundances, we might measure soil characteristics, topographic descriptors, and so on. We will refer to these as the *environmental data matrix*. In landscape ecology, we often also tally the *locations* of the samples, in terms of latitude and longitude or other spatial coordinates. There are other ecological data sets, of course. Life-history *traits* mediate the responses of species to environment, and genetic data (as genetic *markers*) mediate the expression of traits in species. All of these data sets, when analyzed in their native form will be referred to as *primary data matrices*.

For purposes of analysis, a primary data matrix might be processed further into another format, in ways that are specific to particular analyses. These reworked data sets will be referred to as *secondary matrices*. A correlation matrix derived from a primary matrix of environmental variables is a familiar example. Here we consider some basic attributes of these data matrices.

3.2.1 The Primary Data Matrix

A primary data matrix might be reworked only slightly from the format in which the data were actually collected in the field. For example, in the field, one might tally the abundance of various plant species in a sample quadrat, with the resulting field form readily converted into a primary data matrix of n rows (sample quadrats) and m columns (species) (Table 3.1).

Of course, the same data could be coded as the *transpose* of this matrix, in which the rows would be species and the columns, samples. The two matrices are equivalent, and the choice of whether to code the data as shown above or as its transpose is largely a matter of coding convenience—depending on how the data were tallied in the field. It is a trivial matter to transpose either matrix as necessary. Often, in the

Table 3.1 An example of a primary data matrix of species on sample quadrats

	Spp 1	Spp 2	Spp 3	...	Spp *m*
Sample 1	0	16	30	...	42
Sample 2	2	8	11	...	34
Sample 3	8	0	15	...	9
...	x_{ij}	...	...	...	...
Sample *n*	0	11	23	...	28

field species are tallied only when present on a sample, and so constructing the full species matrix entails inserting the zeros that correspond to absences (which typically far out-number the presences).

We will refer to this primary data set in matrix notation, for example, **X**. By common convention: a bold-faced, capitalized letter (**X**) is a matrix. A bold lowercase letter (**x**) is a vector (a *column* of elements; its transpose **x′** is a *row vector*). An italicized lowercase letter (x) is a scalar variable. Element x_{ij} of **X** corresponds to row i and column j of that matrix.

3.2.2 Ancillary Data Matrices

In many applications, we will analyze more than one primary data matrix at the same time. In particular, we often will be interested in how the variables in **X** (species) are related to those in **Y** (environment). In landscape ecology, the ancillary variables often include location (i.e., geographic coordinates), which allows the special cases of asking how species abundances or environmental variables vary according to location or proximity (distance apart), and which broaches a whole new realm of ecological questions and statistical inference. Indeed, much of Chap. 6 is concerned with inferences about the relationships among species, environment, and location.

Ancillary data matrices must take the same form and have the same number of rows as the focal primary matrix, although the number of ancillary variables need not match. Thus, if the primary species matrix has n rows and m columns, then the ancillary (environmental) matrix must be n x p (read "*n by p*") for p variables matched by sample locations. Likewise, if the primary matrix is coded m x n the ancillary matrix must be p x n.

There are other possibilities for ecological data matrices, of course. These include trait data (species x trait), genetic markers (individual x marker), and so on. An important task in data analyses is to reconcile the sample units and dimensions of these data matrices. For example, we expect species to sort along environmental gradients based on their life-history traits (e.g., shade tolerance, drought tolerance, nutrient requirements) or behaviors. These traits are "packaged" only as species (they cannot be separated from the species that carry them), and so inferences about gradient response are made at the species level or in terms of the interaction between environmental variables, species, and traits (the so-called fourth-corner problem, because the inference of interest is in an unmeasured "corner" defined by the other three matrices, Dray and Legendre 2008). Similarly, inferences about genetic influences on species distributions depend on genetic markers extracted from individuals, of particular species, and which are tallied at specific locations.

3.2.3 Data Transformations

The primary matrix is often transformed before further analysis. The goal of data transformations is, in general, to reduce numerical problems associated with ecological data. For example, data are frequently transformed so that their distribution more closely meets the assumptions of normality typical of many parametric analyses. But ecological data have quirks beyond distributional issues. Some of these problems include the preponderance of zeros in the species matrix (i.e., most species do not occur in most samples), the fact that species abundances often vary over orders of magnitude (effectively trivializing rare species in many analyses), and so on. While we will consider some common data transformations in conjunction with specific analyses, it is worth anticipating that transformations can have a profound effect on subsequent analyses.

For example, one common transformation of species-compositional data relativizes species values by the maximum value recorded for that species on any sample (i.e., relativizing by column maximum). This effectively treats all species equitably: all species values vary on [0,1] (or [0,100] percent). This means that abundant species do not dominate the analysis; rare species are allowed to contribute equally. Another common data transformation relativizes all elements x_{ij} by their respective row (sample) totals, yielding proportions (or percentages, if multiplied by 100) as new data elements. If this is done for vegetation samples, any subsequent analysis would emphasize the relative composition of the samples as their main feature; the analysis would be blind to gross differences among samples in terms of their actual biomass or stature. The *Wisconsin double* relativization (Bray and Curtis 1957), popular with many plant ecologists, first relativizes by column maximum, then by row sum, to yield values that are proportions of relative abundances.

Another common transformation for species abundance is to take the logarithm to compress a huge natural range of abundance values. This might use $\log_2$, natural log *ln*, or $\log_{10}$ depending on the range of abundance values. (By convention, we would add an appropriately tiny number to each 0 in the data matrix, to avoid taking the log of 0.) While untransformed data would produce results highly biased by the dominant species, log-transformed data down-weight abundant species so they do not overwhelm the influence of less common species. This has the same aim but it less extreme than relativizing species data. The extreme case would be to convert all species abundances to presence/absence (1/0).

For example, consider the (fake) species data in Table 3.2. In terms of abundances, species *B* outweighs the others, and unless the data are relativized (i.e., by

Table 3.2 An example of species data and the implications of relativizing or not

	Spp A	B	C	D	E	*Total*
Smp 1	1	200	5	0	0	206
2	2	400	10	0	0	412
3	0	1	0	10	5	16
4	0	2	0	20	10	32
Max	2	400	10	20	10	

column maximum), this species will dominate numerical analyses while species *A* would have very little influence. Samples 1 and 2 are identical in relative composition but sample 2 has twice the overall abundances. Samples 3 and 4 are also the same, relatively, but they differ in absolute abundance from each other and from samples 1 and 2.

Importantly, there is no right and wrong approach here: Various transformations might make sense for particular applications. But decisions about data transformations cannot be made lightly, as they affect both the numerical results of an analysis and its ecological interpretation.

Issues about data transformations also apply to ancillary data, particularly if these variables are in different measurement units. For example, ancillary environmental variables might include elevation (m), soil pH (log of H ion concentration), percent clay in the B soil horizon, and exchangeable cations in μeqs. If these variables are to be used separately, transformation might not be an issue. If they are to be combined in an analysis, then they *must* be transformed into consistent units. Converting all variables to percentages (relative to the maximum for each variable) or standardizing to *z*-scores are common approaches to reconciling disparate measurement units. Converting to commensurate units, of course, still invites further transformations to deal with distributional issues (normality, skewness).

3.2.4 Secondary Data Matrices

There are two ways in which a primary data matrix can be processed for subsequent analyses. One approach focuses on relationships among the variables (e.g., species or environmental factors), while the alternative approach focuses on relationships among the samples. Many multivariate methods proceed from either one or the other of these *secondary* matrices. It should be clear that these two modes of analysis are complementary approaches to looking at the same data. Because relationships among variables are often indexed as correlation coefficients (elements *r* of matrix **R**), approaches that emphasize patterns among variables are sometimes termed *R-mode* analyses. By comparison, approaches that emphasize relationships among samples (quadrats) are termed *Q-mode.* Legendre and Legendre (2012, Chapter 7) discuss additional variations on these approaches, including cases where the data include repeated measures (i.e., samples x variables x time).

3.2.4.1 Relationships Among Variables

While relationships among variables are often summarized in terms of correlations, these are not the only approaches. Here we consider a few options.

Covariance and Correlation Matrices A familiar processing of a primary data matrix **X** is to summarize it in terms of associations among the measured variables. A common form of this is the variance-covariance matrix **C**. The matrix **C** is computed from a few basic calculations on the primary data matrix. Recall that the mean for any column j of the n x p matrix above is:

$$\overline{x}_j = \frac{1}{n} \sum_{i=1}^{n} x_{ij} \tag{3.1}$$

for variable x_j measured over n samples. Similarly, the sample variance for the jth variable is:

$$s_j^2 = \frac{1}{(n-1)} \sum_{i=1}^{n} \left(x_{ij} - \overline{x}_j\right)^2 \tag{3.2}$$

and the covariance between two columns j and k is:

$$c_{jk} = \frac{1}{(n-1)} \sum_{i=1}^{n} \left(x_{ij} - \overline{x}_j\right)\left(x_{ik} - \overline{x}_k\right). \tag{3.3}$$

The variance-covariance matrix **C**, known as the sum-of-squares and cross-products (SSCP) matrix before division by $(n - 1)$, is a p x p matrix in which the elements along the diagonal are the sample variances for each variable and the off-diagonal elements are covariances between pairs of variables (Table 3.3).

This matrix is often further transformed by dividing the elements by their respective pooled variances:

Table 3.3 An example of a covariance matrix **C**

	Var 1	Var 2	Var 3	...	Var p
Var 1	s_1^2	c_{12}	c_{13}	...	c_{1p}
Var 2	c_{21}	s_2^2	c_{23}	...	c_{2p}
Var 3	c_{31}	c_{32}	s_3^2	...	c_{3p}
...	c_{jk}	...	...	s_j^2	...
Var p	c_{m1}	c_{m2}	c_{m3}	...	s_p^2

Table 3.4 A tally matrix from which to derive indices of species association

Species association		**Species B**		Row totals
		Present	*Absent*	
Species A	*Present*	a	b	m = a + b
	Absent	c	d	n = c + d
Column totals		r = a + c	s = b + d	N = m + n

$$r_{jk} = \frac{c_{jk}}{s_j s_k} \tag{3.4}$$

yielding a correlation matrix **R**. In this, the diagonals are all 1.0 and the off-diagonals *rjk* are product-moment correlations on [−1,1]. Note that if the original variables have different measurement units, the matrix **C** might be dominated by the variable (s) with the largest values based on their measurement *units* rather than the greatest inherent variability; using the standardized matrix **R** treats all variables equitably.

Principal components analysis (PCA) and factor analysis (FA), two very common multivariate techniques (see Chap. 4), proceed from the secondary matrices **C** or **R**. Note that the correlations are linear, and so PCA and FA are linear models. These are especially common in analyzing environmental variables, less so for species abundances because the pairwise relationships among species are often not linear (and see below).

Indices of Association Among Species There is, of course, no reason why the association among species (or other variables) needs to be described by covariance or correlation. There are other indices of association that might be more appropriate for certain applications. In particular, for primary data matrices where the variables are species, it might be appropriate to summarize species associations in terms of their tendency to co-occur on the same sites (or conversely, to *not* occur on the same sites). There are dozens of indices of interspecific *association* (Ludwig and Reynolds 1988). Most are based on various combinations of basic tallies of presence/absence from the primary data matrix (Table 3.4). To be clear, these are tallies for pairs of species *A* and *B* as they occur over all samples.

From the tallies *a*, *b*, *c*, and *d*, a wide variety of association indices have been devised. For example, the Jaccard Index summarizes the proportion of samples where two species co-occur relative to the proportion of samples where at least one species occurs:

$$J = \frac{a}{(a + b + c)}. \tag{3.5}$$

Many statistical ecology textbooks review a variety of association indices (e.g., Ludwig and Reynolds 1988; McCune and Grace 2002; Legendre and Legendre 2012). Much of the proliferation of these indices stems from a concern among ecologists as to the meaning of joint occurrences as compared to *joint absences*. Clearly, two species occurring together on a site implies that they share *something* ecologically; perhaps they have similar environmental tolerances or niches. Conversely, their *not* occurring on the same site does not imply the same level of information. We cannot know why a species does not occur on a site, and consequently, we cannot assume that its absence tells us anything with confidence. To anticipate, this nebulous meaning of joint absences will arise as a complication in a number of analyses. Computationally, a common response to this issue is simply to ignore joint absences (note that term d is not included in Eq. 3.5).

Given an appropriate index of species association, a secondary matrix can be constructed in a manner analogous to matrix **C** or **R** (above): an m x m matrix **A** summarizing the pairwise associations among all m species in the primary matrix.

3.2.4.2 Relationships Among Samples: Distance Matrices

An alternative approach to summarizing the primary data matrix in terms of the variables is to focus on the relationships among the samples. In this, we describe the correspondence among pairs of samples in much the same way we produced the covariance matrix **C**, correlation matrix **R**, or association matrix **A** (above)—except that we now focus on rows instead of columns in the primary data matrix.

There are two complementary ways of summarizing relationships among pairs of samples, by focusing on their similarities or, reciprocally, their differences. In general, the former approach gives rise to a *resemblance* or *similarity matrix* while the latter yields a *dissimilarity* or *distance matrix*. A similarity index is typically the additive inverse or complement of a distance index, and so the approaches are equivalent (with some exceptions that we revisit later).

One familiar measure of ecological distance is simply the multivariate form of the familiar Euclidean distance metric:

$$ED_{ij} = \sqrt{\sum_{k=1}^{p} \left(x_{ik} - x_{jk}\right)^2} \quad (3.6)$$

for all variables p on the two samples i and j. This procedure, conducted for all pairs of samples, produces an n x n matrix **D** summarizing the pairwise ecological distances between samples. (Note here that the subscripts i and j still index rows and columns of the primary matrix, but that both items now refer to samples in the secondary matrix.)

For a data set with many correlated variables, Euclidean distances overestimate distances. One way to address this is to compute Mahalanobis distances, which are essentially Euclidean distances corrected for the covariance among variables (Goslee and Urban 2007).

One common instance of a distance matrix is actual distances: pairwise distances (typically Euclidean) between samples based on their geographic coordinates. Geographic distance matrices will feature prominently in Chap. 6.

For species data, there are various similarity indices that are analogous to the species association indices described above. For example, there is a Jaccard analog of Eq. (3.5) computed between samples i and j, with the presence/absence tallies comparing two samples as summed over all species (i.e., essentially the same index but on the transpose of the primary data matrix). Computed this way, Jaccard's sample similarity indexes the proportion of species presences that are shared in common by two samples. This version of Jaccard's index is sometimes called the *coefficient of community*.

Similarly, the familiar Bray-Curtis (aka Sorenson's) index of sample similarity is:

$$S = 2\frac{W}{(A + B)} \tag{3.7}$$

where W is the number of species shared in common and A and B are the numbers of species on each of the two samples. This index compares the shared species to the average richness of the two samples. Subtracting this from 1.0 converts it to a dissimilarity index.

A more quantitatively sensitive index that considers actual abundances rather than presence/absence is the Bray-Curtis (1957) dissimilarity index, also known as *percent difference*:

$$PD = \frac{\sum_{k=1}^{m} \left| x_{ik} - x_{jk} \right|}{\sum_{k=1}^{m} \left(x_{ik} + x_{jk} \right)}. \tag{3.8}$$

An analogous coefficient computed as *percent similarity* uses the minimum value of x_{ij} or x_{jk} as the numerator of Eq. (3.8), which emphasizes shared abundances. Both are very commonly used in community analysis (see Chaps. 4, 5, and 6).

Indices such as the Bray-Curtis forms are often computed from data that have been transformed, e.g., relativized by column maximum, row sum, or both. While any of these transformations can be justifiable on ecological grounds, it is worth emphasizing again that they will likely yield different results from the same subsequent analysis.

As was the case with species association indices, there is a huge variety of distance (or similarity) indices used by ecologists (Legendre and Legendre 2012 review more than two dozen indices). The indices tend to vary according to a number of numerical issues that plague ecological interpretation (e.g., the joint-absences issue). Beyond this, there is some concern over whether the distance measure is *metric*. A metric distance measure satisfies the geometric condition that, for three samples a, b, and c:

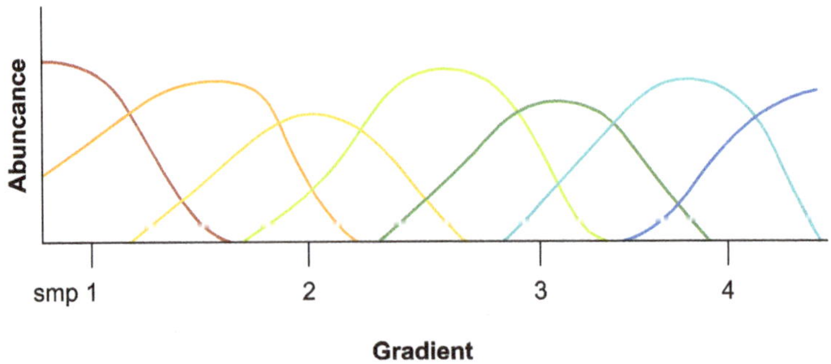

Fig. 3.1 A case where species (colors) sort along a long environmental gradient. Samples 1 and 2 share species and so have a distance $d_{12} < 1.0$. Distances d_{13} and d_{14} are both 1.0, though sample 4 is ecologically farther from 1 than sample 3 is. One solution is to *extend* d_{14} as the sum of d_{12}, d_{23}, and d_{34}

$$d_{ac} \leq (d_{ab} + d_{bc}). \tag{3.9}$$

A metric index also satisfies three additional properties: that if $a = b$, then $d_{ab} = 0$; if $a \neq b$, then $d_{ab} > 0$; and $d_{ab} = d_{ba.}$ Semi-metric or nonmetric indices fail to meet one or more of these conditions. A metric may also satisfy the more stringent condition and be strictly Euclidean (i.e., replacing "less than or equal to" with "equals" in Eq. 3.9). Whether an index is metric has implications in ordination studies because only metric indices can be mapped directly into a Euclidean ordination space (and, by extension, simply drawn on a physical page); we return to this in Chap. 4.

One issue that can confound distance measures is *saturation* of the index. That is, once two samples share no species in common, their distance is 1.0 and cannot reflect any further ecological dissimilarities. This is often an issue with samples collected over long environmental gradients, i.e., systems with high *beta*-diversity. One solution to this problem is to estimate long distances ($d > 1.0$) as the sum of distances between pairs of samples that are closer together ($d < 1.0$). These are termed *extended* distances, and they are typically estimated as the shortest stepping-stone path between two samples that have $d = 1.0$, using as stepping-stones only samples for which the distances are <1.0 (Bradfield and Kenkel 1987; De'Ath 1999) (Fig. 3.1).We will revisit some of these indices as we turn to ordination methods, some popular versions of which proceed from distance matrices. Principal coordinates analysis and nonmetric multidimensional scaling (Chap. 4) proceed from a sample *x* sample distance or dissimilarity matrix. Distance matrices are also used in some popular classification techniques (Chap. 5) and in Mantel tests (Chap. 6).

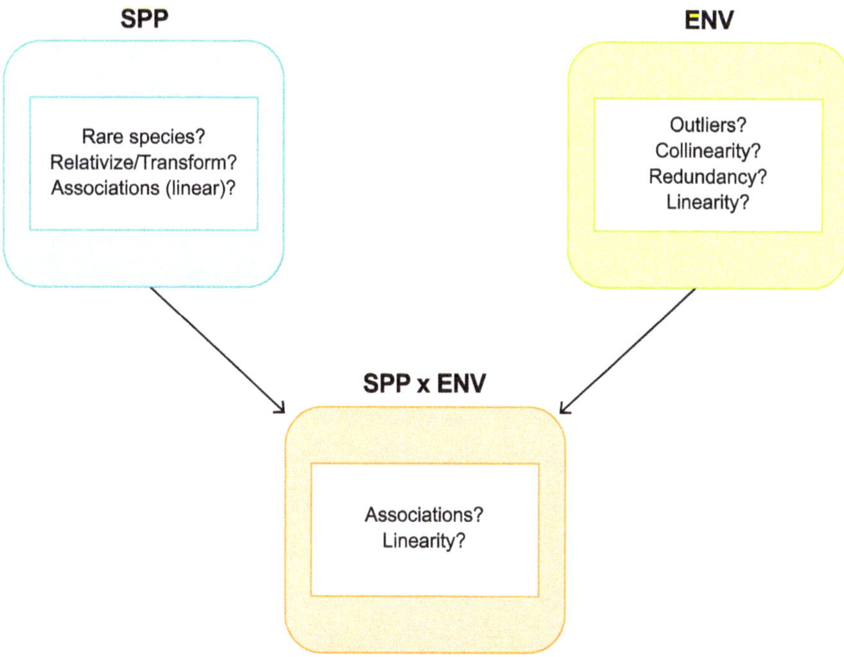

Fig. 3.2 Workflow for exploratory data analysis (EDA) for among species (SPP), among environmental (ENV), and between species and environmental variables. Details, of course, depend on the particular data sets

3.3 Exploratory Data Analysis

An important part of any analysis of ecological data is to preview the data in a very simple and straightforward fashion, by *looking at it.* There are a number of ways to display data, and these "looks" convey several different perspectives on the data. These preliminary inspections should then inform subsequent analyses.

As noted, the primary data sets in many ecological studies are matrices of species and environmental variables measured over the same samples. Preliminary inspection would focus on each of these matrices separately, as well as any relationships between the matrices, that is, *among* species, *among* environmental variables, and *between* species and environmental variables (Fig. 3.2).

The following illustrations are examples of methods for the preliminary inspection of these data sets. The data are measures of abundance of forest trees and environmental factors, collected in Sequoia National Park in the Sierra Nevada of California, USA (Urban et al. 2002). The data sets include species abundances, environmental factors (see below), and geographic coordinates (UTM easting and northing, m). The 99 samples were distributed in clusters of 3–4 quadrats (each 20 x 20 m), with the clusters distributed over a long elevation gradient (recall sampling designs in Chap. 1, Sect. 1.3.3 and Fig. 1.7) (Fig. 3.3). We will return to

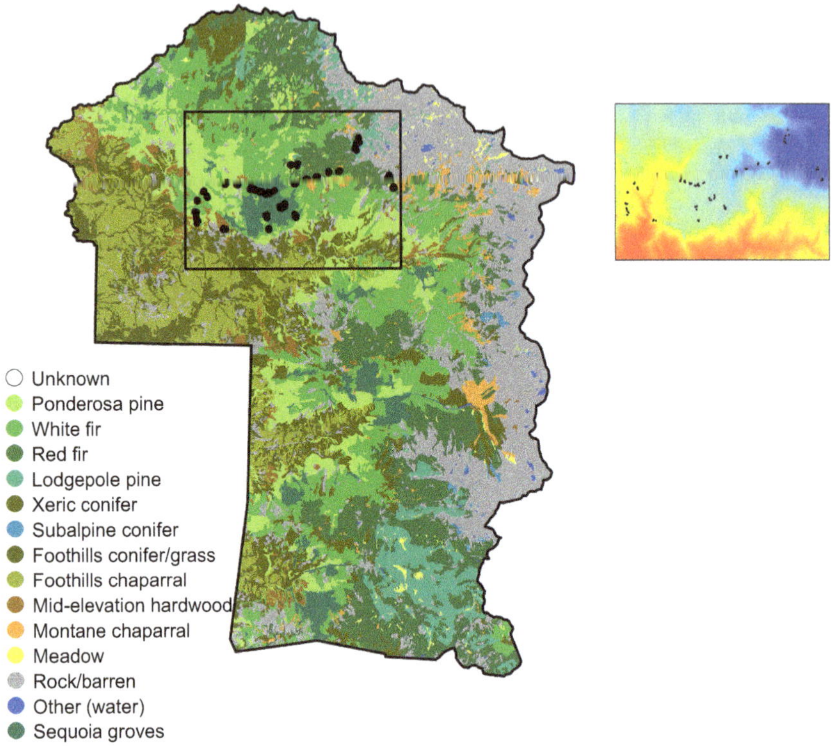

Fig. 3.3 Distribution of 99 samples in clusters distributed over an elevation gradient in Sequoia National Park, California (USA). The inset map is a digital elevation model, ranging from ~800 (red: warm and dry) to ~3300 m (blue/violet, cool and wet). The pattern of samples roughly follows roads and trails. Vegetation map courtesy Sequoia-Kings Canyon National Park. (Reproduced with permission from Urban (2023); permission conveyed via Copyright Clearance Center, Inc.)

the spatial structure in Chap. 6; here we focus on the species and environmental variables.

3.3.1 Inspection of Species Data

The species data set has 17 tree species, each tallied in terms of basal area (the summed cross-sectional area of all trees per species, in m^2ha^{-1}) (Table 3.5). There are a number of useful and informative methods for visually inspecting species data. In particular, a variety of graphical approaches focus on relative species abundances.

One useful summary is the familiar dominance-diversity curve in which species abundance is plotted against rank species sequence (i.e., species ranked in order of decreasing abundance) (Fig. 3.4). Such curves often show a steeply decreasing initial slope, indicating that much of the importance is attributable to a very few species.

Table 3.5 Tree species sampled in the Sierran data set

Code	Scientific name	Common name
ABco	*Abies concolor*	White fir
ABma	*Abies magnifica*	Red fir
ACma*	*Acer macrophyllum*	Bigleaf maple
ARvi*	*Arctostaphylos viscida*	Whiteleaf manzanita
CAde	*Calocedrus decurrens*	Incense cedar
CEin*	*Ceanothus integerrimus*	Deer brush
COnu	*Cornus nuttallii*	Pacific dogwood
PIco	*Pinus contorta*	Lodgepole pine
PIje	*Pinus jeffreyi*	Jeffrey pine
PIla	*Pinus lambertiana*	Sugar pine
PImo	*Pinus monticola*	Western white pine
PIpo	*Pinus ponderosa*	Ponderosa pine
QUch*	*Quercus chrysolepis*	Canyon live oak
QUke	*Quercus kelloggii*	California black oak
SEgi	*Sequoiadendron giganteum*	Giant sequoia
TOca*	*Torreya californica*	California torreya
UMca*	*Umbellularia californica*	California bay

Six uncommon species (frequency <0.05) were deleted (tagged *)

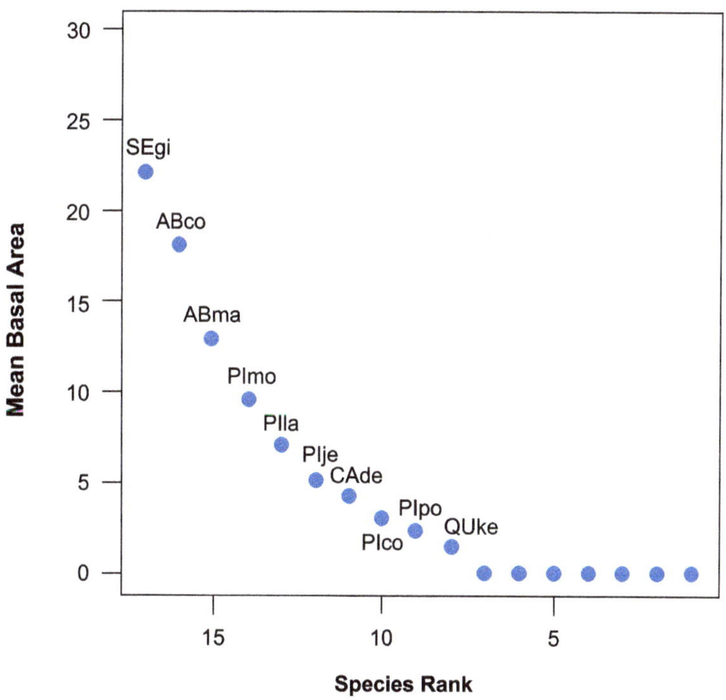

Fig. 3.4 Dominance-diversity curve for 17 tree species sampled in Sequoia National Park in California (data from Urban et al., 2002). Species are ordered on the abscissa according to their relative basal area (see species codes in Table 3.5)

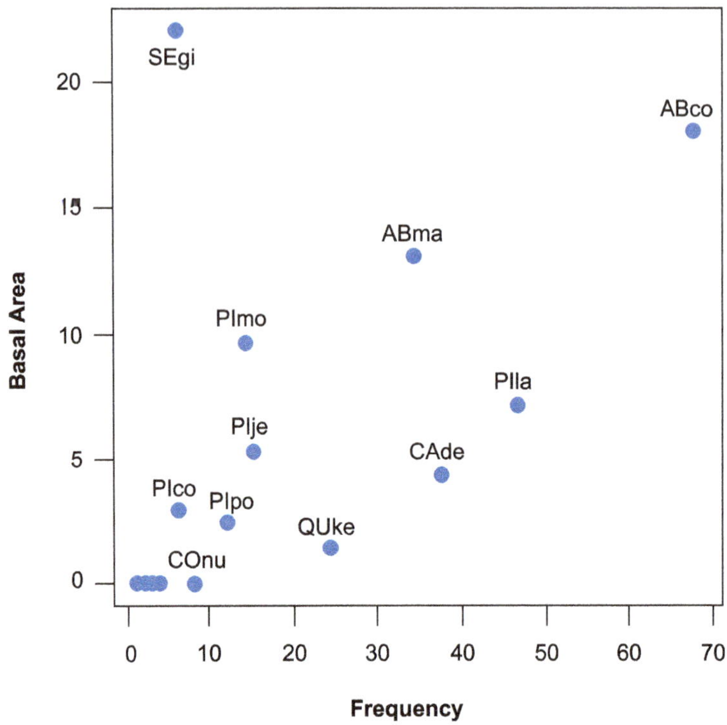

Fig. 3.5 Plot of species in diagnostic space defined by frequency (% of samples on which the species occurred) and dominance (as basal area), for tree species in Sequoia National Park. (Data from Urban et al. 2002; species codes in Table 3.5)

More importantly, the long flat tail of this curve confirms that *most species are uncommon or rare*. The presence of a large number of rare species in the data set might be interesting ecologically, but it can degrade multivariate techniques that are plagued by the prevalence of zeros in the matrix. Indeed, one frustration in ecological analysis is that species that are compelling for conservation purposes precisely *because* they are rare are also hard to analyze for that same reason.

There is no strict rule for dealing with uncommon species, but a common response is to delete species that occur in fewer than 5% of the samples. In the Sierran data set, this removes six species—the last six in the flat tail of the dominance-diversity curve.

Another useful summary entails plotting species positions in a two-space defined by frequency (percent of samples occupied) and some measure of abundance (e.g., basal area, percent cover, density—but be aware that these measures might produce different results). The resulting graph (Fig. 3.5) highlights species that are widespread versus locally distributed, and rare versus abundant (for trees, reflecting their density and stature). Beyond offering a concise visual summary of community structure, the graph also can indicate species that might bias subsequent multivariate analyses. For example, a species that is widespread and abundant will dominate

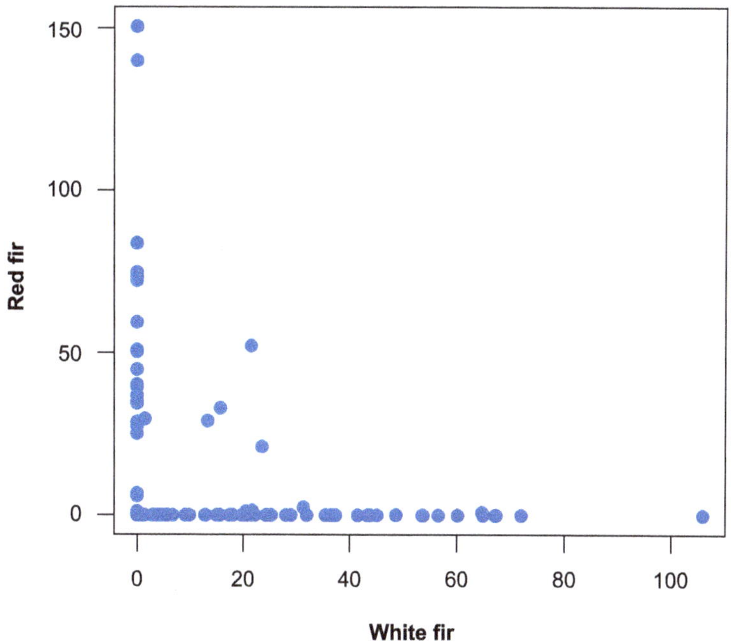

Fig. 3.6 A plot of the abundances (m^2ha^{-1} basal area) of two common species in the Sierran data set, red and white fir, emphasizing their nonlinear relationship

nearly any multivariate analysis; if that species is similarly abundant everywhere, it is not very interesting ecologically. By contrast, a large number of species that are quite locally distributed and also rare will confound some techniques, primarily because these columns of the primary data matrix will contain mostly zeros; in the limit, very rare species are merely noise. Ideally, species that occur with moderate frequency (few zeros) but variable abundance would be handled best by most multivariate techniques.

Of the two graphs, the plot of relative abundance versus frequency would seem to provide more information, as it separates "commonness" into two components. For example, giant sequoias (SEgi in Fig. 3.5) are rather uncommon in the data set (low frequency) but are enormous where they do occur (high basal area). Again, the species in the lower left corner of this figure are rare and small; these were deleted for subsequent analyses.

We touched on the issue of linear versus nonlinear relationships among variables when we considered indices of species association (Sect. 3.2.4). It is worth emphasizing this here, by showing the pairwise relationship between two tree species in the Sierran data set (Fig. 3.6). If the two species replaced each other in a zero-sum sense, we might expect one to increase when the other decreases, and vice versa. This would yield a negative and perhaps nearly linear relationship between the two species. But the reality is that most species do not occur in most places, so that many of the samples have neither species (the crowd of points in the lower-left

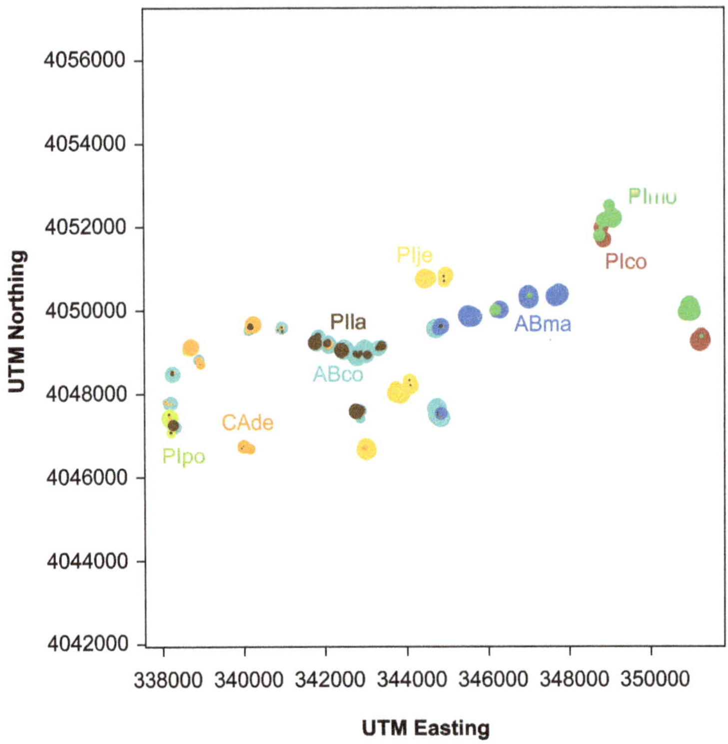

Fig. 3.7 Species abundances mapped into the study area in Sequoia National Park, CA. Symbol sizes are proportional to relative basal area per species (see species codes in Table 3.5)

corner of the figure). This yields a strongly nonlinear relationship between the species. Extended to three or more species, the distribution shows "spikes" for each common species on its own axis, with an accumulation of samples in the lower corner, near the origin. McCune and Grace (2002) refer to this as a "dust bunny" distribution (samples accumulate in the lower corner), and it underscores why linear models are often not a good fit for species data.

Finally, it is always useful to examine a plot of species occurrences or abundances in geographic space, a map. With species abundances, one way to display this is with a color bubble plot, in which symbols are colored by species and sized proportional to abundances (Fig. 3.7).

3.3.2 Preview of Environmental Patterns

The environmental data set for the Sierran case study has 23 variables, including a variety of topographic and soil factors (Table 3.6).

Table 3.6 Environmental variables used in the Sierran case study

Variable	Description	Note
Elevation	Elevation (m)	From GPS, DEM
Slope	Maximum slope (°)	In field
TAspect	Transformed aspect[1]	−1 × COS(45-aspect)
TSI	Terrain shape index[2]	<1: dome; >1: cove
xDepth	Mean soil depth (cm)	To maximum of 100 cm
sDepth	Standard deviation of depth	From 30 measurements
pH	pH	
C	Soil carbon (%)	Total carbon
N	Nitrogen (%)	Total nitrogen
C:N	Carbon:nitrogen ratio	
P	Phosphorus (μg/g)	Total exchangeable
Ca	Calcium (cmol(+)/kg)	
Mg	Magnesium (cmol(+)/kg)	
K	Potassium (cmol(+)/kg)	
Ac	Acidity (cmol(+)/kg)	Total exchangeable
ECEC	Cation exchange capacity	Σ(Ca, Mg, K, Ac)
BS	Base saturation (%)	Σ(Ca, Mg, K)/ECEC
xLitter	Mean litter depth (cm)	Depth to mineral soil
Clay	Clay (%)	Particles <2 μm
Silt	Silt (%) 3	Particles 2–50 μm
Sand	Sand (%)	Particles >50 μm

Described more fully in Urban et al. (2002)

Preliminary inspection of relationships among environmental variables can be conducted through simple correlation. A correlation matrix highlights pairs of variables that are either positively or negatively associated and so provides an obvious foundation for subsequent analyses via principal components or factor analysis—techniques based on covariance analysis.

One caveat to bear in mind, however, is that correlation analysis is linear and so cannot reflect accurately any relationships that are nonlinear. The safest way to guard against misinterpreting nonlinear relationships is to simply graph variables against one another in a pairwise fashion (Fig. 3.8). Several commercial statistics packages have graphical utilities to produce pages of "thumbnail" graphics that, while tiny, are more than adequate for preliminary data inspection. Manly (2004) refers to these plots as a *draftsman's display* of the data; they are also known as *pairs plots.*

3.3.2.1 Outliers

Graphical displays provide an easy means of identifying unusual samples (outliers) in data sets, whether these are species or environmental variables. As in any statistical analysis, outliers can bias or confound results by lending undue influence

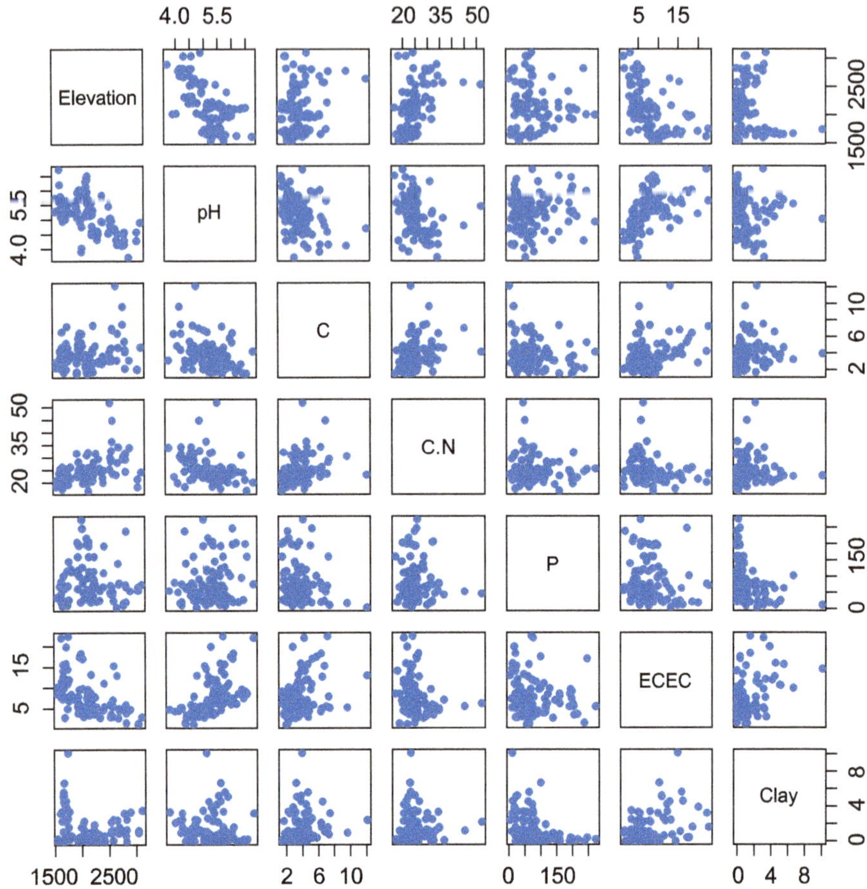

Fig. 3.8 Draftsman's display (pairs plot) of a set of environmental factors measured in Sequoia National Park (see variable codes in Table 3.6)

to the statistical estimate or fit. Most would agree that extreme outliers should be removed from the data sets before further analysis.

In other instances, numerical approaches to finding outliers are easier to use. For example, one might compute summary statistics on the raw data or a secondary matrix and look for samples that are much more than three standard deviations beyond the pooled sample mean. (Be aware that the inclusion of outliers in the data might confound the estimate of the standard deviation.)

3.3.2.2 Collinearity

Sometimes variables are so highly correlated that they are essentially duplicate versions of the same variable. Collinearity is a vexing problem in statistical analyses because it confounds the interpretation of individual variables—as we discovered in fitting habitat models as multiple regressions in Chap. 2 (Sect. 2.4.2). In analyses that combine variables into synthetic new variates, such as principal components analysis (Chap. 4), the inclusion of collinear variables inflates the relative importance of the synthetic variate by loading it with arbitrarily redundant information.

One solution to that issue, in part, is to screen variables so that none are correlated too strongly. Dormann et al. (2013) suggested $|0.70|$ as a cutoff value for correlations. This is a practical recommendation based on empirical trials (not statistical theory). Equivalently, some analysts use *variance inflation factors* (VIFs) as a means to screen redundant variables. VIFs are often computed in post-processing a regression analysis, a procedure outlined by Legendre and Legendre (2012, their Chapter 10). An alternative approach that can be done in preprocessing or exploratory data analysis is based on the inverse of the correlation matrix **R**. The inverse of a matrix ($\mathbf{X}^{-1}$) is such that **X** matrix-multiplied by $\mathbf{X}^{-1}$ yields the identity matrix **I** (a diagonal matrix with 1's in the diagonal and 0's elsewhere). In this instance, the diagonal of the inverse matrix $\mathbf{R}^{-1}$ provides the VIFs from the correlation matrix (Legendre and Legendre 2012). As with correlations themselves, there is not a formal rule about how large is too large for a VIF; but general guidance would be to remove variables with VIFs much larger than 5–10.

As a rule, collinear variables should be edited from a primary data matrix. While there is no strict guidance on this, if two variables are redundant, one of the pairs should be removed. As practical guidance, one might adopt a few decision rules for this:

1. Of a collinear pair, choose to retain the variable with the fewest or lowest correlations with other variables.
2. In a multiple regression model, retain the variable with the most readily interpreted (in a narrative sense) relationship to the dependent variable (and see below).
3. For variables that are devised as alternative measures of essentially the same thing, choose the one that has the highest univariate correlation with the dependent variable or the strongest correlation with synthetic variates in exploratory analyses. This last case would include variables measured at alternative scales—common in landscape ecology.

Collinearity can also arise in sets of variables that are mutually redundant. For example, soil texture is typically reported in terms of sand, silt, and clay fractions—which fractions sum to 100%. That is, one of the variables is a linear combination of the others. Thus, the three variables are collinear even if none of the pairwise correlations is quite strong; only two of the three need to be used.

In the Sierra case, many variables were strongly correlated and several redundant variables were removed. This resulted in a set of 13 environmental variables over 99 samples.

3.3.3 Species-Environment Relationships

Many of the same approaches outlined above can be applied to relationships between species abundances and environmental variables. Simple correlations between species and environment can provide a quick look at these relationships. Of course, there is the concern that species responses to environmental variables might not be linear; indeed, much of the work in plant community ecology explicitly assumes that these relationships are nonlinear (ter Braak and Prentice 1988, Austin and Smith 1989).

Species distributional data from the Sierra Nevada illustrate this point quite nicely. Simple correlation shows that four common species are each correlated, though weakly, with elevation (Table 3.7).

But, as a graph more clearly shows, the species sort out rather nicely along an elevation gradient (Fig. 3.9). Importantly, in this case, a linear model of species response would be an inappropriate choice for a descriptive model, and, similarly, a multivariate analysis based on linear relationships would be misleading if not completely confounded. The issue of linear as compared to nonlinear also depends, in part, on the length of the gradient: even if species response is nonlinear, over a short gradient, a linear model might fit (we will revisit this in Chap. 4.)

Table 3.7 Linear correlations between elevation (m) and basal area ($m^2\ ha^{-1}$) in Sequoia National Park in California (species names in Table 3.5)

	Ponderosa pine	White fir	Red fir	Western white pine
Elevation	−0.273	−0.412	0.486	0.526

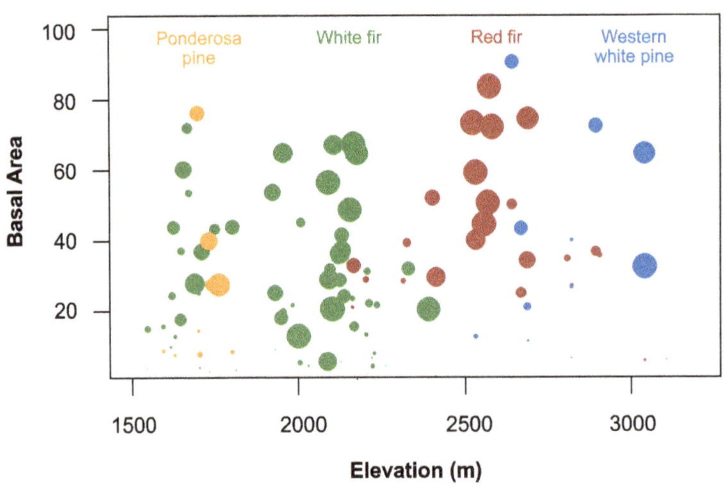

Fig. 3.9 Distribution of four common species in Sequoia National Park, with respect to elevation (m), illustrating the nonlinear trends that compromise many multivariate methods (data are the same as in Table 3.7; symbols are sized proportional to abundance)

The growing availability of user-friendly software packages makes it easy to simply "point and click" to conduct increasingly sophisticated multivariate analyses. Unfortunately, this also makes it increasingly tempting to skip the preliminary work and to dive right into the fun stuff. Remember, time spent in preliminaries will be repaid many times over in time saved interpreting subsequent, more complicated analyses. *Take time to look at the data!*

3.4 Reporting

In technical writing, data are typically described as a subsection of a Methods section, before the detailed description of the Analyses. The Data description should include details on the primary data set(s), any editing or transformation of the original data, and any insights from exploratory data analysis that would inform or constrain subsequent analyses.

☑ Summary description of the rows and columns (names, units, codes; probably in tables)
☑ Sample sizes (rows and columns, species and environmental factors)
☑ Any editing of the rows or columns (e.g., removing redundant/collinear environmental variables, uncommon species, or removing outlier samples)
☑ Any relativizations or other transformations of the original data, with an explanation of why these were done
☑ Insights from EDA that would influence subsequent analyses, including which analyses might be used or how the analyses would be conducted (these would be detailed separately)

We will revisit some of these details as we delve into subsequent analyses in the next few chapters.

3.5 Further Reading

Any text on multivariate analysis in ecology will include some preliminary discussion of the nature of ecological data in an introductory chapter. Among the more focused for ecologists are McCune and Grace (2002), Legendre and Legendre (2012), and Dale and Fortin (2014). Other (sometimes older) texts are still useful (e.g., Gauch 1982; Pielou 1984; Digby and Kempton 1987; Ludwig and Reynolds 1988; Tabachnick and Fidell 1996; Manly 2004). The first chapters of these books are the right entry points in most cases.

For those who are not comfortable with the notation and terminology of matrices, both Manly (2004) and Legendre and Legendre (2012) include overviews of matrix algebra in their second chapters; McCune and Grace (2002) lead with this in their first chapter, while Digby and Kempton (1987) include this material as an appendix.

Most of these texts include some discussion of data transformations and the calculation of secondary data matrices. Legendre and Legendre (2012, Chapter 7) is especially thorough.

A number of basic texts include chapters dedicated to data preview, inspection, or visualization: Digby and Kempton (1987, their Chapter 2), Tabachnick and Fidell (1996, Chapter 3), McCune and Grace (2002, Chapter 2), and Manly (2004, Chapter 3).

Beyond this, the utilities for data inspection are somewhat software-specific. R (R Core Team 2022) takes some pride in its graphics and exploratory procedures, and many of the illustrations in this chapter were produced in R. Borcard et al. (2011) use R to cover many topics in EDA (their Chapter 2) and secondary data matrices (their Chapter 3). Plant (2012) covers EDA for spatial data in Chap. 6.

But nearly any statistics or graphics package can provide the basic tools for this crucial part of multivariate analysis. For general insights and recommendations for the graphing and display of scientific data, see the classics such as Cleveland (1985, 1993) or Tufte (1983, 1990).

3.6 Summary and Prospectus

Ecological data sets include tallies of species, environmental factors, and other variables measured over samples that are typically also indexed by location. These data sets have characteristic features that can have profound influences on numerical analyses. For species data, a preponderance of zeroes (absences) and nonlinear relationships among species or between species and environmental factors can confound many analyses. With environmental factors, natural correlations lead to redundancies or collinear relationships that can be similarly vexing in multivariate analyses.

Ecologists have devised approaches to deal with these problematic issues with their data. These include rule-of-thumb guidance on removing uncommon species or redundant variables and approaches for relativizing or transforming variables prior to other analyses.

Exploratory data analyses (EDA) can be hugely informative in identifying problematic data issues so they can be dealt with appropriately and to reveal patterns in the data that might inform the choice of techniques in subsequent analyses. In the next few chapters, we will use the results of EDA as a point of entry into more focused analyses.

References

Austin MP, Smith TM (1989) A new model for the continuum concept. Vegetatio 83:35–47

Borcard D, Gillet F, Legendre P (2011) Numerical ecology with R. Springer, New York

Bradfield GE, Kenkel NC (1987) Nonlinear ordination using flexible shortest path adjustment of ecological distances. Ecology 68:750–753

Bray JR, Curtis JT (1957) An ordination of the upland forest communities of southern Wisconsin. Ecol Monogr 27:325–349

Cleveland WS (1985) The elements of graphing data. Wadsworth, Monterey

Cleveland WS (1993) Visualizing data. Hobart Press, Summit

Dale MRT, Fortin M-J (2014) Spatial analysis: a guide for ecologists, 2nd edn. Cambridge Univ. Press, Cambridge

De'ath G (1999) Extended dissimilarity: a method of robust estimation of ecological distances from high beta diversity data. Plant Ecol 144:191–199

Digby PGN, Kempton RA (1987) Multivariate analysis of ecological communities. Chapman and Hall, London

Dormann CF, Elith J, Bacher S, Buchmann C, Carl G, Carré G, García Marquéz JR, Gruber B, Lafourcade B, Leitão PJ, Münkemüller T, McClean C, Osborne PE, Reineking B, Schröder B, Skidmore AK, Zurell D, Lautenbach S (2013) Collinearity: a review of methods to deal with it and a simulation study evaluating their performance. Ecography 36:27–46

Dray S, Legendre P (2008) Testing the species traits-environment relationships: the fourth-corner problem revisited. Ecology 89:3400–3412

Gauch HG (1982) Multivariate analysis in community ecology. Cambridge University Press, Cambridge

Goslee SC, Urban DL (2007) The ecodist package for dissimilarity-based analyses of ecological data. J Stat Softw 22:7

Legendre P, Legendre L (2012) Numerical ecology, 3rd English edition. Elsevier, Amsterdam

Ludwig JA, Reynolds JF (1988) Statistical ecology: a primer on methods and computing. Wiley and Sons, New York

Manly BFJ (2004) Multivariate statistical methods: a primer, 3rd edn. Chapman and Hall, London

McCune B, Grace JB (2002) Analysis of ecological communities. MjM Software Design, Gleneden Beach

Pielou EC (1984) The interpretation of ecological data: a primer on classification and ordination. Wiley, New York

Plant RE (2012) Spatial data analysis in ecology and agriculture using R. CRC Press, Boca Raton

R Core Team (2022) R: a language and environment for statistical computing. R Foundation for Statistical Computing, Vienna, Austria. URL https://www.R-proj-ect.org

Tabachnick BG, Fidell LS (1996) Using multivariate statistics, 3rd edn. Harper-Collins, New York

ter Braak CJF, Prentice IC (1988) A theory of gradient analysis. Adv Ecol Res 18:271–313

Tufte ER (1983) The visual display of quantitative information. Graphic Press, New Haven

Tufte ER (1990) Envisioning information. Graphic Press, New Haven

Urban DL (2023) Agents and implications of landscape pattern: working models for landscape ecology. Springer-Nature, Cham

Urban D, Goslee S, Pierce K, Lookingbill T (2002) Extending community ecology to landscapes. Ecoscience 9:200–212

Chapter 4
Ordination

Abstract Ordination refers to a collection of tools for summarizing continuous trends in multivariate data sets. These are venerable tools that have informed community ecology for many decades and which are now being adopted more generally in "big data" applications. There are three general lineages, corresponding to underlying assumptions about the data (e.g., linear versus nonlinear models). The focus here is on indirect ordinations as tools for summarizing and communicating trends in species or environmental data sets, an application that would arise naturally from an inventory and monitoring program. The workflow for ordination is illustrated using principal components analysis with environmental variables and, in parallel, nonmetric multidimensional scaling with species abundance data. In both illustrations, the process flows from exploratory data analysis to generating the ordination, to post-processing the ordination for interpretation and reporting. In the following chapter, we develop the complementary application of classifying discrete groups in data sets. We return to ordination in Chap. 10, where it provides a framework for ecological assessment.

4.1 Introduction

In the previous chapters, we learned that ecological data are multivariate, redundant (correlated), spatially structured, and noisy. Here we begin a set of chapters aimed at resolving some of these difficulties. In this chapter, we focus on data compression and summary, with tools that reveal the larger trends in big data sets. The aim here is to deal directly with the multivariate, redundant, and noisy aspects of ecological data. In the next chapters, we will then look at discrete groups in the data, an alternative and complementary approach to summarizing trends (Chap. 5); statistical inferences on multivariate data—especially spatial inferences (Chap. 6); and a

---

Supplementary Information The online version contains supplementary material available at https://doi.org/10.1007/978-3-031-72251-6_4.

D. L. Urban, *Landscape Ecology: A Task-Oriented Perspective*,
https://doi.org/10.1007/978-3-031-72251-6_4

structured approach to multivariate explanatory (causal) models, which attempts to explain *why* variables should be correlated (Chap. 7).

Ordination is the general term for a collection of techniques aimed at ordering data to emphasize underlying trends or patterns. The result of an ordination is a rendering of the major trends that captures the redundancy in the data (i.e., correlations among variables, associations among samples) while suppressing the noise. The result is a new set of variables—ordination axes—that represent the trends. Usually, there are only a few main axes and so the ordination reduces the dimensionality of the data: a data set with dozens to hundreds of species might be summarized in two or three main compositional axes. Often, a welcome outcome of this reduction in dimensionality is that the ordination can be depicted on a physical page (e.g., in two dimensions).

The development of ordination techniques is closely tied to vegetation science or community ecology in general and gradient analysis in particular. Gradient analysis has a long history in plant ecology, beginning with the classic studies by Whittaker (1956, 1967) and Curtis (Curtis and McIntosh 1951; Bray and Curtis 1957) (see Gauch (1982) for more history). There are several popular techniques, in three main lineages that have evolved over the past several decades. Some techniques have only recently come into more common usage, largely as a reflection of faster computers and more widely accessible software. Many of these techniques are emerging anew as tools for "big data"—especially for visualizing large, multivariate data sets. In this, new users are embracing the same benefits that led community ecologists to these tools long ago.

The choice of technique is to some extent dictated by the kind of data available for the analysis. As we might anticipate, each technique has its strengths for particular kinds of data or applications. Choices also depend on the specific goals of the application and the extent to which any particular technique meets those objectives explicitly. And so, a user needs to appreciate the options in order to choose an appropriate tool for a task at hand. In this chapter, we develop a craftsman's appreciation for the tools for ordination. Subsequent chapters will build on this foundation.

In terms of the overall workflow outlined in the Preface to this book, ordination will serve us initially as a tool for exploratory data analysis, summary, and communication of highlights of data collected as part of inventory and monitoring programs. Later, we will use ordinations as a framework for more targeted applications aimed at multi-scale inference (Chap. 6) and ecological assessment (Chap. 10).

4.2 An Overview of Ordinations

The three ordination lineages correspond to three underlying models. These models are (1) *linear*, (2) *nonlinear* and unimodal, and (3) a model that makes no explicit assumption (here, termed *agnostic*). These response models are directly related to the computational details of the techniques. The linear/nonlinear distinction can be applied in two ways. In one case, the assumption refers explicitly to relationships among variables: principal components analysis assumes that the variables covary in a linear fashion. In a second case, correspondence analysis and related techniques assume that species respond to underlying gradients in a nonlinear and unimodal fashion—without explicitly assuming that the species covary among themselves in a nonlinear way. These relationships can be evaluated through exploratory data analysis (Chap. 3).

Rather than focusing on the models underlying the analyses, we might also categorize approaches according to the extent to which the researcher wants to influence or constrain the analysis. There are three approaches:

1. *Direct ordination* orders samples (e.g., of species abundances) in direct relationship to one or more ancillary (e.g., environmental) variables selected by the user. Common examples include plotting species abundances relative to elevation or hillslope position.
2. *Indirect ordination* focuses only on relationships among the variables in a single (primary) data matrix. This might seek trends in species composition or the main patterns in a set of environmental variables.
3. *Constrained ordination* forces the ordination of one (primary) matrix to be expressed in terms of variables from an ancillary matrix. Typically, this would aim to capture trends in species composition that are related to user-selected environmental factors.

The choice of direct, indirect, or constrained ordination depends on the data available for the analysis or, in some cases, the extent to which the researcher chooses to *use* these data explicitly. All ordination techniques aimed at exploring species-compositional patterns require species abundance (or presence/absence) data: a samples *x* species primary data matrix. Direct and constrained ordinations also require an ancillary matrix, typically of environmental descriptors such as elevation, topography, soils, or climate variables. Indirect ordinations, of course, may be performed simply by ignoring the existence of the ancillary matrix for the ordination and addressing any relationships between ancillary variables and the ordination afterward.

4.2.1 The Ordination Toolkit

The set of common ordinations can be sorted by response model and degree of constraint (Table 4.1).

Principal components analysis (PCA) is a linear model, an unconstrained (indirect) ordination. PCA is covered in any multivariate text, including several written from an ecological perspective (e.g., Pielou 1984; Manly 2004; McCune and Grace 2002; Legendre and Legendre 2012). A related technique, *factor analysis* (FA; Tabachnick and Fidell 1996), is also linear and indirect. Factor analysis is much invested in labeling ordination axes, as *latent factors* identified via *indicator*

Table 4.1 A catalog of ordination techniques (see text for details)

	Indirect	Constrained
Linear	PCA, FA	RDA
Nonlinear	CA (RA), DCA	CCA
Agnostic	NMS, PCoA	dbRDA

Direct ordination is not included here

variables. The constrained version of PCA is *redundancy analysis* (RDA; Legendre and Legendre 2012). There is no constrained version of factor analysis.

Correspondence analysis (CA), also known as reciprocal averaging (RA; Hill 1973), is based on a nonlinear unimodal response model that assumes species sort along environmental gradients in something akin to bell-shaped curves. An extension to CA, detrended correspondence analysis (DCA; Hill and Gauch 1980) addressed some numerical issues with CA. CA and DCA are indirect ordinations. The constrained version is *canonical correspondence analysis*, CCA (ter Braak 1986, 1987).

The original agnostic ordination was based on ecological distances (Chap. 3, Sect. 3.2.3) and was known as polar ordination (PO), an indirect ordination (Bray and Curtis 1957). PO has largely been subsumed by two alternatives: *nonmetric multidimensional scaling* (NMS; Kruskal 1964) and *principal coordinates analysis* (PCoA; Anderson and Willis 2003). NMS is a numerical algorithm based on ecological distances, while PCoA is a PCA of ecological distances; both are indirect ordinations. There is no constrained version of NMS. The constrained version of PCoA is *distance-based redundancy analysis*, dbRDA (Legendre and Anderson 1999).

To these several techniques, we can add direct ordination, which is a user-constrained approach in which the analyst chooses the ancillary (environmental) variables by which to sort the primary variables (species). This yields three response models and indirect versus constrained techniques: six approaches plus direct ordinations. All of these approaches are in common usage today.

The full set of ordination techniques is detailed elsewhere (see especially McCune and Grace (2002) and Legendre and Legendre (2012)) and in the digital Supplement (S4) to this chapter. Supplement S4 also includes some guidance on choosing among the tools. In the following sections, we work through a typical application of ordination, focusing on the logical workflow and decision points. We do this for two complementary approaches, principal components analysis (PCA) and nonmetric multidimensional scaling (NMS), to recognize some nuances in the details while emphasizing the overall similarity of the workflows.

The first example is a PCA of environmental factors, while the second illustrates species composition using NMS. Both examples use the data sets from Sequoia National Park in the southern Sierra Nevada of California (Urban et al. 2002), introduced in the previous chapter (Sect. 3.3).

4.3 Ordination Workflow

In this section, we first develop an application of PCA. In the following section, we develop a parallel application using NMS. We then extract the commonalities of these two illustrations, to generalize the approach.

An ordination analysis can be rather involved, as there are several steps along the way and a deliberate approach is helpful in working through the steps. In general, the aim is to arrive at a sample ordination, in which the samples are sorted along a few axes that represent the main trends in the data set. These trends might be in terms of species composition or environmental factors initially, but it will often be useful to add species to an environmental ordination or, reciprocally, to add environmental factors to an ordination of species compositional trends. However this might unfold, the net result is similar, with a similar set of decision points and steps (Fig. 4.1).

In any application, there will be details that matter depending on the data and the analytic approach. But applications share many issues in common, as we illustrate with the following examples. The illustrations presume that exploratory data analysis (Chap. 3) has already been completed.

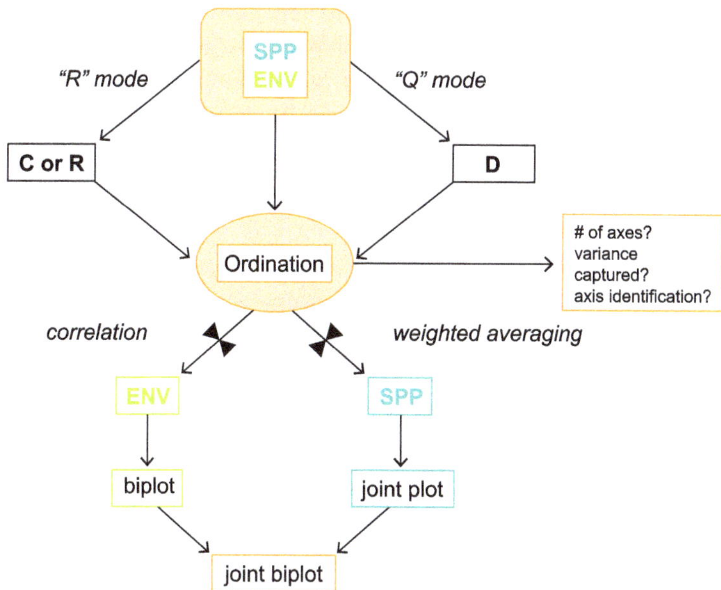

Fig. 4.1 Workflow for an ordination application. Analysis might begin with either a species (SPP) or environmental (ENV) (or other!) data matrix and proceed via either R mode (based on a covariance matrix **C** or correlation matrix **R**, or Q mode (based on a distance matrix **D**), or via the raw data matrix (central route). The ordination is then evaluated in terms of a few queries (right-side box), and perhaps then further post-processed with information from either the SPP or ENV data sets, or both (see text for details)

4.3.1 *Principal Components Analysis*

Principal components analysis is a linear model and thus appropriate for data sets in which relationships among variables are approximately linear (the technique is reasonably robust to minor violations of this assumption). Linearity comes into this because the analysis proceeds not from the primary data matrix but from a secondary matrix (Chap. 3, Sect. 3.2.3). The secondary matrix is either a covariance matrix or a correlation matrix; both are linear models.

Unlike some of the tools we will consider, PCA has an exact analytic solution. The solution is developed along several perspectives elsewhere (especially Pielou 1984; McCune and Grace 2002; see also Supplement S4), and here it is presented in its essentials only.

4.3.1.1 Data Preparation

PCA would normally be preceded by exploratory data analysis (EDA, Chap. 3). In this, issues related to sample sizes, normality, outliers, and so on would be discovered and dealt with appropriately. The main decision point at this step is to base the analysis on either a covariance matrix (**C**) or a correlation matrix (**R**). Correlations are standardized from the covariances, and so treat all variables equally (each is in the same units, standard deviations). By contrast, covariances reflect the measurement units and ranges of values of the raw data. If the variables are in the same units (e.g., species abundances), then covariances will be biased toward the variables with the largest values (e.g., the most abundant species) ... and this might be perfectly reasonable. Alternatively, the data might be relativized or transformed to downweight the most abundant species so rare species can contribute to the analysis (again, see Chap. 3).

If the variables are in different units (as would be the case for many environmental factors), then variables must be relativized or standardized or the covariances would reflect the measurement units (which would be silly). Using a correlation matrix solves this issue neatly; relativizing the data would also reconcile disparate units.

In practice, data on species abundances are often difficult to reconcile with a linear model (Sect. 3.3.1, Fig. 3.6). Environmental factors are more likely to fit a linear model. In the illustration presented here, PCA is applied to a correlation matrix computed from a set of environmental factors collected in Sequoia National Park in the Sierra Nevada of California, USA (Fig. 4.2, data from Urban et al. (2002), previewed in Chap. 3).

4.3.1.2 The Analysis

PCA aims to capture the main trends in the data by constructing new variables (i.e., the principal components) that represent common patterns in the way that the raw

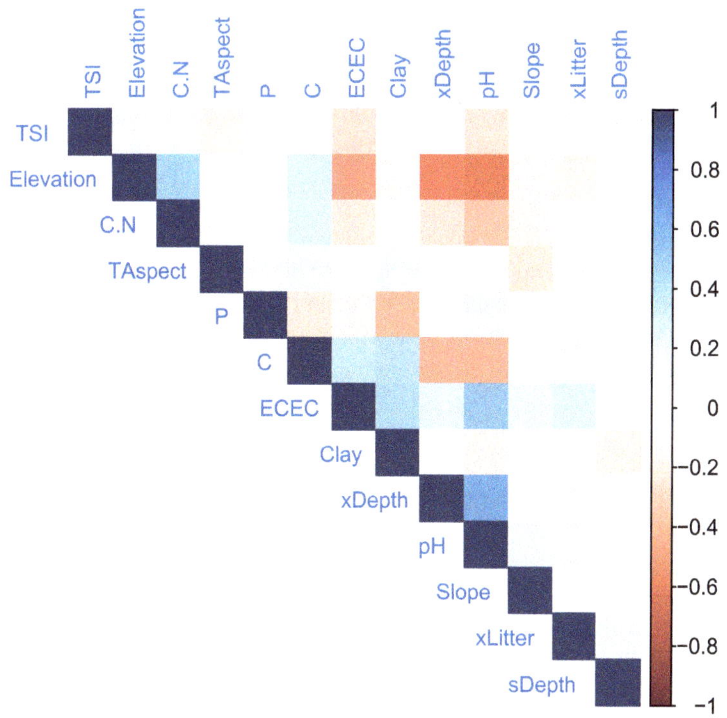

Fig. 4.2 Graphical representation of the correlation matrix for environmental factors in the Sierran data set. Variables are arranged to emphasize patterns of correlations. See Table 3.6 for description of variables. Plot constructed using the *corrplot* package (Wei and Simko 2021) in R (R Core Team 2021)

variables covary. A subset of several variables that tend to covary strongly (either positively or negatively) would be extracted as a principal component representing those variables. Another PC would be extracted for a different set of variables that covary, and so on. The resulting set of principal components provides a new coordinate system for the data, a set of just a few main axes in which the samples can be projected. This new coordinate system reduces the dimensionality of the data, suppresses noise, and reveals the main trends in the data—precisely what an ordination is intended to do.

The analytic solution to PCA is provided by *eigenanalysis* of the covariance or correlation matrix. For the correlation matrix **R**, the solution is to find two new matrices **U** and **Λ** (lambda) such that **RU** = Λ**U**. Here, **Λ** is a diagonal matrix of the *eigenvalues* of **R** and the matrix **U** holds the *eigenvectors* as its columns. For a data set of p environmental factors measured over n samples, **R** is p x p (variable x variable), **Λ** is p x p (PC x PC), and **U** is p x p (variable x PC).

The matrix **Λ** is diagonal, with *eigenvalues* in its diagonal and 0's otherwise. The eigenvalues λ indicate the amount of variance from the input matrix that is represented on each of the principal components. the most on PC 1, the next most on PC 2, and so on. Typically, most of the variance is captured on the first few PCs, with the last few PCs of minor importance and interpreted (and disregarded) as noise. Note that because there are p PCs computed, the analysis captures *all* of the variance in the input data matrix, but it repackages this variance into just a few main axes and so provides a compact summary of the data.

The matrix **U** holds *eigenvectors* as its columns, one column for each PC. The eigenvectors are used to reproject the samples into PC space. These are regression coefficients: the first eigenvector (first column of **U**), multiplied by the original variables for the first sample (i.e., the first row of the n x p primary matrix), yields that sample's position on the first PC. The second eigenvector locates the samples on the second PC, and so on. This is done efficiently, by matrix multiplication, for all samples and all PCs. A plot of samples into a low-dimensional PC space provides the sample ordination.

The raw result of PCA is the summary matrices $\mathbf{\Lambda}$ and **U**. Note that these two summary matrices completely describe the input data ... but there is still a lot of post-processing and interpretation to do.

4.3.1.3 Post-processing and Evaluation

A principal components analysis is perhaps typical of many ecological analyses, in that the solution is quick and tidy, but there are still several post-processing steps (recall the extensive post-processing of the species distribution models in Chap. 2!). These steps are developed below, and illustrated with the environmental data set from Sequoia National Park.

How much variance do the PCs capture? How many axes to retain? For this Sierran data set, there are 13 environmental variables, and so there are 13 principal components. The variance on each can be visualized with a *scree plot*, a histogram of variances over the PCs (Fig. 4.3). These scree plots typically are rather steep, illustrating that most of the variance is summarized on the first few PCs.

The decision about how many axes to retain is ultimately subjective, but there are some guidelines. On a purely pragmatic basis, if the application is to be plotted on a physical page, then 2 is the right number of axes to retain—or 3, though these would then require either a 3D graphic or a set of pairwise plots (1 vs 2, 1 vs 3, 2 vs 3). Perhaps more typically, the scree plot is examined for a natural break that might reveal the main components.

For a PCA computed on a correlation matrix, the total variance is equal to the number of variables, as each variable contributes unit variance. In this case, a PC that has an eigenvalue greater than 1.0 contributes to data summary, while a PC with an eigenvalue less than 1.0 provides less information than a raw variable. From this, it is typical to interpret PCs with eigenvalues larger than 1.0. (For an analysis from a covariance matrix, there are randomization tests that can support a similar decision process.)

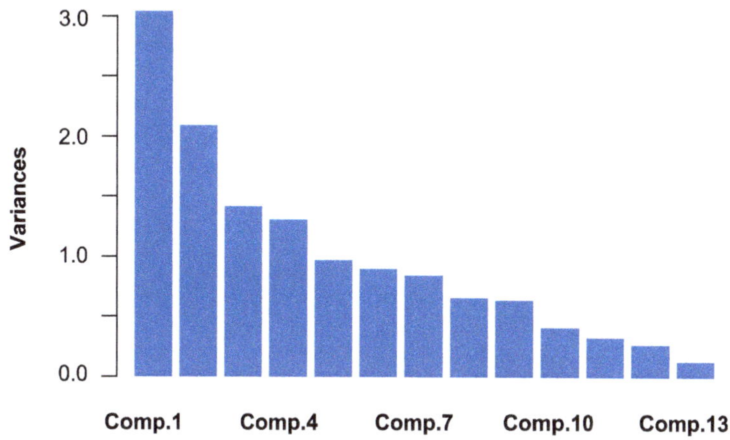

Fig. 4.3 Scree plot of the eigenvalues (as variances) from a PCA of the environmental variables in the Sierran data set. Constructed using the native plot function for a PCA object in R

In the Sierran case, the first four PCs have eigenvalues larger than 1.0 (the fifth is marginal), but the first two are clearly the most important. In what follows, four are interpreted in tabular summaries while only the first two are plotted graphically.

What *are* the axes? The PCs represent the main trends in the data, but what does this mean? To identify the axes ecologically, it helps to relate these to the original input variables. The eigenvectors can suggest how each variable relates to each PC (these are the *loadings* on the PCs), but it is tidier to compute correlations. This generates a table that has p rows for p input variables and a column for each retained PC. It is convenient to add a couple rows to this table, to include the variance on each PC and the cumulative variance over all retained PCs (Table 4.2).

Tabular correlations are precise but not always easy to interpret at a glance. Another way to interpret these is to plot them as correlation vectors. A correlation vector in PC (or any ordination) space graphically shows the relationship between an input variable and each plotted PC: a vector nearly parallel to a PC is very strongly correlated with it (positively to the right or up, negatively to the left or down). The length of the vector indicates the magnitude of the correlation. A vector that lies "in-between" two PCs is correlated with both (and probably not strongly). The angle is the arccosine of the correlation itself. (We can verify this relationship by recalling that we draw Cartesian (X,Y) coordinates at 90° to each other to show their independence: acos(0) = 90.) In computing correlation vectors, it is typical to also estimate a test of significance of the correlations, with only significant correlation vectors plotted in graphics (here, using package *ecodist* in R, Goslee and Urban 2007). A plot of sample scores on selected PCs, with the correlation vectors overlaid, is an ordination *biplot* (Fig. 4.4). This display makes it easier to interpret PCs as "bundles" of covarying input variables.

Table 4.2 Correlations between environmental factors and the first four principal components (blanks are not significant, $P > 0.05$), along with eigenvalues and proportion of total variance (raw and cumulative), for the Sierran data set (variables described in Table 3.6)

Variable	PC 1	PC 2	PC 3	PC 4
Eigenvalue	3.037	1.987	1.470	1.420
Variance	0.234	0.153	0.113	0.109
Cumulative	0.234	0.387	0.500	0.609
Elevation	0.844			
Slope	−0.188	−0.275	0.244	−0.578
TAspect	−0.180	−0.205	0.709	
TSI	0.306	0.240	−0.480	
xLitter	−0.244	−0.262	0.511	0.396
xDepth	−0.719	0.281		
sDepth	−0.174	0.725		
pH	−0.842	0.186		
C	0.307	−0.783	0.197	
C.N	0.553	−0.173	0.173	0.224
P	0.525	0.325	0.352	
ECEC	−0.691	−0.584		
Clay	−0.180	−0.629	−0.533	

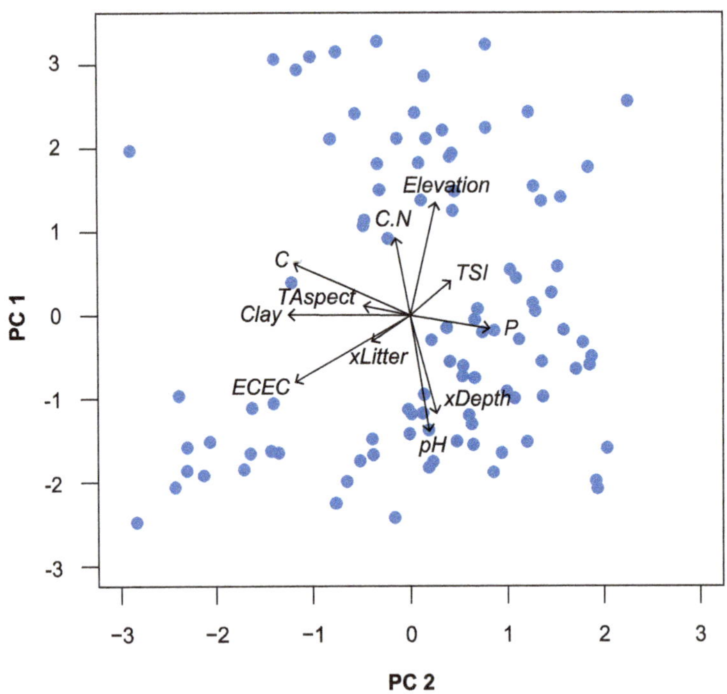

Fig. 4.4 A biplot of the PCA of the Sierran environmental data, showing correlation vectors for the environmental factors (variable codes in Table 3.6). Note that this figure is plotted so that the "elevation gradient" is on the vertical axis, a purely cosmetic decision. Constructed using the *ecodist* package in R (Goslee and Urban 2007)

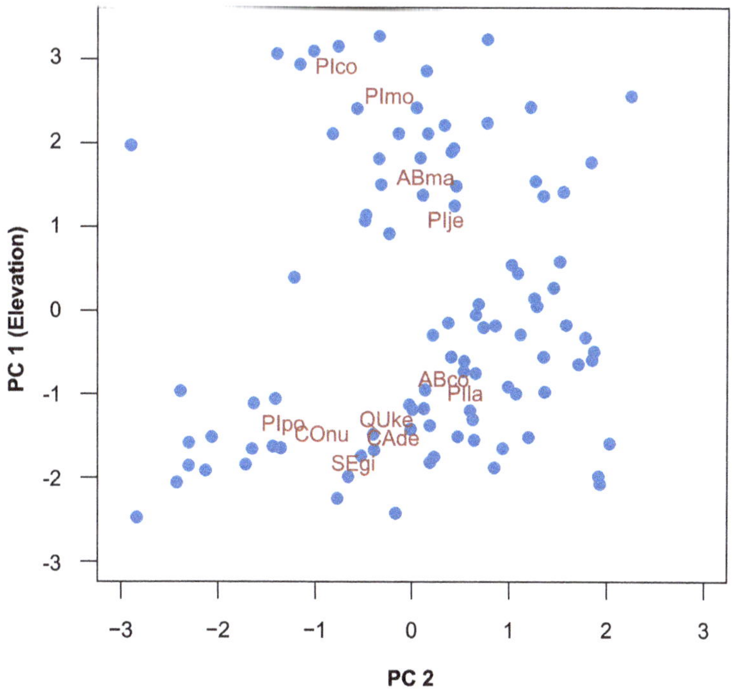

Fig. 4.5 A joint plot of the environmental PCA from the Sierran data set, displaying sample plots (points) as well as species at their weighted-average positions along the first two PC axes (species codes in Table 3.5). Note that the "elevation gradient" has been displayed on the vertical axis, a cosmetic decision

Adding species into the sample ordination This ordination was conducted on environmental variables, but it is a straightforward task to add species to the ordination. In this, we calculate the average ordination score, on each axis, for each species. This average is weighted by the relative abundance of each species on each sample, so that samples where the species is most abundant contribute the most to the average, while samples on which the species is absent do not contribute at all. While this *weighted-averaging* is a focus on one ordination technique (correspondence analysis), species can be added to any sample ordination after the fact.

Note that the weighted averaging assumes that a species shows a unimodal response to the ordination axis, so that it makes sense to summarize its distribution along the axis as an average. This would not work, by contrast, if a species showed a distinctly bimodal distribution along an axis. Weighted averaging also is more appropriate than computing correlations between species abundances and ordination axes, if the species responses are unimodal (i.e., nonlinear). A plot that includes both samples and species positions is a *joint plot* (Fig. 4.5), and this overlaid with environmental correlation vectors is a *joint biplot*. Depending on the number of environmental factors and species, these can be a bit busy

4.3.1.4 Reporting

The following information communicates how the analysis was conducted and key decision points along the way:

☑ Data description (what the variables are, sample sizes); any editings or transformations (from EDA); whether **R** or **C** was used for the analysis
☑ Software used to perform the analysis (should not matter for PCA, but some implementations vary in the details)
☑ How many axes were retained, and why; variance captured on each axis and cumulative (as in Table 4.2)
☑ Axis identification: correlations with environmental factors (table and/or biplot)
☑ If the analysis was of environmental factors, how species were added (joint plot); if species were analyzed, how environmental factors were added (as biplot)
☑ Optional: joint biplot

4.3.2 Nonmetric Multidimensional Scaling

In this section, we develop an ordination workflow to parallel the application of PCA. This example uses nonmentric multidimensional scaling (NMS), and some of the details are quite different from PCA ... but the basic workflow is the same.

NMS is an indirect ordination based on dissimilarities or ecological distances among samples. As it is based on distances, it is agnostic about any underlying responses of species to environment (it can work with linear or nonlinear data). Unlike PCA, which has an exact analytic solution, NMS is a numerical algorithm: it finds a solution by successive approximation. We illustrate NMS here as a contrast to PCA and because NMS has been shown to perform well with a variety of ecological data sets. (It is also featured in many other applications beyond ecology.) We will come back to NMS in Chap. 10.

4.3.2.1 Data Preparation

Data for NMS are typically screened and explored as with other ecological data sets (Chap. 3). The analysis proceeds from a distance matrix that contains the dissimilarities among all pairs of samples, an n x n matrix **D**. This matrix is symmetic and holds 0's in its diagonal. For computational simplicity, the matrix is typically stored as a vector, with elements corresponding to the lower triangle of **D**: d_{21}, d_{31}, d_{32}, ..., $d_{n,n-1}$.

There are very many alternative distance measures (Sect. 3.2.3), and a major decision point in NMS is choosing a measure appropriate to the data set. For species compositional data, variations of the Bray-Curtis measure are often used (Sect. 3.2.3, Eq. 3.8). Note that this measure will produce different results depending on how the species data are relativized or transformed. For this reason, researchers often repeat the analyses to explore alternative data relativizations or distance measures.

4.3.2.2 The Analysis

NMS aims to find a sample ordination in which the distances among samples in ordination space reflect, as nearly as possible, the ecological distances among samples. For a two-dimensional ordination, this means that the distances apart on the page (a plot of the ordination) are as similar as possible to the m-dimensional ecological distances (e.g., Bray-Curtis distances). This solution is generated by numerical approximation.

NMS requires that the number of ordination axes, k, be decided in advance. Sometimes this is subjective and arbitrary, as in the case where the application requires a 2D solution that can be easily visualized. More typically, the proper number of dimensions is unknown, and so the user computes ordinations of several dimensions and then chooses k afterwards, based on goodness-of-fit or explanatory power.

The numerical algorithm of NMS toggles between two representations of the data. One is the ordination itself, in k dimensions. The other is a plot of ecological distances (e.g., Bray-Curtis distances) among samples and sample separation in ordination space. In the heuristic in Fig. 4.6, the ordination is on the left and ordination distances are simply the Euclidean distances between pairs of samples (dots) in ordination space. On the right is the Shepard diagram (Shepard 1962) of ordination versus ecological distances. The aim is to make this fit as linear as possible.

The ordination is generated in a stepwise fashion:

1. Begin with an arbitrary configuration to the ordination (e.g., random coordinates in k-space), corresponding to the left panel of Fig. 4.6.

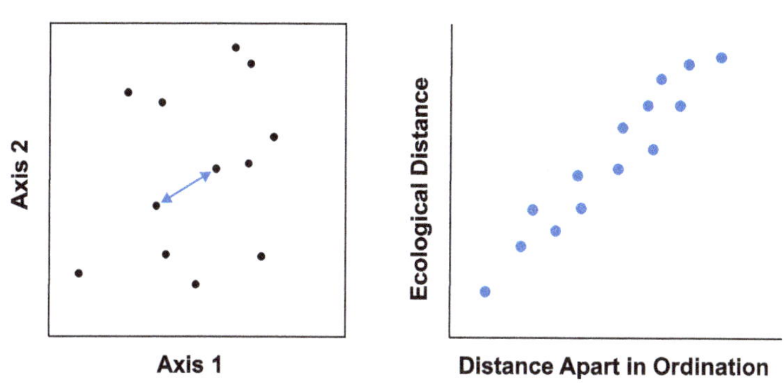

Fig. 4.6 Schematic of how NMS works. Ordination is on the left, Shepard diagram on the right. The aim is to generate an ordination such that pairwise distance relationships in the ordination reflect ecological distances as faithfully as possible

2. Compute pairwise distances among samples in ordination space, and fit these to the ecological distances as in a regression (right side of Fig. 4.6). Compute a "badness-of-fit" index, termed *stress*, which is the complement of a more conventional goodness-of-fit term for a regression.
3. Find samples that fit particularly badly in this regression and move them slightly in ordination space, to try to improve their fit.
4. Repeat (2–3) until the fit cannot be improved further. Save the final sample configuration as the ordination, and save the stress value for this configuration.

Clearly, in the first iterations, the fit will be terrible (i.e., from random numbers), and every iteration will improve the fit. Eventually, the adjustments are quite small as the solution converges. As a numerical algorithm, an exact solution is not guaranteed, and so the convention is to compute several ordinations and then choose the best version (lowest stress) as the final ordination.

When the number of dimensions k is not known in advance, a *step-down procedure* is followed, in which a set of several replicate ordinations is computed for a range of values of k (say, 10 iterations each, for $k = 1, 2, 3, ..., 6$). This is a lot of ordinations, and for large data sets, this can be computationally demanding. (Modern computers make this less of an issue than it used to be.)

From a step-down procedure, a plot of stress versus k helps choose an appropriate number of axes. Like the PCA scree plot, this curve is typically steeply decreasing and the aim is to find a natural break in the curve to choose k. Again, as with PCA, the choice of how many axes to use is ultimately subjective.

Beyond its numerical algorithm, NMS is unusual in that it finds the k-dimensional solution for all axes simultaneously. That is, the second axis is not found from the residuals of the first (as with most other ordinations). The solution is also unusual in that it is "free floating" in k dimensions: the configuration is arbitrary in k-space. For this reason, it is conventional to rotate the initial solution so that the first axis represents most of the variation, the second axis the next most, and so on. A simple way to do this is to perform a PCA on the initial NMS solution and to retain all the principal components. This keeps the configuration exactly as it was but rotates it so that the axes represent successively decreasing proportions of the variance in the data (revisit PCA, above, to verify that this makes sense!).

The end result of this processing is an ordination of samples in k dimensions, a new data set of sample scores on each of the ordination axes.

4.3.2.3 Post-processing and Evaluation

As with any ordination, the initial solution is a starting point for additional interpretation. In the case of NMS, the first decision is to choose an appropriate number of axes, from a step-down procedure (above). For the illustration here, the NMS is of a matrix of Bray-Curtis distances computed from species abundances, relativized to treat all species similarly (a Wisconsin double relativization, see Chap. 3, Sect. 3.

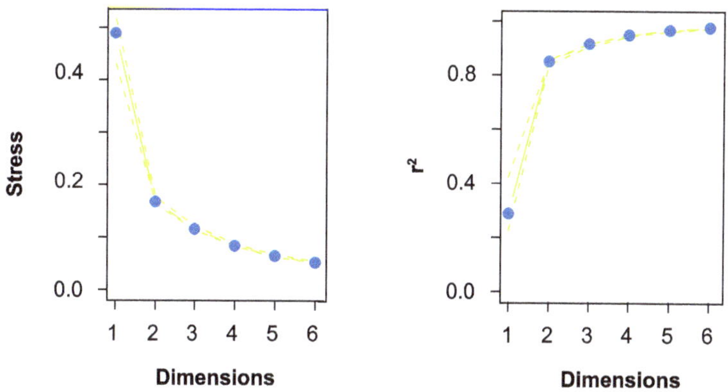

Fig. 4.7 Step-down procedure for the Sierran tree species data set. Left: stress; right, R^2. Constructed using the *ecodist* package in R (Goslee and Urban 2007). Solid lines are means and dashed lines, min/max values from ten replicate ordinations for each dimension 1–6

2.3). There are 11 species in the analysis. The Bray-Curtis distances were extended using the *stepacross* function in the *vegan* package in R (Oksanen et al. 2021).

How many axes? The step-down procedure shows a large reduction in stress from one to two dimensions, and lower reductions in stress beyond that (Fig. 4.7). Correspondingly, the overall goodness-of-fit of the ordination, as the R^2 of the regression in the Shepard diagram, shows a large increase at two dimensions. For these reasons, a two-dimensional solution was selected. The best 2D configuration had a stress of 0.16 and a total R^2 of 0.86. It might be noted here that while the stress values vary on [0,1], what constitutes a suitably low stress value depends on the data set.

It is often informative to plot the final Shepard diagram for an NMS ordination. This clearly reveals the goodness-of-fit. In the Sierran case, it also shows the effect of using extended distances in the analysis: in Fig. 4.8, all of the points above a value of 1.0 on the *Y* axis have been extended; otherwise, all of those values would have been fixed at 1.0, and the overall fit would be much worse (and nonlinear). The takeaway from this figure is that the fit is nearly linear, which means that we can interpret sample location in ordination space as a reasonable reflection of ecological distances. That is, samples that are close together in ordination space are compositionally similar, while samples that are far apart are ecologically different from each other. This interpretation is the explicit aim of an NMS ordination.

How much variance do the axes capture? The ordination fit provides an estimate of the explanatory power of the ordination: the R^2 of the fit between ordination distances and ecological distances. This is an estimate of how much of the compositional variability in the data set is represented in the ordination. In this case, the explanatory power is rather high for an ecological ordination, a result of the low species diversity and strong sorting along a long elevation gradient (of course, we don't know that yet!).

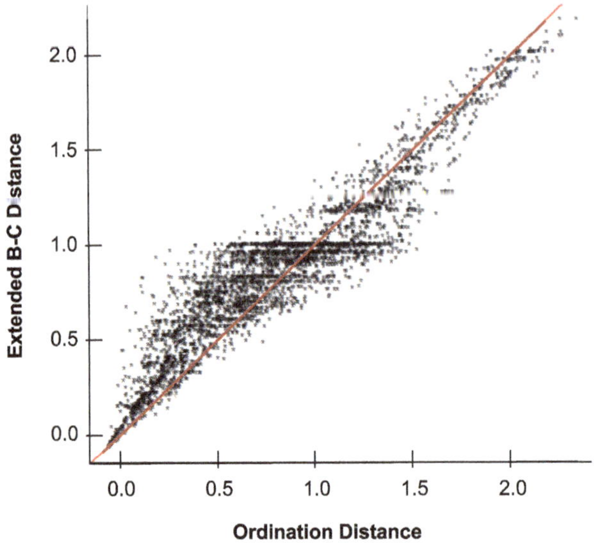

Fig. 4.8 Shepard diagram for the NMS of Sierran tree species. The 1:1 line is in red. Distances above 1.0 were extended in the analysis (see text)

The relationship shown in Fig. 4.8 is between distance matrices, and the correlations behind the R^2 values are Mantel correlations. Mantel tests (Mantel 1967) are based on correlations between distance or dissimilarity matrices. Beyond this example here, we will revisit Mantel tests again in Chaps. 5 and 6.

Assigning the explanatory power of individual NMS axes is complicated by the way they are computed, simultaneously. That is, where a sample falls on axis 2 depends on where it is located on axis 1; the axes are not fitted independently. The variance accounted on axis 1 can be estimated simply, as the regression r^2 between sample separation on axis 1 and Bray-Curtis distances. For axis 2, the variance on that axis is the total R^2 (here, in 2 dimensions) minus the r^2 on axis 1. For higher-dimensional NMS ordinations, subsequent axes are estimate similarly, by differencing. In this case, the first axis explains 72% of the compositional distances, and the second axis explains 14%.

What *are* the axes? As with PCA, we will want to identify the axes, to label them with an ecological interpretation. Because we ordinated species data, we start by looking at how species sort in ordination space. Again, because we suspect that species response might not be linear, we can plot species abundances from the primary data matrix against sample scores on NMS axis 1. This is not shown here, but we do suspect a nonlinear response from EDA (Chap. 3), so computing correlations would be inappropriate. Instead, we can compute weighted-average species scores on the ordination axes. These can be saved in a table, or plotted into the ordination as a joint plot (Fig. 4.9).

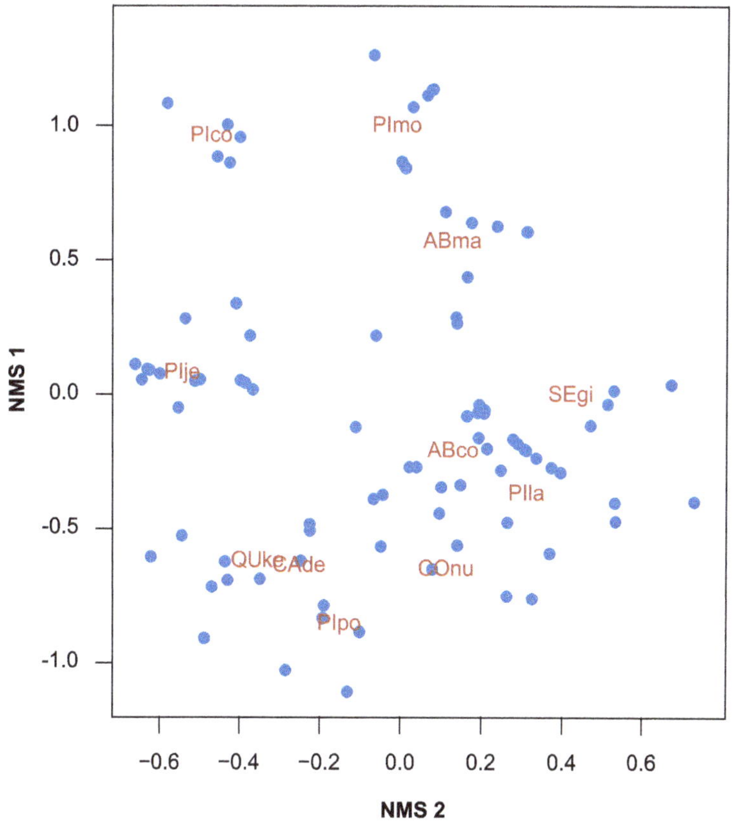

Fig. 4.9 A joint plot of the NMS ordination of compositional distances, with species located at their weighted average positioned on each axis (species codes in Table 3.5)

Adding environmental factors into the ordination The NMS axes here are defined in terms of species composition, but we might also wish to interpret them in terms of environmental factors. To do this, the ordination axes can be correlated with the environmental factors, yielding a tabular summary (Table 4.3).

These correlations can also be visualized by adding correlation vectors to the ordination diagram, a biplot (Fig. 4.10).

Again, the sample locations, species positions, and environmental correlation vectors can all be combined into a joint biplot (Fig. 4.11). This plot can be busy but holds an impressive amount of information.

Note that some of this information is redundant (e.g., tables and figures with species scores or environmental correlations). The user would either choose which to include (table or figure but not both) or would relegate some bits to appendices.

Table 4.3 Summary of the NMS in terms of compositional variance explained and correlations with environmental factors (blank: $P > 0.05$) (environmental variable codes in Table 3.6)

	NMS1	NMS2
R^2	0.72	0.14
Cumulative	0.72	0.86
Variable		
Elevation	0.927	
Slope		
TAspect	0.188	
TSI		
xLitter	0.249	
xDepth	−0.535	0.425
sDepth		
pH	−0.570	0.460
C	0.299	−0.187
C.N	0.439	
P	0.301	
ECEC	−0.409	0.191
Clay		

4.3.3 *Reporting*

A full NMS ordination analysis involves a lot of processing steps and some explanation of how the ordination was constructed. These details need to be communicated:

☑ Data preparation, any relativizations or transformations, and results of EDA
☑ Choice of distance measure and whether the distances were extended because of saturation
☑ Results of step-down procedure and decision about how many axes to use
☑ Stress value and overall R^2 for final configuration
☑ Variance accounted on each axis
☑ Identification of axes in terms of species composition; weighted-average species scores on the axes (table, joint plot)
☑ Optionally, identification of axes in terms of environmental or other factors; correlations with environmental factors (table, biplot)
☑ Optionally, joint biplot

4.3.4 *Collecting Terms*

PCA is an analysis of trends among variables and proceeds to an ordination of relationships among samples. For this reason, it is sometimes referred to as an "R-

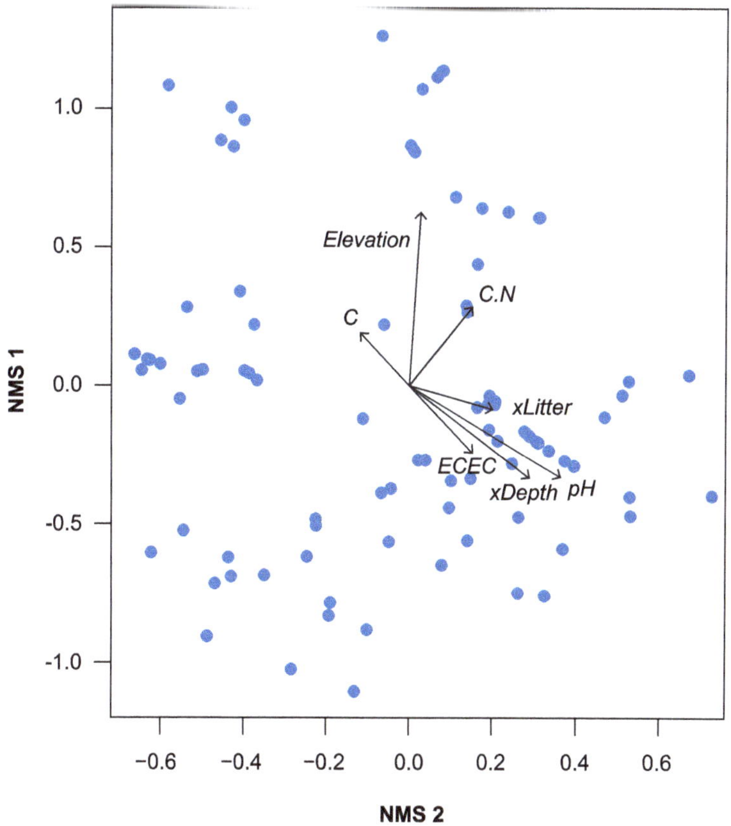

Fig. 4.10 A biplot of the NMS ordination of compositional distances for Sierran tree species (environmental variable codes in Table 3.6)

mode" analysis (think correlations, **R**). By contrast, nonmetic multidimensional scaling is an analysis of trends among samples; this is a "Q-mode" analysis (think sample *q*uadrats). With NMS, we began by focusing on species composition, and then overlaid the environmental factors. Which is to say: R and Q refer to alternative routes to similar endpoints.

Beyond this, we did the PCA using environmental factors and the NMS with species composition. These suggested the choice of techniques: PCA for environmental factors because the assumption of linearity was (nearly) met; NMS for the species because exploratory data analysis suggested nonlinearities. Again, the analyses are parallel in that both ended by incorporating both species and environmental variables (i.e., into a joint biplot).

But these two examples are not interchangeable; they tell us different things. The PCA emphasizes environmental patterns and the species did not contribute to this solution. In a sense, it is simply our good luck that the species tend to sort along the main environmental PCs. Elevation is a major contributor to the first PC, but as PC1

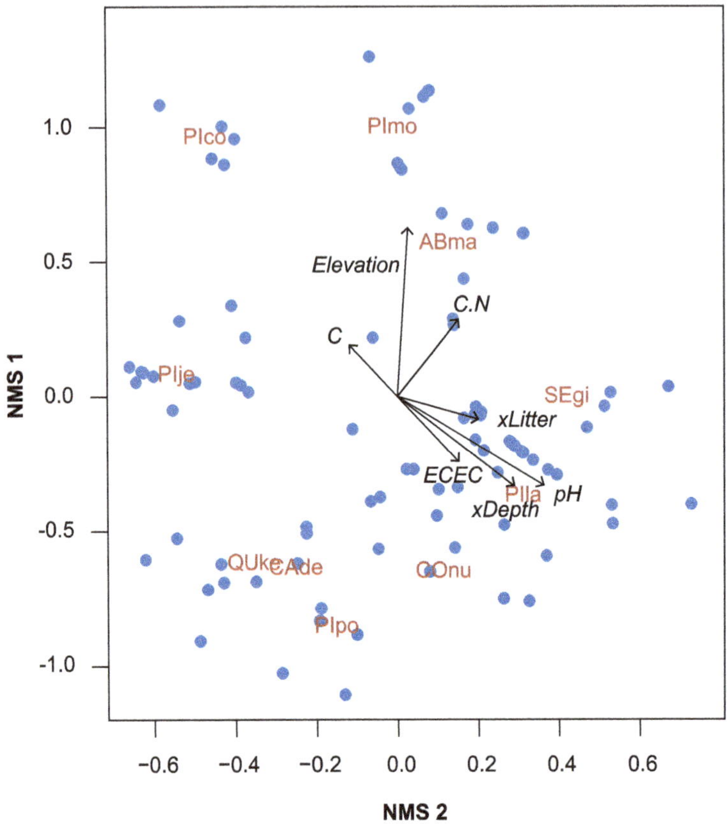

Fig. 4.11 A joint biplot of the Sierran NMS ordination, combining Figs. 4.9 and 4.10 (see Table 3. 5 for species codes, Table 3.6 for environmental variables)

captures only 23% of the total variance, there is clearly a lot of environmental variability in this data set that is unrelated to elevation.

By contrast, the NMS emphasizes compositional trends and is unconstrained by the environmental factors. Again, it is our good luck that these factors are indeed correlated with the compositional trends (or perhaps more charitably, that we chose appropriate candidates when we measured environmental factors). In this case, the elevation gradient expressed on NMS1 is quite dominant: it captures 72% of the compositional variability (as dissimilarities) in the species data set.

The two examples are complementary and provide different but compatible insights into these forests. In practice, it is often useful (and not onerous!) to use complementary techniques on the same data sets.

These illustrations have focused on two popular and robust techniques, but similar workflows could be developed for alternative techniques such as factor analysis, correspondence analysis, or principal coordinates analysis. Some details would vary but the overall flow is the same.

In the following chapter, we turn to classification as an analysis that is complementary to but compatible with ordination. While the examples here have been mostly descriptive and exploratory, we will turn to inferential approaches, using constrained ordinations, in Chap. 6.

Which is to say, ordination can serve as a fundamental tool with ecological data, an exploratory entry point as well as a foundation for subsequent applications.

4.4 Further Reading

Much of the history of ordination and gradient analysis is itself rather dated but still quite relevant. These sources include several books (Whittaker 1978; Orlóci 1978; Gauch 1982; Jongman et al. 1995; Manly 2004) and some in-depth reviews (especially Beals 1984). McCune and Grace (2002) provide a user-friendly presentation, focusing on guidelines for performing and evaluating ordinations. Legendre and Legendre (2012) remains an authoritative resource for ordination and related numerical methods.

Most texts offer some insights and recommendations about the relative strengths and weaknesses of these techniques for various sorts of applications (e.g., McCune and Grace 2002; Legendre and Legendre 2012). Often these recommendations are based on statistical (i.e., theoretical) arguments. While these are certainly valid, it is also important to compare tools from a more pragmatic basis. Supplement S4 for this chapter offers some guidance, along with additional references.

4.5 Summary and Prospectus

Ordination refers to a collection of tools used to summarize the main trends in multivariate data sets. In this, the tools reduce the dimensionality of the data—typically to a few axes that can be depicted easily on a page. The analysis also suppresses noise in the data, by relegating this to minor axes that are discarded.

There is a very long history of the development and application of ordinations in ecology. There are three main lineages based on the underlying response model: linear, nonlinear, and agnostic. Each of these has indirect and constrained versions of the analysis. Each of the resulting approaches is in common use by ecologists.

Ordination serves as a powerful tool for exploratory analysis of complex data sets. Analyses often combine species and environmental data via after-the-fact addition of environmental factors to a species ordination, or vice versa. The summary provided by the analysis can be an end result in itself, but ordinations also can serve as the framework for subsequent analyses. We turn to one complementary analysis, in the following chapter. We return to ordination, and constrained ordination in particular, in Chap. 6 where we explore inferential methods for dissecting species-environment relationships over a range of spatial scales.

References

Anderson MJ, Willis TJ (2003) Canonical analysis of principal coordinates: a useful method of constrained ordination for ecology. Ecology 84:511–525

Beals EW (1984) Bray-Curtis ordination: an effective strategy for analysis of multivariate ecological data. Adv Ecol Res 14:1–55

Bray JR, Curtis JT (1957) An ordination of the upland forest communities of southern Wisconsin. Ecol Monogr 27:325–349

Curtis JT, McIntosh RP (1951) An upland forest continuum in the prairie-forest border region of Wisconsin. Ecology 32:476–496

Gauch HG (1982) Multivariate analysis in community ecology. Cambridge University Press, Cambridge

Goslee SC, Urban DL (2007) The ecodist package for dissimilarity-based analyses of ecological data. J Statistical Software 22:7

Hill MO (1973) Reciprocal averaging: an eigenvector method of ordination. J Ecol 61:237–249

Hill MO, Gauch HG (1980) Detrended correspondence analysis: an improved ordination technique. Vegetatio 42:47–58

Jongman RHG, ter Braak CJF, van Tongeren OFR (1995) Data analysis in community and landscape ecology. Cambridge University Press, Cambridge

Kruskal JB (1964) Multidimensional scaling by optimizing goodness of fit to a non-metric hypothesis. Psychometrika 29:1–27

Legendre P, Anderson MJ (1999) Distance-based redundancy analysis: testing multispecies responses in multifactorial ecological experiments. Ecol Monogr 69:1–24

Legendre P, Legendre L (2012) Numerical ecology, 3rd English edn. Elsevier, Amsterdam

Manly BFJ (2004) Multivariate statistical methods: a primer, 3rd edn. Chapman and Hall, London

Mantel N (1967) The detection of disease clustering and a generalized regression approach. Cancer Res 27:209–220

McCune B, Grace JB (2002) Analysis of ecological communities. MjM Software Design, Gleneden Beach

Oksanen J, Blanchet FG, Kindt R, Legendre P, O'Hara RB, Simpson GL, Solymus P, Henry M, Stevens H, Wagner H (2021) Vegan: community analysis package. R version 1.17-11. http://CRAN.Rproject/org/program=vegan.

Orlóci L (1978) Multivariate analysis in vegetation research, 2nd edn. Junk, The Hague

Pielou EC (1984) The interpretation of ecological data: a primer on classification and ordination. Wiley, New York

R Core Team (2021) (et seq.) R: A language and environment for statistical computing. R Foundation for Statistical Computing, Vienna, Austria. http://www.R-project.org/

Shepard RN (1962) The analysis of proximities: multidimensional scaling with an unknown distance function. Psychometrika 27:125–139

Tabachnick BG, Fidell LS (1996) Using multivariate statistics, 3rd edn. Harper-Collins, New York

ter Braak CJF (1986) Canonical correspondence analysis: a new eigenvector technique for multivariate direct gradient analysis. Ecology 67:1167–1179

ter Braak CJF (1987) The analysis of vegetation-environment relationships by canonical correspondence analysis. Vegetatio 69:69–77

Urban DL, Goslee S, Pierce KB, and Lookingbill TR (2002) Extending community ecology to landscapes. Ecoscience 9:200–212

Wei T, Simko V (2021) R package *corrplot*: visualization of a correlation matrix (version 0.92). Available from https://github.com/taiyun/corrplot

Whittaker RH (1956) Vegetation of the great smoky mountains. Ecol Monogr 26:1–80

Whittaker RH (1967) Gradient analysis of vegetation. Biol Rev 42:207–264

Whittaker RH (1978) Ordination of plant communities. Junk, The Hague

Chapter 5
Classification

Abstract Classification is the complement of ordination (Chap. 4): while ordination summarizes continuous trends in a multivariate data set, classification seeks discrete groups. With species data sets, these groups are community types; with environmental variables, the groups are biophysical settings or habitat types. As with ordination, there are many alternative tools available, with a long history in ecology. Here we begin with perhaps the most common approach, cluster analysis. The workflow begins with exploratory data analysis, then proceeds to post-processing to interpret and communicate the results. This post-processing might summarize the groups in terms of the variables used to define them (e.g., indicator species for compositional types, environmental predictors for habitat types) or it might cross-walk the data (e.g., indicator species for habitat types or environmental predictors for community types). As with ordination, classification provides a powerful summary for the presentation of data sets collected as part of an inventory or monitoring program. Ordination and classification are especially powerful when integrated together.

5.1 Introduction

There are two complementary approaches to summarizing multivariate ecological data sets. With ordination, we focus on continuous trends in the data, such as species-compositional trends along environmental gradients. In classification, we focus on discrete groups in the data. These two perspectives are complementary and not exclusive, as it is not unrealistic to have more-or-less discrete communities that are themselves arrayed along environmental gradients, or trends within more-or-less discrete communities. Because of this, we will come to appreciate how powerful these two approaches can be when applied in tandem.

Supplementary Information The online version contains supplementary material available at https://doi.org/10.1007/978-3-031-72251-6_5.

D. L. Urban, *Landscape Ecology: A Task-Oriented Perspective*,
https://doi.org/10.1007/978-3-031-72251-6_5

We considered ordination in the previous chapter; now we attend classification techniques. In community ecology where the data set consists of species abundances measured on sample locations, the goal is to discover typal species assemblages or communities. This goal presumes that some species co-occur naturally or that groups of species respond similarly to environmental heterogeneity, in contrast to the individualistic model of species response implicit in many ordinations. If the data set comprises environmental variables instead of species, classification will result in typal biophysical settings (e.g., topographic positions, sites with similar soil chemistry, and so on) or habitat types.

Identifying discrete groups in a data set will invite two additional questions. First, we will ask whether the groups are *real* in the sense that the groups really do differ in some way; this will require that we develop a way to test group differences statistically. Second, we will ask *how* the groups differ, that is, which species occur differentially across groups or which environmental variables best account for group differences. This will lead to a variety of after-the-fact analyses of groups. In the end, if we determine that groups identified by classification are not really discrete or ecologically uninterpretable, we would be left with an ordination-based gradient model as the logical alternative. In many applications, both approaches are useful and, as we will see, classified types are readily embedded into ordinations.

In terms of the overall workflow of this book, classification joins ordination as a toolkit for exploring, summarizing, and presenting the main patterns in ecological data sets. This would apply especially to inventory and monitoring data. Sometimes the analysis is an end in itself (i.e., identifying communities), but more typically classification provides a foundation for further analyses such as site prioritization (Chap. 8) and ecological assessment (Chap. 10).

5.2 Overview of Classification

Various classification methods have been devised over the past several decades. These range from rather simple methods that were originally performed by hand to more recent and more computationally demanding methods. Most methods, however, can be categorized according to a few simple criteria. And, once classified, we tend to ask the same questions about the groups no matter how they were created. Here we consider classification in a cursory pass, before delving into more detail on a few popular approaches.

5.2.1 Classification Techniques

First, classification methods can form groups in one of two directions. In *agglomerative* methods, the analysis begins by considering each sample a discrete group and samples are then combined into larger groups ("from the bottom up"). By

Table 5.1 A classification of classification techniques (items in parentheses exist but are deprecated or uncommon; n/a means no method exists)

Direction	Nesting	Monothetic	Polythetic
Agglomerative	*Nested*	n/a	Clustering
	Not nested	n/a	Pooling
Divisive	*Nested*	(Association)	(Partitioning)
	Not nested	(Partitioning)	(Divisive analysis)

contrast, *divisive* methods begin by considering all the samples to represent a single group. This group is then partitioned into smaller groups ("from the top down").

Second, classification can result in *hierarchical* or nonhierarchical groups. In the former, groups are nested so that each group contains smaller groups and is itself contained in a larger group. Nonhierarchical groups are not nested.

Finally, these methods can be sorted according to the number of variables used to define or drive the classification. In *monothetic* methods, the classification depends on a single variable or criterion (typically, single variables applied sequentially at each step). By contrast, *polythetic* methods consider multiple criterion variables at each step of the classification.

These sorting criteria would seem to yield eight possible classification methods, but this is not the case. There are at least two polythetic ways to perform agglomeration and no way to perform a monothetic agglomeration. There are two ways, monothetic and polythetic, to perform a divisive classification. The following sections outline the common classification approaches used in ecology. We then develop a workflow for applications, beginning with what is probably the most commonly used technique. This workflow can be adapted readily to alternative techniques (Table 5.1).

To be thorough, we might note that species distribution modeling (Chap. 2) is a trivially simple monothetic divisive technique: we divide the data into one group of samples where the species occurred and another group where the species did not occur. Thus, the *identification* of the groups is not very interesting, and we focus rightly on whether and how these two groups differ—an after-the-fact *discrimination* of the groups. This discrimination is common to all classification applications.

5.2.1.1 Hierarchical Agglomerative Classification

By far, the most common classification method is polythetic hierarchical agglomerative classification, or *clustering*. In this, the analysis begins by considering each sample to be a separate group. The two most similar groups are then joined into a new group. The next most similar groups are then joined, and so on. Early in the clustering, the joinings are between pairs of individual samples. Later, the joinings are of pairs of groups, each group including many samples. Ultimately, all samples are included in a single group. Because of the joining algorithm, each group contains smaller groups and is itself contained in a larger group, so the procedure is *hierarchical*. The result is typically summarized in a *dendrogram,* an inverted tree that traces the sequence of joinings. Clustering is intuitive and superficially simple, which explains its popularity. Decisions must be made, however, about how to compute ecological similarity between two samples and especially about how to compute similarities between groups that contain more than one sample. The user must also decide how many groups to retain; the dendrogram allows many levels. These decisions give rise to several alternative approaches to clustering, to which we will return.

5.2.1.2 Nonhierarchical Agglomerative Classification

An alternative to clustering is a polythetic agglomeration that is not hierarchical. In such a solution, the groups are not contained by larger groups, nor do they contain smaller groups. The simplest algorithm begins by choosing a number of *kernels* or centers in the multivariate space defined by the primary data matrix (e.g., sample points in m-dimensional species space). Each sample is then assigned to the nearest kernel, that is, the kernel to which the sample is most similar, based on Euclidean distance. The result is "pools" of similar samples in the multivariate space, and so Pielou (1984) coined the term *pooling* for this approach. This same method is often called *K-means partitioning* because it creates K clusters in the sample space, with the clusters as internally homogeneous as possible. Importantly, the number of clusters (K) must be specified in advance by the user. *Unsupervised classification* of remotely sensed images is typically conducted by K-means partitions of the spectral bands (variables) that comprise the images (e.g., Lillesand et al. 2015).

Pooling is often recommended as the technique of choice for very large data sets. In these cases, an initial pooling is used to provide a smaller number of more homogeneous groups, which are then themselves used as aggregate (averaged) samples for subsequent analysis. This further analysis might be hierarchical clustering or ordination. K-means clustering also can be used to partition an ordination space into discrete regions (groups of samples), as we will consider later.

5.2.1.3 Monothetic Divisive Classification

The most common (perhaps the *only* common) monothetic divisive method is *association analysis* (Williams and Lambert 1959). The procedure was designed for species presence/absence data but could be used on any binary data matrix. In the analysis, the samples are partitioned into two groups according to the presence or absence of a single species. That species is then removed from further consideration and the procedure repeats, choosing a second species on which to partition the samples. The result is a divisive hierarchical classification in which each group can be "keyed out" according to the presence or absence of the chosen indicator species at each iteration (i.e., each branch of the resulting decision tree).

5.2.1.4 Polythetic Divisive Classification

A logical extension of association analysis, at least conceptually, would be to use actual abundances (instead of binary data) and to allow for multiple criterion species at each level of the classification. This approach would be polythetic and divisive. Legendre and Legendre (2012) discussed possible algorithms but noted the computational challenge. The task is to find the "best" partition of samples in a multivariate space, and the only sure solution to this task is examine every possible partition of the data—a daunting task if the number of samples is large. Kaufman and Rousseeuw (1990) provide an algorithm for a polythetic divisive classification based on the method of Macnaughton-Smith et al. (1964).

One method that has proven popular is a hybrid technique that partitions an ordination divisively. This reduces the task above into the much simpler task of finding the optimum partitioning of a single ordination axis (i.e., a one-dimensional problem rather than a multidimensional one). One method, called two-way indicator species analysis (TWINSPAN, Hill 1979) uses a reciprocal averaging ordination as its basis. This ordination axis is "split" into two groups, and species that show a preference for one side or the other of this axis are identified as *indicator species*. The algorithm is actually rather complicated and is not considered further here (but see Supplement S5.3). The net result is an ordered table which summarizes the ordination, the groups derived from the ordination, and the species that show high indicator value for each group. As with association analysis, TWINSPAN also provides a key via which any sample can be assigned to a group according to its species composition.

5.2.2 *Issues in Classification*

While classification techniques often seem beguilingly simple, a number of issues can complicate the analyses. Many of the techniques, especially hierarchical clustering, depend on an ecological distance measure (recall our previous discussion of secondary data matrices in Chaps. 3 and 4), and the choice of distance metric can substantially affect the classification results.

Classification methods also vary in the geometry or other attributes of the groups they produce. Some produce compact groups, some diffuse; some work best for small data sets as compared to larger ones; and some are more or less vulnerable to noisy data or other data issues. We will address these as we consider the more popular classification methods used by ecologists.

5.2.3 *Working with Classified Groups*

Having defined groups—whether arbitrarily or through a statistical classification technique—there are a few questions that arise naturally about the groups. First, we will want to know whether the among-group differences are statistically significant: Are the groups *real*? Next, we will probably want to know *how* the groups differ. Finally, we might want to be able to predict group membership for new samples (i.e., those not used to define the groups). These are exactly the questions we addressed while evaluating species distribution models, when the number of groups was two ("habitat" or "not"); now, we will allow more groups. Further, because we might create groups from either species compositional or environmental variables, we will broach more options in how we interpret the groups. Because these questions apply generally to groups—no matter how there were defined—we return to these questions after we consider techniques for creating groups.

5.2.4 *Workflow*

An application in classification invites a straightforward workflow that unfolds in several steps (Fig. 5.1). As with other analyses, the main step—creating the groups—is important but might not be the most involved bit of the workflow. Post-processing the groups can be more involved. In particular, we will want evaluate the groups in terms of whether they are statistically different and how

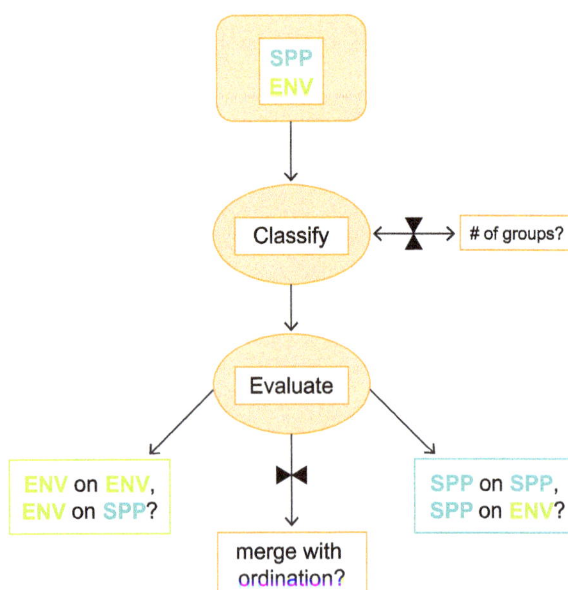

Fig. 5.1 Workflow for an application of classification. Regardless of the classification technique, the workflow unfolds in similar steps: deciding how many groups to use and then evaluating the groups in terms of how they differ. In this, groups might be defined on species (SPP) or environmental factors (ENV) and evaluated in terms of either SPP or ENV or (typically) both. A classification can also be merged with an ordination (Chap. 4)

they differ. The classification can be of either species or environmental (or other ancillary) data. The post-processing would naturally evaluate the groups in terms of the variables that formed them (i.e., species groups in terms of species, environmental groups in terms of environmental factors). But the complementary assessments are also interesting and often an aim of the analysis: evaluating species groups in terms of environmental factors or environmental groups (habitat types) in terms of species response to these.

In the following sections, we develop two illustrations of a classification workflow. The first demonstrates cluster analysis, and then we compare this to a nonhierarchical pooling. These two starting points converge in a parallel workflow in evaluating group differences.

Additional details on classification techniques, as well as some alternative approaches to evaluating group differences, are provided in a digital supplement to this chapter (Supplement S5).

5.3 Cluster Analysis

Clustering is a hierarchical agglomerative method for identifying groups of samples in a multivariate data set. If the data are species abundances measured on sample plots, then the groups are compositionally similar species assemblages or community types. If the data are environmental variables, then the groups are similar biophysical settings or habitat types. While conceptually intuitive and reasonably straightforward as an algorithm, there are subtleties to clustering that present decision points in applications. Nonetheless, clustering remains by far the most popular classification technique used today. The technique is intuitive, well understood by statisticians as well as practioners, and (perhaps most importantly!) it performs quite well in most applications.

5.3.1 Hierarchical Agglomeration

Clustering is *agglomerative* in that it begins by treating each sample unit as a group and then proceeds to combine samples into increasingly larger groups. It is *hierarchical* in that the joining method produces groups that contain smaller groups while being contained by larger groups. Clustering is *polythetic* in that joinings are based on between-group (dis)similarities computed from a set of several measured variables (in the case illustrated here, species).

The algorithm begins with n individual samples and then chooses the two most similar samples, joining them into a new group. The next pair of most-similar samples are then joined, and so on. Early in the process, the groups joined tend to be individual samples, while later the groups are larger clusters themselves. After $n - 1$ joinings there is a single group. Because the solution is hierarchical, groups can be defined at various levels ranging from a few high-level groups to a larger number of low-level groups. The decision about how many groups to retain or interpret is left to the user (and see below).

Clustering, while quite simple in general outline, depends critically on two decisions about how the clusters are defined. These decisions are (1) the choice of distance measure that is used to define the (dis)similarity of groups and (2) the criteria by which groups are joined when there are multiple samples within a group.

5.3.1.1 Distance Measures

Clustering requires a distance measure to define the pairwise dissimilarity of groups in the analysis. Any distance measure can be used, although there are preferred indices for certain kinds of data. For species data, indices such as Bray-Curtis and similar indices are used commonly. Many of these measures are similar in that they are concerned with the treatment of joint absences ("double zeroes") of species on samples: we might not want to treat two samples as ecologically similar because neither supports a particular species. Various transformations or relativizations can substantially influence the behavior of these distance measures.

For other kinds of variables such as environmental factors, Euclidean or Mahalanobis distances might be used. Again, issues of data transformations or relativizations must be considered when selecting an appropriate index.

We considered distance measures in Chap. 3 (Sect. 3.2.4) and again with distance-based ordinations (Chap. 4, Sect. 4.3.2). Classifications and ordinations based on the same distance measures can be combined readily, as we will see below.

5.3.1.2 Joining Criteria

To begin, the clustering algorithm computes a sample x sample dissimilarity matrix and scans this to find the pair of samples that are most similar (least dissimilar). These samples are then joined into a new group. The process then repeats. A crucial point of the analysis lies in deciding how to compute the ecological distance between two groups when each group itself consists of several sample units. This decision about *joining rules* defines a variety of clustering algorithms.

A number of joining rules have been proposed and codified in cluster analysis. Three of these can serve to frame the general issues in joining groups of sample units. The three methods are *nearest neighbor*, *farthest neighbor*, and *centroid linkage* (Fig. 5.2). Additional joining rules follow from these conceptual starting points. McCune and Grace (2002) provide practical guidance on these alternatives.

Nearest-Neighbor Linkage The nearest-neighbor joining rule, also known as *single-linkage*, computes the distance between two groups of samples as the minimum sample-to-sample distance for any pair of samples i and j such that these two samples are in different groups g and h. The minimum distance is found simply by

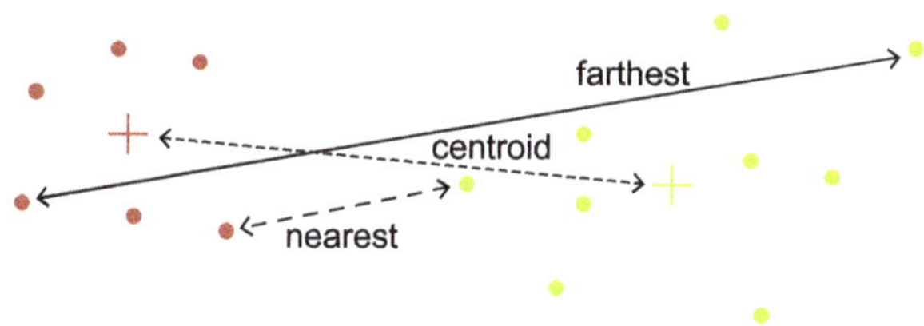

Fig. 5.2 Schematic of three alternative methods for joining groups that consist of multiple samples in each group. Here, red and green samples are the two groups. The effective distance between the two groups might be based on the nearest-neighbor sample pair, the farthest neighbors, or the average distance between samples in each group (i.e., the distance between their centroids, indicated by "+" signs)

computing the distance between every pair of samples and retaining the minimum of these values.

Nearest-neighbor linkages tend to produce groups that are "loose" because they can be joined if any two samples are ecologically close. In particular, this method often results in *chaining*, or the sequential addition of single samples to an existing group (see also Fig. S5.2). In this, a single group might grow by accretion, adding single samples to the "edges" of the growing cluster. In effect, this contracts the space around the clusters, making them appear closer together than they actually are in the original multivariate space.

Farthest-Neighbor Linkage By contrast to nearest-neighbor joining rules, farthest-neighbor (also known as *complete*) linkage joins groups by finding the maximum pairwise distance between two samples, i in group g and j in group h. Again, this maximum distance is found by scanning all pairwise distances for the two groups.

Not surprisingly, farthest-neighbor linkage tends to produce very compact clusters of samples. This method does not tend to chain as it joins groups. Farthest-neighbor linkage tends to produce groups that are similar in size. This also tends to make the groups appear more different than they are, in effect expanding the space around the clusters.

Centroid Linkage Single- and complete-linkage joinings are defined by single pairs of samples (the closest or farthest). It would seem more appropriate to use information about *all* of the samples in each group to define the joinings. One way to do that is to find the centroids of each group and then base joinings on the distances between group centroids. For species data, the group centroid is the average species composition of the group and so this makes intuitive sense as a basis for a joining rule. As might be expected, centroid linkage produces clusters that are intermediate in size and shape as compared to single- or complete-linkage methods. This approach is space-conserving, in that the joining rule does not distort the apparent distances among groups.

Intermediate Linkage An alternative intermediate case entails considering all of the pairwise distances for samples in two groups and joining the groups if some percentage of the pairs meet some threshold distance value. This is known as intermediate linkage, and in some instances (i.e., for some threshold values), this is also referred to as *average linkage*. (This is an unfortunate label, in that the method does *not* average the actual links between members of each group.)

Alternative Linkage Methods Sneath and Sokal (1973) presented a two-way contingency table of methods based on averaging distances between members of groups to be joined. This contingency defines four *average linkage* methods, typically referenced by their rather awkward acronyms UPGMA, WPGMA, UPGMC, and WPGMC. The base method, often referred to simply as *average linkage*, is unweighted, arithmetic average clustering (i.e., *u*nweighted *p*air-*g*roup *m*ethod, *a*rithmetic average). Its counterpart UPGMC is based on distances between centroids of groups, rather than average distances (as in centroid linkage, above). In the unweighted methods, each sample contributes equally in the distance calculations. The weighted versions of each method further adjust the group joinings to give equal weight to each group being joined—in essence, down-weighting individual members of the larger group.

An important algorithmic consideration in clustering is whether the distances between higher-level groups (i.e., clusters of the original samples) can be computed "on the fly" or whether these needed to be recomputed from the original data at each level of clustering. Methods in which the new distances can be computed are *combinatorial* methods (Lance and Williams 1967a, c), which clearly have some advantages in terms of algorithmic efficiency. Many clustering algorithms can be implemented in terms of the same generic formula for combining distances among three groups: parameters can be chosen to generate single-linkage, complete-linkage, average-linkage, and many other intermediate forms (McCune and Grace 2002; Legendre and Legendre 2012). This is the algorithm used by most computer packages. (See Supplement S5 for more on this.)

5.3.1.3 Presentation and Interpretation of Results

Most clustering programs provide two summaries of the process. The first is tabular and recounts which groups were joined at each cycle of the process. Because there are $n - 1$ cycles required to join n samples, this can be a rather intimidating volume of output. The second, and more intelligible, summary is the *dendrogram*. This tree diagram joins the samples hierarchically, beginning with n individual samples at the bottom of the diagram (sometimes drawn rotated, the "bottom" to the left) and a single group at the top (right, if rotated). The height at which each joining is drawn indicates the ecological distance between the groups. Thus, the initial joinings are at low heights (small distances) and the last joining is at maximum height.

A key step in the interpretation of clustering results is to decide how many groups there are. This is clearly a subjective decision, because the groups can be selected at several levels. A simple and graphic way to identify groups is to draw a line across the dendrogram, which "cuts the branches of the tree" to identify groups at that level

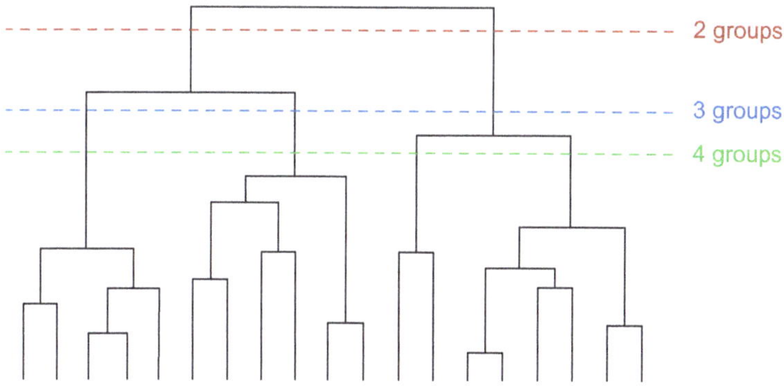

Fig. 5.3 Dendrogram of a clustering; at the distance levels indicated by the arrows, there are 4, 3, or 2 groups

(Fig. 5.3). For example, the arrows in Fig. 5.3 shows that there may be 2, 3, or 4 groups in the data, depending on where the tree is cut. One way to make this decision more objective is to insist that the groups be statistically different from each other. We will return to these statistical tests later, because they apply to other classifications as well.

Another way to choose the number of groups is to retain those that are ecologically interpretable (experiments suggest that groups might be statistically different well beyond any ecological significance). Clusters can be interpreted by examining the sample units in a particular cluster or the species or other variables that characterize it. Again, because this assessment can apply to any classification, we return to this later.

5.3.1.4 Complications and Confounding Issues

Clustering, for all its intuitive appeal, can be confounded by a number of issues. An obvious source of confusion is that the results might change dramatically if one selects different joining rules. Choice of distance measure can also have an effect, though usually not as dramatic as the joining rule. There is no "correct" joining rule, and so this decision may hinge on utilitarian issues in most cases. For example, single-linkage clusters tend to chain, which does not provide tight clusters. If chaining results from other linkage methods beyond single-linkage, this probably means that the data do not form natural groups (i.e., a naturally continuous gradient would chain under nearly any clustering method).

If clusters are to be interpreted in terms of ecological space (e.g., plotted into an ordination), then it is important to choose a joining rule that preserves this space as much as possible. Average linkage methods (including methods approximating this) are space-preserving; single- and complete-linkage methods are space-distorting.

Because it is hierarchically agglomerative, replicate samples that are essentially equivalent pose a problem for initial joinings in clustering. In the case of ties (more than one pair of samples that are equivalently distant), some arbitrary decision must be made. One option is to join random pairs of equivalent samples. One consequence of these initial decisions is that they propagate to higher-level groups; thus, arbitrary decisions with no real information content are still retained in the highest-level groups. Fortunately, samples that are very similar should naturally end up in the same group, so this is not typically an issue.

Clustering is sensitive to outliers but these appear readily in the dendrogram as long branches with few leaves, very small groups, or single samples that join the rest of the dendrogram at very high levels. Cutting the tree at a high level then produces a few "good groups" and a few outliers that are merely distracting (and which can be discarded). With noisy data, and especially for very large data sets, a nonhierarchical method can often perform better than clustering. For example, pooling (below) not only solves the replicate similarity problem by pooling them together, it also provides a solution to outliers: they are simply ignored (not classified) in many pooling algorithms.

Finally, clustering can be tedious for very large data sets and inefficient because generally only the last few high-level clusters are of interest. In these cases, it would make sense to use a divisive technique instead: that is, starting with one group and stopping the divisive process when the desired number of groups is reached. For example, if we have 1000 samples and want three groups, clustering requires 997 cycles while a divisive technique needs two to reach the goal. A useful alternative here, a compromise, is to first pool the data into small composite samples and then cluster the composites.

5.4 Nonhierarchical Agglomeration

We have considered clustering (polythetic hierarchical agglomeration) as perhaps the most popular classification technique in common usage today. Clustering also provided a conceptual bridge between ordination via nonmetric multidimensional scaling (NMS) and classification, as both NMS and clustering can proceed from the same ecological distance matrix. But there are cases, especially with huge data sets, where clustering is inefficient. In such cases, nonhierarchical methods might be more appropriate than clustering.

There are two basic approaches to nonhierarchical classification. The first and by far most common method produces a *partition* of the samples. A partition of a data set consists of a number of groups (subsets) such that each sample unit belongs to one and only one subset. Note that all nonhierarchical methods share the result that while groups are defined to be as internally homogeneous as possible, there is no information provided about the relationships among groups, as results from clustering.

Here we focus on nonhierarchical polythetic agglomeration, beginning with the most popular approach but diverging to an emerging alternative.

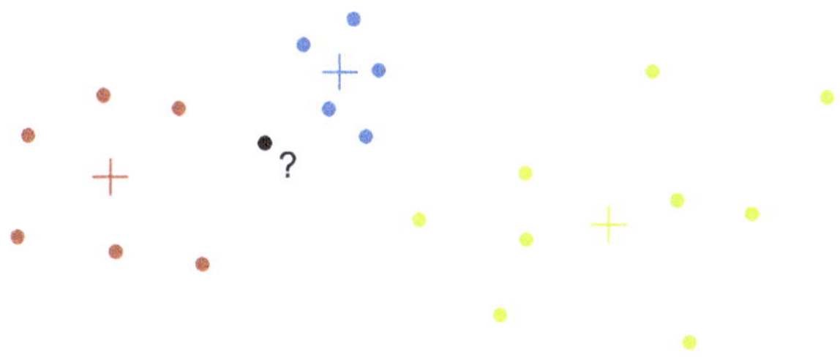

Fig. 5.4 Schematic of *K*-means classification with three groups (their centroids indicated by +'s). In the "build" phase, each sample point is assigned to the closest kernel (groups colored accordingly; here, the black sample indicated by "?" will be assigned to the closest centroid, in blue). In the "clean" phase, the kernels are recomputed as group centroids and probably move. The process then repeats until the kernels stabilize (stop moving)

5.4.1 *Partitioning Methods:* K-*Means Pooling*

The most popular nonhierarchical techniques are partitioning methods known variously as pooling (Pielou 1984) or *K*-means classification (MacQueen 1967). There are in fact several different algorithms, considered below. Most of the methods share at least a few elements in common. The general task is to find *K* groups in multivariate space such that within-group similarities are maximized (Fig. 5.4). The result is groups that are as ecologically distinctive as possible. The number of groups must be specified by the user.

5.4.1.1 Pooling Algorithms

The solution to *K*-means classification is typically a two-stage process. In the first pass, *K* group centroids are identified as reference points in *m*-dimensional ecological space (*m* the number of species or variables). Each sample unit is assigned to the group (centroid) to which it is most similar. This is often referred to as the *build* or *assignment* phase of the algorithm. The centroids of each group are then recalculated as the mean of each variable over all samples in a group. This is the *clean* phase. The samples are then reassigned as necessary, again to the nearest centroid (which will have moved in the *clean* phase). This iteration continues until the centroids (hence group assignments) no longer change.

Note that in assigning each sample to the nearest centroid, this approach uses Euclidean distances. Thus, the objective is to minimize the within-group variances, estimated as ordinary sums of squared deviations (Euclidean distances). This technique thus is especially applicable to data sets for which is makes sense to use Euclidean distances to estimate between-sample dissimilarities.

5.4.2 Partitioning Around Medoids

An alternative to pooling is partitioning around medoids (PAM) (Kaufman and Rousseeuw 1990). This approach is similar to K-means pooling but uses "medoids" instead of centroids (kernels averaged from samples) of the pools to compute distances. A medoid is a sample that is used as the centroid or exemplar for a cluster. In PAM, a set of K of the samples are used as medoids, and all samples are assigned to the nearest medoid. But the objective is different than with K-means pooling. In PAM, the objective is to minimize the average pairwise dissimilarity from sample to medoid within each group (and hence, over all groups). This is the *build* phase. In the *swap* phase, each sample in a group is swapped with that group's medoid, to see if the within-group distances can be improved (minimized). That is, might another sample in that group serve as a better medoid? The *build* and *swap* phases repeat until the groups stabilize.

The use of actual samples (medoids) as compared to the (synthetic) group centroids of K-means pooling is not the most important distinction here. What is important is that PAM can use *any* distance or dissimilarity measure, not just Euclidean distance. In particular, it can use distances more appropriate for species data (e.g., Bray-Curtis) or for highly correlated environmental factors (e.g., as Mahalanobis distances). This aligns PAM more readily with NMS ordinations or clustering, in their ability to use more nuanced ecological distances.

5.4.2.1 How Many Groups?

Again, with pooling (K-means or PAM), the user must decide in advance how many groups to create. Because the pooling is not hierarchical, the groups cannot be compared directly from analyses creating different numbers of groups. That is, the groups in a six-group solution are not directly comparable to those of a five- or seven-group solution. We turn to the decision on group number in the next section.

5.5 Working with Groups

We have considered a few common techniques for identifying or creating discrete groups of samples from a heterogeneous data set. Defining groups broaches the natural question: *How valid are these groups?* That is, are the groups actually significantly different statistically? This question leads quite naturally to the further question, if the groups differ: *How are the groups different,* or more specifically, *Which variables best distinguish or discriminate among the groups?* Here we will begin with the first issue of testing *whether* groups differ and then proceed to consider *how* they differ.

Note that these questions do not require that we have previously generated the groups statistically—the groups might have been defined by other methods, or administratively, or for management purposes. In any case, it makes sense to assess the groups statistically.

Note again that these questions are the same as those we asked in the initial application in this book, species distribution modeling. In that case, we had only two groups (habitat or not). Here it will become obvious that the same questions apply to the multiple-group case, but that there are additional questions we might ask of groups defined from multivariate ecological data.

5.5.1 Are *the Groups Different?*

The basic question we wish to answer is: *Are these groups different?* This is a vague question that needs to be rephrased more specifically for testing. One way to rephrase the question is: *Are samples within the same group more similar than samples from different groups?* This is equivalent to contrasting the among-group to within-group variability in the data set. This test is essentially a multivariate *F*-test, a contrast of among- to within-group variances. Indeed, a MANOVA *F*-test would be a reasonable solution to the question of group differences—if we could meet the assumptions of the parametric test. This is difficult with ecological data, in general. In the cases developed here, the groups are based on distances or dissimilarities, and we know that these are not independent (Chap. 3) and so cannot be assessed with parametric tests.

Alternative nonparametric methods are available. Perhaps the most straightforward is simply a nonparametric version of MANOVA, with the test statistic based on permutation (Anderson 2001; Oksanen et al. 2021). Even this approach suffers from imbalanced groups (e.g., from very different samples sizes over the groups), an issue with analysis of variance in general. Other approaches differ in how they compute among- versus within-group homogeneity (see Supplement S5). Often, these tests area based on ecological dissimilarities or distances. The approaches vary in the details of the test statistic, but each is concerned with among- as compared to within-group dissimilarities. One general approach is developed below.

5.5.1.1 Mantel's Test

Mantel's test is a correlation between distance matrices (Mantel 1967). The Mantel's test is quite flexible and can be used here as a general tool for testing group differences.

In this instance, a dissimilarity matrix can be defined as a contrast (or design) matrix for groups, coding distance as "1" if two samples are in different groups or "0" if they are in the same group. Mantel's test conducted between this group contrast matrix and an ecological distance matrix (e.g., the Bray-Curtis index) tests the hypothesis that among-group distances are large relative to within-group distances. In this case, the one-tailed test is of interest because larger among-group differences are the expectation. The test statistic is evaluated by permutation (Goslee and Urban 2007), as with the permutation version of MANOVA. In this application, the test would be repeated for multiple numbers of groups—different cut levels of a cluster dendrogram or separate classifications using pooling.

5.5.1.2 Statistical and Ecological Significance

The classifications we have just examined here all create groups by maximizing among-group differences or (equivalently) minimizing within-group differences. It should come as no great surprise, then, that these groups are usually statistically different from one another. And so, this test of significance, while important as a box to be checked, is typically not a difficult test to pass. The question is, rather, of several alternative classifications—all of which are statistically significant—what is the appropriate number of groups to use?

One way to address this is to repeat the statistical analysis for a range of numbers of groups. This might be multiple cut levels of a cluster dendrogram, or a set of poolings with different numbers of groups. Plotting the test statistic (here, the Mantel correlation, but any comparable statistic would also work) against number of groups would typically reveal a curve that might suggest a level that works best for that data set, corresponding to a peak in the test statistic (Goslee and Urban 2007; Borcard et al. 2011).

In a best-case scenario, this curve is monotonic and peaks at an intermediate number of groups. This can be interpreted intuitively relative to the among- versus within-group contrast of the test statistic. With too few groups, the internal variability increases so the within-group dissimilarity is too high, while with too many groups, the among-group dissimilarity is too low; both cases reduce the among- to within-group contrast. The best case is an intermediate number of groups. A PAM classification of the Sierran tree data featured in Chaps. 3 and 4 exhibits this relationship (Fig. 5.5). In this case, the species data were analyzed using the same extended Bray-Curtis distances used previously in the NMS ordination (Sect. 4.3.2), and the pooling was conducted for a range of 2–10 groups. Note that at all levels, the tests are highly significant ($P < 0.001$) and the test statistics are actually quite similar; the "best" number of groups is six but any choice from two to seven would be reasonable.

In practice, more groups will often be better (statistically) than fewer groups, but too many groups can degrade the usefulness of the solution. In this, this decision is similar conceptually to choosing an appropriate number of axes for an ordination (recall Chap. 4, Sect. 4.3, especially Figs. 4.3 and 4.7): More axes will generally be better statistically but less useful practically. The aim is to find a workable and justifiable compromise.

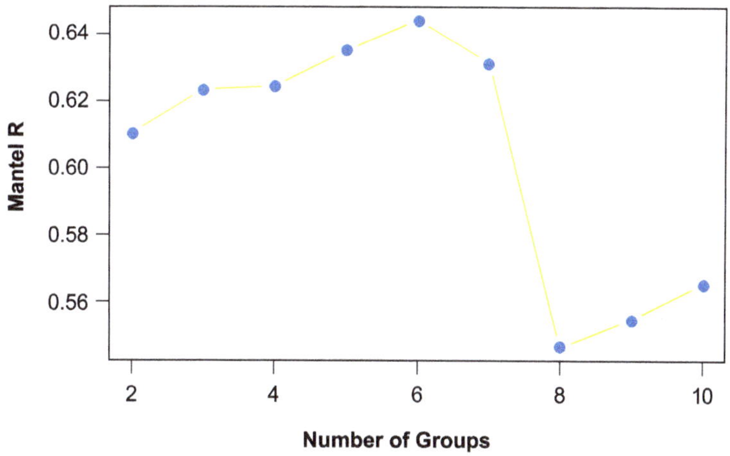

Fig. 5.5 Mantel's test (see text) for a classification using PAM, illustrating the among- to within-group contrast in extended Bray-Curtis distances used in the classification of tree species abundances for samples in Sequoia National Park in the Sierra Nevada in California. A six-group solution is indicated

5.5.2 How *Are the Groups Different?*

This question of among-group differences can be phrased in several ways, depending on the nature of the data. In particular, we can adopt two different perspectives on group differences: (1) *Given groups defined on a set of variables, which of these variables actually distinguish the groups?* For example, if we cluster typal communities based on species composition, which species are most characteristic of each community? Similarly, if we define habitat types based on biophysical environmental variables, which of these variables actually defines the various types? Alternatively: (2) *Given groups defined on a set of variables, which of another set of variables best captures these differences?* For example, if we define community types based on composition, which environmental variables best account for these communities? Reciprocally, if we define habitat types on biophysical variables, are there species that show affinities for each of the various types? These two applications lead to different methods, reflecting the numerical vagaries of species abundances as compared to environmental scalars.

Any evaluation of groups should begin with simple descriptive statistics: summarizing the average species composition of the groups (Table 5.2) or their averages on environmental factors (Table 5.3). This applies for groups defined on either species composition or environmental factors. Inspection of these tabular summaries should suggest an initial interpretation of the groups ecologically.

As might be expected, there are many techniques for comparing groups. Some of these are detailed in Supplement S5. In the next sections, we illustrate two such approaches, using the Sierran forest community types classified using PAM. In the

Table 5.2 Average species composition for six groups classified using PAM (see text) (species codes in Table 3.5)

Species	Group 1	2	3	4	5	6
ABco	38.16	11.94	2.77	0.00	8.43	0.00
ABma	0.64	0.00	67.03	15.71	7.77	17.24
CAde	1.39	16.52	0.00	0.00	0.64	0.00
COnu	0.05	0.15	0.00	0.00	0.00	0.00
PIco	0.00	0.00	0.00	2.87	0.00	52.10
PIje	0.41	0.00	0.00	0.00	37.28	0.00
PIla	13.39	9.09	0.39	0.00	0.88	0.00
PImo	0.00	0.00	3.28	106.36	0.00	16.82
PIpo	1.03	9.68	0.00	0.00	0.00	0.00
QUke	0.35	6.02	0.00	0.00	0.01	0.00
SEgi	60.86	0.00	0.00	0.00	0.00	0.00

Tabled values are average basal area (m^2 ha^{-1}) per species. Species codes are in Table 3.5

Table 5.3 Average environmental conditions for the six forest communities (see text) (environmental variables in Table 3.6)

Factor	Group 1	2	3	4	5	6
Elevation	2003.1	1696.9	2523.2	2867.0	2189.9	2817.7
Slope	18.14	23.91	15.86	25.13	15.54	13.00
TAspect	0.08	−0.04	−0.19	0.01	0.28	0.82
TSI	0.00	0.02	0.02	0.02	0.01	0.01
xLitter	8.47	8.03	4.68	6.35	5.82	4.68
xDepth	81.48	78.04	50.97	24.46	40.89	36.64
sDepth	23.37	20.96	30.55	20.23	18.78	18.68
pH	5.56	5.45	4.91	4.34	4.67	4.37
C	3.28	3.36	4.28	5.54	4.02	4.02
C.N	24.15	22.18	31.11	26.03	24.54	29.15
P	109.59	58.90	69.44	53.50	71.13	100.83
ECEC	8.99	10.17	6.28	4.53	4.70	4.09
Clay	1.06	2.22	1.16	2.06	0.98	1.10

Environmental variables are described in Table 3.6

first case we ask, which species best characterize each of the groups? The approach is indicator species analysis. In the second illustration, we ask which environmental variables best distinguish these forest types. In this, we use a classification and regression tree (CART), exactly as developed as a species distribution model in Chap. 2 (Appendix A2.2).

5.5.2.1 Merging Classification and Ordination

Classified groups are generated without any reference an interpretative framework beyond the other groups. Summarizing groups in terms of the classifying or ancillary data can help interpret the groups. But it can be quite revealing to display a classification in an ordination. This can be done with any classification and any ordination, simply by color- or symbol-coding the samples by group in a plot of the ordination. But this is especially compelling with classifications done by clustering or PAM, embedded in an NMS ordination. This is because the analyses all proceed from the same distance matrices.

In the case of the Sierran forest community types, plotting these into the corresponding NMS ordination (Figs. 4.9, 4.10, and 4.11) emphasizes the strong elevation gradient while also suggesting the separation of forest types at lower and mid-elevations (Fig. 5.6).

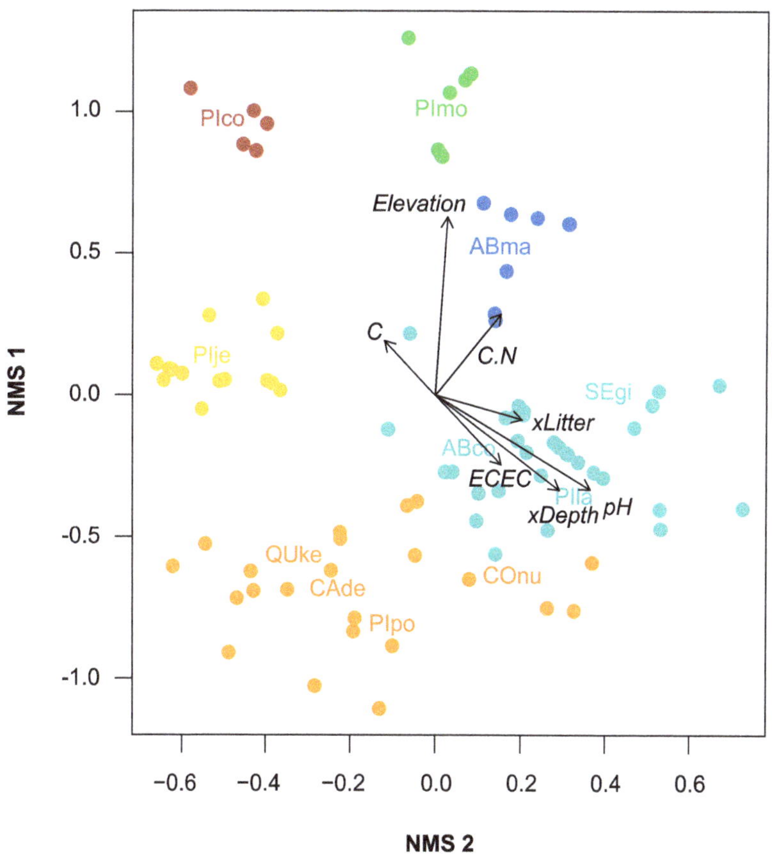

Fig. 5.6 A joint biplot of an NMS ordination (the same as Fig. 4.11), with groups classified via PAM overlaid. Species names are colored according to the group in which they reach their maximum abundance (see Table 3.5 for species codes)

While this illustration features the PAM partitioning, it can be especially useful to explore a clustering in an ordination. Because the clustering is hierarchical (nested), viewing the clusters corresponding to various cut levels in the dendrogram shows precisely how each group is merged or divided from lower or higher levels in the tree. For example, a clustering of the Sierran forests reveals the relationships among two groups (lower/mid and higher elevation, Fig. 5.7a) and three (Fig. 5.7b, splitting the Jeffrey pine community (PIje) off at middle elevations). At four groups, the lodgepole pine (PIco) group separates from other higher-elevation types (Fig. 5.7c), and at five groups the lower-elevation group splits the white fir (ABco) group from the lower elevation and drier forests characterized by Ponderosa pine (PIpo) and incense-cedar (CAde) (Fig. 5.7d). Finally, at level 6, the red fir (ABma) group splits off at upper elevations (Fig. 5.7e). At six groups, the classification is very similar to the PAM groups illustrated in Fig. 5.6.

5.5.2.2 Identifying Compositional Groups: Indicator Species

One approach to distinguishing among groups is based on the notion of *indicator species* (Dufrêne and Legendre 1997). Their method is designed for species compositional data. The goal is to identify those species that have shown high fidelity to a particular group and, as such, can serve as indicators for that group. A good indicator species would occur with high relative abundance and high frequency in its group while simultaneously *not* occurring in other groups.

Dufrêne and Legendre's (1997) method combines the notions of local abundance and frequency into a single indicator value (see Supplement S5 for details). They compute the relative abundance of each species in each group and then also the relative frequency of each species in each group. The *indicator value* for each species for each group is the product of its relative abundance and relative frequency. Indicator values range from 0 to 100, with 100 representing a perfect indicator. A perfect indicator species would be abundant in *every* sample of its type (group), and *never* occur in a sample from a different group.

The indicator values are tested for statistical significance by permutations of the group membership assignments in the original data. This test of significance is important, because rare species will often be appealing as indicators—but their low overall frequency tends to render them unreliable (not significant) as indicators.

Indicator species can be a powerful aid to interpreting compositional groups, as they are the species that best highlight among-group differences. Indicators also can be used to predict group membership for new samples, often providing for efficient keys in the field.

For the Sierran forest communities classified using PAM, ISA finds significant indicators for all six groups (Table 5.4). This analysis is via the implementation of ISA in R (de Caceres and Legendre 2009). In the table, elements A and B are strict interpretations of the general notion of site fidelity. For species k and sample i in group g, A is the conditional probability that sample i is in the indicated group g, given that the focal species k is present on the sample. B is the conditional probability

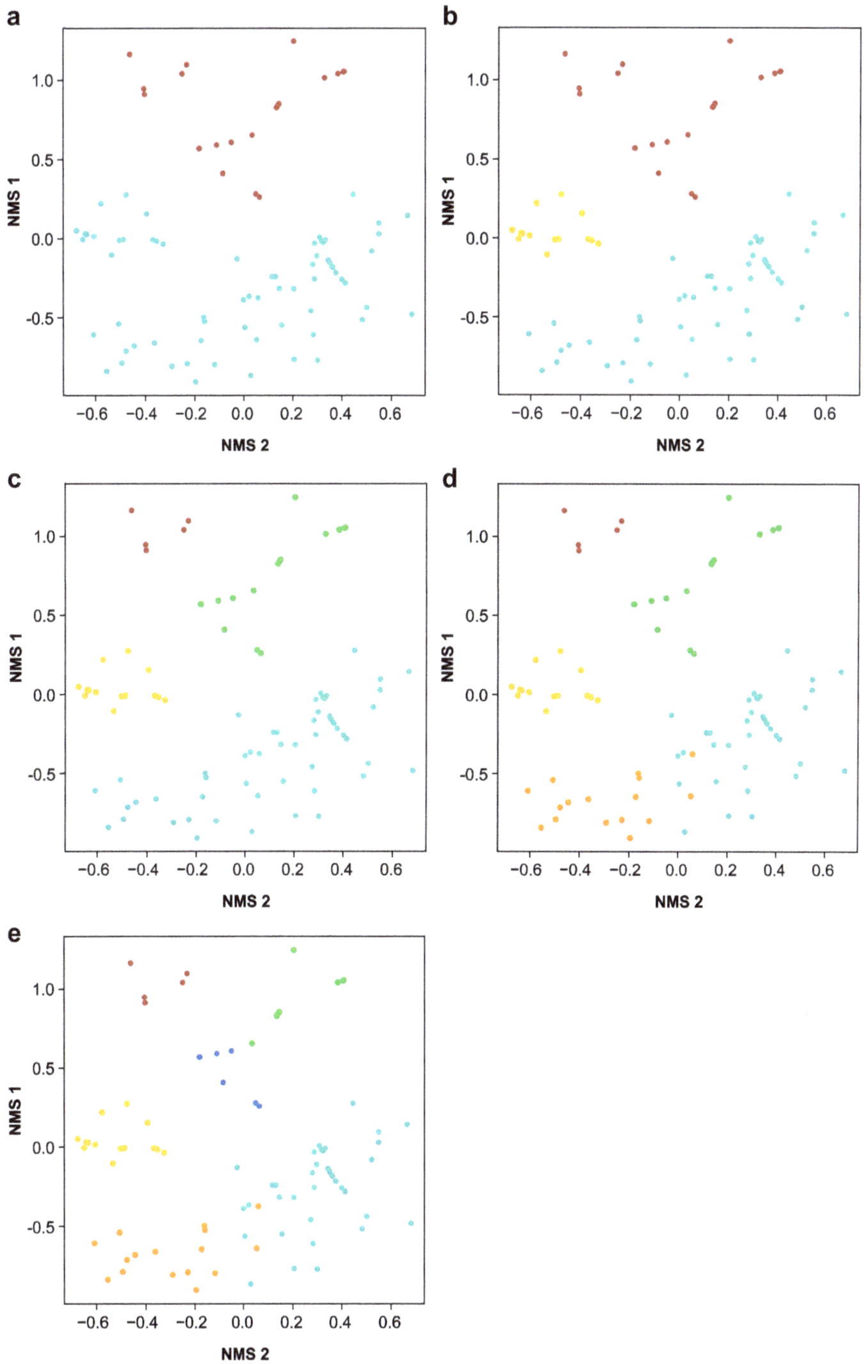

Fig. 5.7 Illustrations of a sequence of clusterings of the Sierran forest data, from two to six groups (**a**–**e**). Successive splits are shown as new symbol colors; see text for further explanation. The six-group solution is very similar to the PAM classification shown in Fig. 5.6

Table 5.4 ISA results for the six forest communities identified using PAM (species codes in Table 3.5)

Species	Group	A	B	IV	$P < \sim$
ABco	1	0.63	1.00	0.79	0.0001
ABma	2	0.62	1.00	0.79	0.0001
CAde	6	0.89	1.00	0.94	0.0001
COnu	6	ns	ns	0.37	0.2
PIco	3	0.95	1.00	0.97	0.0001
PIje	4	0.99	1.00	0.99	0.0001
PIla	1	0.56	0.83	0.90	0.01
PImo	5	0.84	1.00	0.92	0.005
PIpo	6	0.90	0.45	0.64	0.005
QUke	6	0.94	0.86	0.90	0.0001
SEgi	1	ns	ns	0.41	0.1

See text for an explanation of table elements

Table 5.5 Back classification of the six Sierran community types using indicators from ISA (species codes in Table 3.5)

	ABco	ABma	PIco	PIje	PImo	PIpo
ABco	0.639	0.214	0.000	0.000	0.000	0.000
ABma	0.000	0.500	0.000	0.000	0.000	0.000
PIco	0.000	0.000	1.000	0.000	0.125	0.000
PIje	0.028	0.000	0.000	1.000	0.000	0.000
PImo	0.000	0.286	0.000	0.000	0.875	0.000
PIpo	0.333	0.000	0.000	0.000	0.000	1.000

Table values are proportions correct. Rows are predictions; columns are the training data. Group names correspond to species labels in Fig. 5.6

that species k is present in sample i, given that the sample is a member of that group g.

This is a rather simple system and some of the groups are nearly monospecific types, so it is perhaps unsurprising that some of the indicators are quite strong (e.g., lodgepole pine, PIco, for the "lodgepole pine community," group 3). By contrast, Ponderosa pine, PIpo, is not a perfect indicator for what many think of as the "Ponderosa pine community," because it is not reliably present in its type. Pacific dogwood (COnu) and giant sequoia (SEgi) are not common enough to serve as reliable indicators for any group (even though almost all of their occurrences are in group 1).

The indicators can be used to predict group membership. In this instance, they were used to back-classify the training data (a model verification, as these are not independent data). The indicators predict group membership with 78% accuracy overall, with some groups predicted perfectly while others had misclassifications to ecologically similar groups (Table 5.5).

5.5.2.3 Identifying Compositional Groups: Environmental Predictors

Having interpreted groups in terms of the species that generated them, we might also ask whether species compositional groups can be distinguished in terms of environmental predictors. This is essentially the multigroup extension of habitat or species distribution models (Chap. 2), and some of these modeling techniques are readily extended to the multiple-group case. Because we have already considered each of these approaches as habitat models, we will focus here on their multinomial extensions. In this instance, we focus on a classification and regression tree (CART) model (see Appendix A2.2). Supplement S5 also describes a couple other tools for this task.

A classification tree is a special case of a class of models, regression trees, that attempt to partition a data set by recursively explaining subsets of the data (Breiman et al. 1984). In a classification tree, the response variable is categorical, and in this case, the variable assigns group membership. The predictors may be continuous (interval scale) or categorical variables. CART analysis of these data attempts to separate the groups by subsetting the groups and accounting for among-group differences in terms of the predictor variables. As we have already considered CART models in some detail, it is appropriate here to consider only the extension of these models to the multinomial case.

In the case of multiple groups, the procedure of CART is exactly the same as in the binomial case. The aim is to produce "pure" terminal leaves, and each of these will be labeled as a single group. As with classification trees, multiple branches might lead to the same group, allowing for alternative pathways and contingencies.

In this instance, the classification tree (using R package *rpart*, Therneau and Atkinson 2022) produces a relatively simple tree under tenfold cross-validation (Fig. 5.8). Even so, the tree suggests a variety of environmental settings that support the white fir forest type (ABco), while other forest types are more localized environmentally.

The CART model can be used to verify its classification success by back-classifying the data used to create the model. This generates a confusion matrix analogous to Table 5.5 but this time summarizing the environmental associations of the groups (Table 5.6).

The overall classification success (the average of the diagonals of Table 5.6) is 92.8% correct. This underscores the strong environmental sorting of the tree species in this system: the compositional types are actually more readily predicted from environmental factors than by the species data that created the groups.

5.5.3 Reporting

Clearly, classification can generate a rather intimidating volume of information, including a wealth of figures and tables. Some of these will depend on the particular technique, but all applications will share a common standard of reporting. These should include:

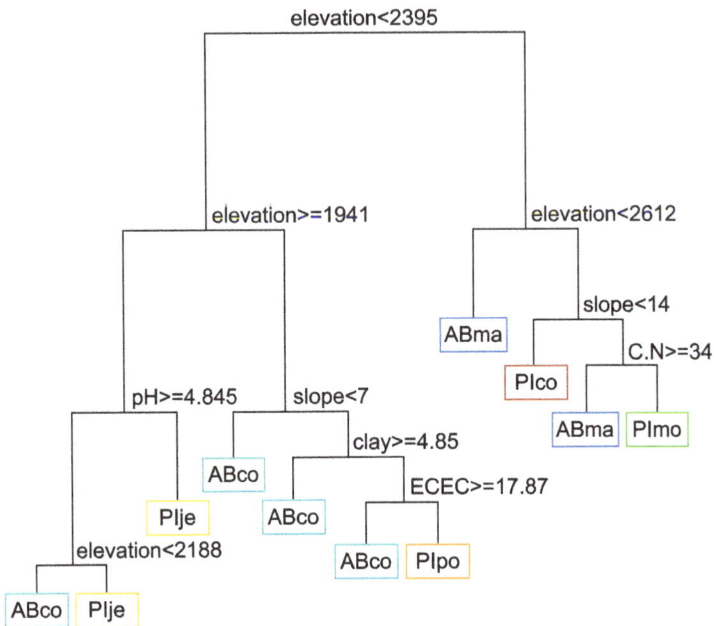

Fig. 5.8 Classification tree for six forest community types identified using PAM, predicted in terms of environmental factors. In each split, the "true" case splits to the left. Group labels, as species names, correspond to the labeling in Fig. 5.6 (species codes in Table 3.5; environmental variables in Table 3.6)

Table 5.6 Confusion matrix for a CART model of six forest types created using PAM, as predicted using environmental factors (species codes in Table 3.5)

	ABco	ABma	PIco	PIje	PImo	PIpo
ABco	0.917	0.071	0.000	0.000	0.000	0.091
ABma	0.000	0.929	0.000	0.000	0.000	0.000
PIco	0.000	0.000	0.750	0.000	0.000	0.000
PIje	0.056	0.000	0.000	1.000	0.000	0.000
PImo	0.000	0.000	0.250	0.000	1.000	0.000
PIpo	0.028	0.000	0.000	0.000	0.000	0.909

Rows are model predictions; columns are data. Group names correspond to coloring and labels in Fig. 5.6

☑ Data preparation, any relativizations or transformations, and results of EDA For a classification based on ecological distances:

☑ Choice of distance measure and whether the distances were extended because of saturation

For any classification:

☑ How many groups were used, and how this was decided (including test statistics as appropriate)
☑ Descriptive statistics on the groups (e.g., mean species composition, mean environmental conditions per group)
☑ Details and results from any further tests among groups, e.g., indicator species analysis, CART, or other model of environmental discrimination among groups

If the classification is merged with an ordination:

☑ Details on the ordination (see Chap. 4, Sect. 4.3.3)
☑ Biplot, joint plot, or joint biplot with classified groups identified by color or symbol codes

5.6 Further Reading

Classification, in general, and clustering, in particular, are venerable techniques and much of the definitive literature is now rather old but still relevant (e.g., Lance and Williams 1967a, b, c, d; Gower 1967; Goodall 1973; Gauch 1982; Pielou 1984; Romesburg 1984). Because these methods are in common use, any textbook on multivariate methods will cover most of the techniques mentioned in this chapter (e.g., Manly 2004). As a very brief introduction to this very large literature, McCune and Grace (2002) cover some of these techniques but with an emphasis on clustering and a focus on ecological decision points and interpretation. Legendre and Legendre (2012) cover most of the techniques discussed here (and some that are not), with a more comprehensive perspective on the statistics and ecology of classification. Kaufman and Rousseeuw (1990) took a fresh look at classification and devised new techniques for many of the available tools. Their software (package *cluster* in R, Maechler et al. 2022) includes functions and support for clustering, partitioning around medoids, polythetic divisive analysis, fuzzy classification, association analysis, and a multistage method for classifying huge data sets such as those generated via remote sensing

References

Anderson MJ (2001) A new method for non-parametric multivariate analysis of variance. Austral J Ecol 26:32–46

Breiman L, Friedman JH, Olshen RA, Stone CJ (1984) Classification and regression trees. Wadsworth and Brooks/Cole, Monterey

Borcard D, Gillet F, Legendre P (2011) Numerical ecology with R. Springer, New York

De Caceres M, Legendre P (2009) Associations between species and groups of sites: indices and statistical inference. Ecology 90:3566–3574

Dufrêne M, Legendre P (1997) Species assemblages and indicator species: the need for a flexible asymmetrical approach. Ecol Monogr 67:345–366
Gauch HG (1982) Multivariate analysis in community ecology. Cambridge University Press, Cambridge
Goodall DW (1973) Numerical classification. In: Handbook of vegetation science, vol 5. W. Junk, Hague, pp 575–615
Goslee SC, Urban DL (2007) The ecodist package for dissimilarity-based analyses of ecological data. J Stat Softw 22:7
Gower JC (1967) A comparison of some methods of cluster analysis. Biometrics 23:623–637
Hill MO (1979) TWINSPAN—a FORTRAN program for arranging multivariate data in an ordered two-way table by classification of the individuals and attributes. Section of Ecology and Systematics, Cornell University, Ithaca
Kaufman L, Rousseeuw PJ (1990) Finding groups in data: an introduction to cluster analysis. Wiley, New York
Lance GN, Williams WT (1967a) Mixed-data classificatory programs. I. Agglomerative systems. Aust Comput J 1:15–20
Lance GN, Williams WT (1967b) Mixed-data classificatory programs. I. Divisive systems. Aust Comput J 1:82–85
Lance GN, Williams WT (1967c) A general theory of classificatory sorting strategies. I. Hierarchical systems. Comput J 9:373–380
Lance GN, Williams WT (1967d) A general theory of classificatory sorting strategies. II. Clustering systems. Comput J 10:271–277
Legendre P, Legendre L (2012) Numerical ecology, 3rd English edn. Elsevier, Amsterdam
Lillesand TM, Kiefer RW, Chipman J (2015) Remote sensing and image interpretation, 7th edn. Wiley, New York
Macnaughton-Smith P, Williams WT, Dale MB, Mockett LG (1964) Dissimilarity analysis: a new technique of hierarchical sub-division. Nature 202:1034–1035
MacQueen J (1967) Some methods for classification and analysis of multivariate observations. In: Le Cam LM, Neyman J (eds) Proceedings of the fifth Berkeley symposium on mathematical statistics and probability, vol 1. University of California Press, Berkeley, pp 281–297
Maechler M, Rousseeuw P, Struyf A, Hubert M, Hornik K (2022) *cluster*: cluster analysis basics and extensions. R package version 2.1.4. https://CRAN.R-project.org/package=cluster
Manly BFJ (2004) Multivariate statistical methods: a primer, 2nd edn. Chapman and Hall, London
Mantel N (1967) The detection of disease clustering and a generalized regression approach. Cancer Res 27:209–220
McCune B, Grace JB (2002) Analysis of ecological communities. MjM Software Design, Gleneden Beach
Oksanen J, Blanchet FG, Kindt R, Legendre P, O'Hara RB, Simpson GL, Solymus P, Henry M, Stevens H, Wagner H (2021) *vegan*: community analysis package. R version 1.17-11. http://CRAN.R-project/org/program=vegan
Pielou EC (1984) The interpretation of ecological data: a primer on classification and ordination. Wiley, New York
Romesburg HC (1984) Cluster analysis for researchers. Lifetime Learning Pubs (Wadsworth), Belmont
Sneath PHA, Sokal RR (1973) Numerical taxonomy: the principles and practice of numerical classification. W.H. Freeman, San Francisco
Therneau T, Atkinson B (2022) rpart: recursive partitioning and regression trees. R package, version 4.1.19. https://CRAN.R-project.org/package=rpart
Williams WT, Lambert JM (1959) Multivariate methods in plant ecology. I. Association analysis in plant communities. J Ecol 47:83–101

Chapter 6
Inferences on Spatial Data

Abstract Ecological data typically are spatially structured, or autocorrelated. Autocorrelation might be due to spatially structured environmental constraints (e.g., topographic controls on soil moisture). Autocorrelation can also arise from local interspecific interactions (e.g., predator/prey relations, competition), or spatial processes such as dispersal. Spatial structure might also be a legacy of past spatial events (e.g., disturbances); this history is often unobserved. There are two approaches to dealing with autocorrelation in inferential models. The first is to avoid it, by sampling deliberately (Chap. 1). Here we adopt the alternative approach, of embracing autocorrelation as a feature of interest in ecological data. The focus here is on multivariate regression, in which we model species abundances in terms of multivariate as well as spatial predictors. The workflow includes pre-processing of spatial data, the analysis itself, and post-processing the results for interpretation and communication. A general summary of the results partitions the variation in species abundances into that explained by the environmental variables as compared to the spatial predictors while also accounting for the spatial structure in the environmental variables themselves.

6.1 Introduction

In previous chapters we have collected landscape-scale data, conducted exploratory data analysis, and summarized the general patterns in the data using ordination and classification as descriptive techniques. In this chapter we extend the descriptive techniques toward explanation, to attempt to explain *why* we see the patterns observed in the data.

In particular, we will wish to estimate how much of the variation in patterns in species abundance can be explained by environmental predictors. As with species distribution modeling (Chap. 2), this effort will be confounded by the correlations among environmental variables. To this, we will add the complication that environmental variables will be spatially structured. In landscape-scale studies, we also will have locations for the samples, as geospatial coordinates, and so can insert explicit

D. L. Urban, *Landscape Ecology: A Task-Oriented Perspective*,
https://doi.org/10.1007/978-3-031-72251-6_6

spatial information into the analyses. This invites analyses that can attend spatial autocorrelation.

There are two possible responses to autocorrelation (Legendre 1993; recall Chap. 1, Sect. 1.3.3). One is to avoid it, as autocorrelation violates the assumption of sample independence that underlies parametric statistics. In this sense, the effect of autocorrelation is to bias the degrees of freedom to be overly generous, because each (autocorrelated) sample does not really represent a "whole" observation to the analysis. The simple solution to this is to sample in such a way as to avoid autocorrelation, by spacing samples to beyond the range of autocorrelation as estimated in a pilot study. An after-the-fact solution entrails adjusting the degrees of freedom downward to account for the partial dependences of samples (Legendre and Legendre 2012; Dale and Fortin 2014). The second response to autocorrelation is to accept it as an interesting feature of the data—to *embrace space*. This latter approach is a hallmark of landscape ecology and is adopted in this chapter.

In this chapter, we infuse spatial considerations into some analyses we have already considered while also introducing some new tools. We begin with an overview of spatial autocorrelation, to make sure we are grounded in the basic issue. We then revisit two approaches from previous chapters, introducing explicitly spatial perspectives. First, we revisit Mantel's tests as a spatial analysis. Then, we introduce spatial predictors into constrained ordinations, using explicitly spatial constructs derived from a matrix of distances among samples. It will become obvious, perhaps, that this adds a new layer of nuance to ecological analyses. How to best do this has been an active topic of methodological innovation and debate.

In terms of the overall workflow that frames this book (Preface, Fig. 2), the task of modeling spatial structure is on the more research-oriented end of things. But this information can be important in less research-intensive applications in landscape management. We all appreciate that nature is patchy, and conservation practice often prioritizes places recognized as special. The question here is "*Why* are these places special?". Is it due to unusual environmental conditions? Legacies of past events? A confluence of migratory or dispersal corridors? Spatial analysis can help sort through the possibilities.

The possibility that a site might be compelling because of local environmental conditions as compared to (or in addition to!) local dispersal or connectivity will invite us to assess the relative importance of these. This, in turn, will force us to decide how to reconcile these features when they conflict as conservation targets—that is, when we need to decide which is more important, habitat quality or connectivity. We turn to this decision process as part of site prioritization in Chap. 8.

6.2 Context: Spatial Autocorrelation

Embracing spatial structure broaches what can be a vexing analytic and inferential problem for ecologists. Spatial autocorrelation[1] can be caused in a few general ways (Wagner and Fortin 2005; Legendre and Legendre 2012; Dale and Fortin 2014). The first is spatial *environmental dependency*, as would be induced by an environmental variable that itself shows strong spatial structure. For example, a plant species that is associated with higher soil moisture available at low slope positions will be spatially structured by the natural scaling of terrain (Urban 2023, Chapters 1 and 4). The second cause is purely *spatial process*, such as local biotic interactions (competition, predator/prey dynamics), dispersal, or contagious disturbance (e.g., fire) or stressors (e.g., pests, disease). For example, primary seed dispersal by plants tends to occur at distances that are a few multiples of plant height, and animal dispersal is also characteristically scaled by body size and mobility; so we might expect dispersal to influence species distributional patterns at characteristic scales. Finally, spatial structure might be observed as the *legacy* of previous events such as disturbances, pest outbreaks, land use practices, the creation (or removal) of dispersal barriers, or past climate.

The inferential problem arises because we rarely measure spatial processes directly, and long-ago spatial events are typically unobserved. For example, we might like to infer the influence of dispersal (or other spatial process) as spatial structure that is residual after accounting for spatial dependencies induced by environmental variables. But we cannot do this logically, because the residual spatial structure might be due to an unmeasured environmental variable or unwitnessed event. We can never truly resolve this logical uncertainty.

In this section, we begin with the conceptual problem of describing or summarizing spatial structure in a single measured variable. This invites a few analytic approaches, which we consider only as a means to frame the general issue. From the univariate case, we will move on to multivariate approaches.

6.2.1 *Descriptive Models of Spatial Structure*

The field of geostatistics (Haining 2003) often is concerned with interpolating information from known (measured) locations to other points, typically to fill in (interpolate) a mapped surface. Geostatistical theory targets *regionalized variables*,

---

[1]Legendre and many colleagues reserve the term "auocorrelation" for local biotic processes and spatial legacies of these, as compared to environmental dependencies. They prefer "spatial correlation" as the general term. While I do not disagree with this, I suspect that "autocorrelation" is used quite generally by many of us. I will use this term to refer to spatial structure in general, but I will try to be explicit about the presumed causes (environmental dependency, spatial process, historical legacy).

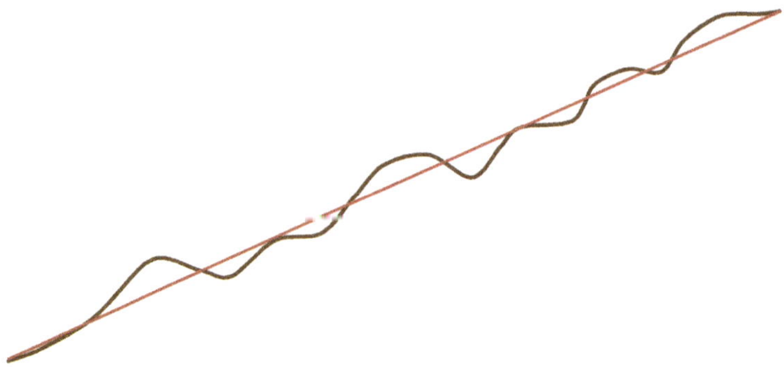

Fig. 6.1 A variable (brown line) as a combination of smooth trend (red line) and local deviations from this trend

which are measurements over surfaces. A working definition of a regionalized variable y measured at location i is:

$$y(i) = f(i) + s(i) + \varepsilon \tag{6.1}$$

where f is a *forcing* or trend, s is local spatial *structure*, and ε is residual error (presumed normal and independent). We distinguish trend and local spatial structure on the basis of spatial scale: trend is larger-scale, while local structure is finer-grained. Importantly, this approach does not distinguish environmental dependencies (e.g., ecophysiological constraints) from purely spatial processes (e.g., dispersal or contagious disturbances); the task is merely to describe the structure. Trends are often fitted as simple linear regressions on spatial coordinates (i.e., lat/lon or other coordinate systems), while local structure is estimated using a correlogram (in spatial statistics) or semivariogram (in geostatistics).

As an example, consider again the illustration of local topography nested in a larger-scale elevation gradient (Chap. 1, Fig. 1.8). While this is presented in ecological terms, a description of the pattern would see only the larger trend and local spatial structure: a straight line and local deviations from this (Fig. 6.1).

Our task here is to capture that structure in quantitative terms.

6.2.1.1 Statistical Models of Spatial Structure

Analytically, the spatial structure illustrated in Fig. 6.1 can be captured in either of two common forms. From spatial statistics (again, recall Chap. 1, Sect. 1.3.3), a common estimate of autocorrelation takes the form:

$$I(d) = \frac{\frac{1}{W}\sum_{i}\sum_{j} w_{ij} z_i z_j}{\frac{1}{(n-1)}\sum_{i} z_i^2} \tag{6.2}$$

for samples $i \neq j$. This is the estimator for Moran's I (Moran 1950; Legendre 1993; Legendre and Fortin 1989). Here, the measurements at locations i and j are converted into z-scores (deviation from the mean, divided by the standard deviation) to rescale the measurements. The w term is an *indicator variable* or weight that takes on a value of 1 if two samples are within some specified range of distances apart (i.e., in that distance class); else it takes on a value of 0. Through this indexing, the formula provides an estimate of autocorrelation for each distance class d. The term W is the sum of the indicator weights (sample size) in each distance class, which along with the overall sample size n rescales the index to vary on the range [−1,1], just as the familiar Pearson correlation coefficient. A plot of autocorrelation versus distance class is a *correlogram*.

The field of geostatistics (mostly at home in engineering) uses an approach to index scaling in terms of the *dis*similarity of measurements as a function of separation distance. Translating from the somewhat different notation of geostatistics into a format consistent with autocorrelation (above), semivariance (*gamma*) is estimated:

$$\gamma(d) = \frac{1}{2W}\sum_{i}\sum_{j} w_{ij}\left(x_i - x_j\right)^2 \tag{6.3}$$

where the indicator variable w acts as in Moran's I (Eq. 6.2) to subset sample pairs by distance class and W is the number of sample pairs in distance class d. Dividing by two rescales the index so that it converges on simple variance as autocorrelation decreases to 0.0. If semivariance is divided by simple variance, γ converges on a value of 1.0 as autocorrelation decreases to 0—a convenient way to convert estimates from varying measurement scales to a common basis. A *semivariogram* (or, simply, *variogram*) plots dissimilarity as a function of separation distance.

With ecological data, correlograms and variograms tend to be monotonic and often steeply so, indicating spatial structure over a narrow range of separation distances. The two approaches, correlograms and variograms, essentially mirror each other: a correlogram (a measure of similarity) decreases with increasing distance, while a variogram (a measure of dissimilarity) increases with distance.

In a simple world, local spatial structure (term s in Eq. 6.1) would take the form of a correlogram or variogram with relatively localized, fine-grained structure, while the forcing or trend (term f) would be much larger scale or even (ideally) linear. It is easy to make this example (Fig. 6.1 as represented in Eq. 6.1) arbitrarily complicated by choosing various forms for the trend and local structure. If we fit the trend as linear but it is in reality nonlinear, then we essentially reassign some of the trend to the local structure. Reciprocally, if we model the trend with a nonlinear polynomial, we might capture some of the local structure in the forcing f.

Finally, if we ignore the local spatial structure and model only the trend (e.g., as a linear forcing), we would expect finer-scale spatial structure in the residuals of the model (transferring information from s to the error term ε in Eq. 6.1). Of these options, only the last one—the case with autocorrelated residuals—is a "bad" regression model because its residual errors are not independent. By convention, we would want to account for local structure in order to estimate a model with independent errors. (It also is legitimate to fit a nonspatial model with autocorrelated errors—as long as the spatial structure is accounted *somewhere*.) Any other approach, in which variability is transferred between the forcing and local structure depending on the shape and scale of the model terms, might be equivalently useful as a description of the data.

In the limit, what we would like to have is a set of model terms that are scale-specific, in effect decomposing the spatial structure of the response variable into variability explained at multiple component scales:

$$y(i) = s_1(i) + s_2(i) + s_3(i) + s_4(i) + \ldots + \varepsilon \tag{6.4}$$

where the structure functions s are identified at a sequence of discrete scales and the residual errors are random.

One method that does explicitly this is spectral analysis (Renshaw and Ford 1984). In this, the data are summarized in terms of sinusoidal forms of varying wavelengths (or, reciprocally, frequencies)—that is, scales. The analysis estimates the proportion of total variance in the data that is expressed at different scales. Wavelet analysis (Bradshaw and Spies 1992; Keitt and Urban 2005; Dale and Fortin 2014) is another method that provides this scale-specific decomposition of the data. Again, these models are typically descriptive in that they merely describe the scales over which the response varies (but see Keitt and Urban (2005) and Dale and Fortin (2014) for inferential applications). We will return to this notion of scale-specific deconstructions of data later in this chapter.

6.2.2 Spatial Explanatory Models with Environmental Predictors

Complications arise when we decide to interpret the spatial structure in terms of explanatory agents. For example, we might expect plant species distribution to respond to a set of environmental constraints that are themselves characteristically scaled, from large-scale gradients such as elevation to fine-scale factors such as those governed by local topography or soils.

Correspondingly, empirical examples of correlograms and variograms show a range of scales of autocorrelation, ranging from truly local structure observed over meters to tens of meters to large-scale gradients that manifest as linear trends

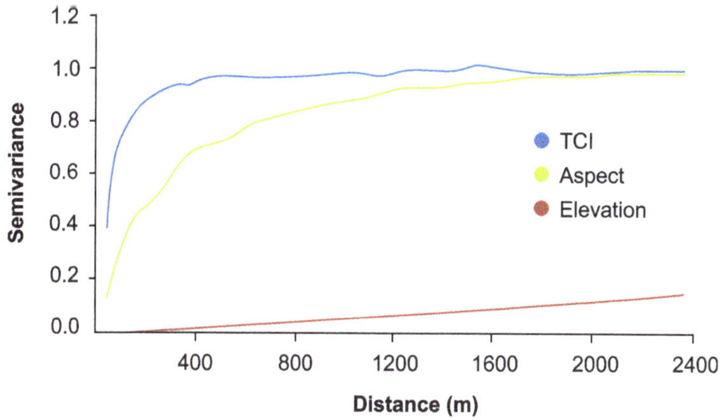

Fig. 6.2 Variograms for elevation, (transformed) slope aspect, and a topographic convergence index (TCI) for Sequoia National Park, in the Sierra Nevada of California. (Modified from Urban (2023), itself redrawn from Urban et al. (2000); permission conveyed via Copyright Clearance Center, Inc.)

(Legendre and Fortin 1989; Urban et al. 2000; Legendre and Legendre 2012; Urban 2023, Chapter 4) (Fig. 6.2).

This approach would result in a familiar multiple regression model, including environmental predictors (elevation, slope, hillslope position, topographic convergence, soil properties, and so on) for the model terms:

$$y(i) = b_0 + b_1x_1(i) + b_2x_2(i) + b_3x_3(i) + \ldots + \varepsilon \tag{6.5}$$

where the x's are environmental predictors and the b's are fitted coefficients. This model emphasizes the explanatory power of environmental constraints or dependencies.

One approach to partitioning environmental constraints or dependencies is to model these effects as with Eq. (6.5) and to look for spatial structure in the residuals of the model. Autocorrelated residuals suggest spatial information in the dependent variable that is unrelated to the measured environmental variables. One way this might arise is due to a pure and unmeasured spatial process such as dispersal or contagious disturbance. That is, one inference we might make about spatial process is that it is manifest in the residuals of an explanatory model invoking environment: a partial regression model.

Unfortunately, again, spatial process is not the only way to generate a spatial residual from a model based on environmental constraints. It is also possible that we simply omitted an important environmental constraint or missed a historical legacy, and so the residuals reflect this omission. A crucial logical limitation in ecology is that we cannot know what causes a spatial residual because the cause is unobserved.

What we *can* do, however, is to model the environmental constraints to the best of our ability and then to describe the spatial structure in the residual variation as precisely as possible in terms of the scale(s) at which this is expressed (Levin 1992; McIntire and Fajardo 2009). Another problem arises, of course, in that some of the environmental predictors in Eq. (6.5) might well correspond to the scales of the spatial descriptions in Eq. (6.4). The solution is to merge these two approaches, which we will do later in this chapter.

6.2.3 Univariate Models with Spatial Structure

Models that incorporate spatial structure along with environmental predictors are very well developed in several disciplines. These include various forms of *autoregression* as used in spatial statistics and the family of *kriging* methods used in geostatistics.

In spatial statistics, various forms of autoregression models pose the expected value at a location i, $y(i)$, in terms of the response variable in the neighborhood of i. That is, the response variable is presumed to be autocorrelated. By contrast, in a *conditional* autoregression (CAR), the value predicted at location i depends on the value of y at i but also on a set of environmental predictors at that location: the effect of the environment is conditional on the state of the response variable (Lichstein et al. 2002).

In the field of geostatistics, the corresponding models are forms of *kriging*. In the simplest case, ordinary kriging, the model is essentially that suggested by Eq. (6.1) but fitted in two steps. The trend is fitted first, and residuals are extracted from this fit. The local spatial structure is then estimated for the detrended residuals. In universal kriging, the trend and local structure are fitted at the same time. Finally, in universal co-kriging, the trend and local structure are fit as well as the effects of a set of (here, environmental) covariate predictors.

These models are developed in most statistical texts (e.g., Haining 1993, 2003; Cressie 1993; Plant 2012). Dale and Fortin (2014) and Fletcher and Fortin (2018) cover many of these approaches for ecologists (and see Augustin et al. (1996)). Keitt et al. (2002) reviewed several such models in the application of species distribution modeling (recall Chap. 2). They found that adding spatial elements improved the performance of the models—and that the particular form of the spatial element was not as important as its inclusion in *some* form. Dormann et al. (2007), in their review of methods for incorporating autocorrelation into species distribution models, came to similar conclusions while encouraging further comparative studies.

In the remainder of this chapter, we focus on two approaches that incorporate spatial elements into multivariate models of species-environment relationships. In this, we will extend approaches previously introduced in Chap. 4. The aim is to explain the distributions of multiple species, in terms of environmental factors as well as spatial structure.

6.2.4 *Workflow for Spatial Inferential Models*

The workflow for these analyses will begin with species (SPP) and environmental (ENV) data sets, appropriately edited and transformed as part of exploratory data analysis (EDA, Chap. 3) (Fig. 6.3). The analysis, a regression, entails separately estimating the effects (explanatory power, in terms of regression) of a set of environmental predictors and a set of explicitly spatial predictors.

We will consider two approaches. The first is a Mantel's test, a correlation between distance matrices. Here, we adapt this to use three distance matrices: one comprising compositional distances, one of environmental distances, and a third that is explicitly spatial and computed from geographic distances among samples. Relative to the workflow in Fig. 6.3, the estimate of the relative explanatory of environment as compared to geographic distances is assembled from several separate Mantel's tests.

In the second example, we use constrained ordination to fit a multivariate regression of species composition (as abundances) on environmental and spatial predictors. As with Mantel's tests, these are fitted piecewise: environmental effects by themselves, then environmental and spatial effects combined, and then as partial

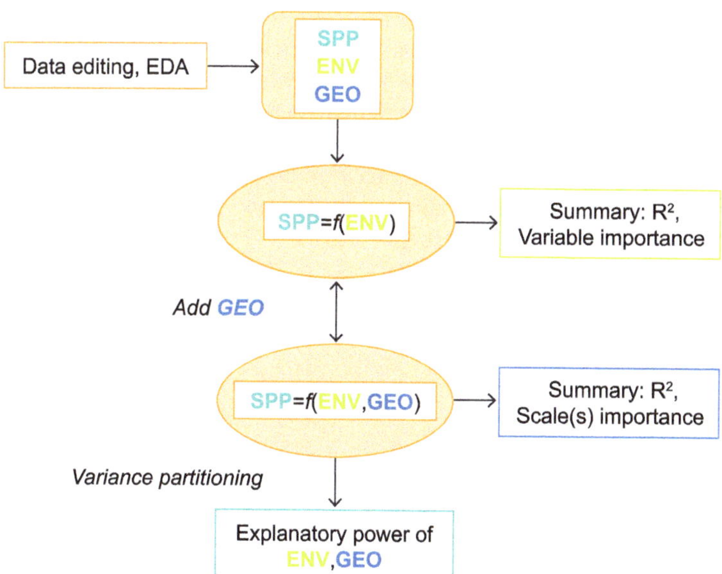

Fig. 6.3 Workflow for inferential models that incorporate environmental predictors (ENV, spatially structured) as well as pure spatial predictors (GEO). The first stage predicts species on environment. The residuals from this model are then extracted, and these residuals are models on spatial predictors. To do the variance partitioning requires a full set of models: SPP~ENV, SPP~GEO, SPP~ENV|GEO, SPP~GEO|ENV, and SPP~ENV + GEO. The results are then pooled and the separate components computed by differencing, yielding a variance partitioning (see text)

regression models in which the spatial effects are fitted after controlling for environment.

In both approaches, the variability explained by environment as compared to space is partitioned afterward into a summary of the modeling (*variance partitioning* in Fig. 6.3).

6.3 Mantel's Test

Here we develop a set of analyses, which collectively can suggest the relative importance of environmental as compared to spatial structure in explaining the relative abundances of a set of species. This addresses two of the issues that plagued our efforts in species distribution modeling: that the environmental predictors are correlated among themselves and that the variables are almost certainly spatially autocorrelated. The approach developed here addresses these two issues by translating the question so that it is framed in terms of distance or dissimilarity matrices.

6.3.1 *The Logic of Mantel's Test*

Mantel's (1967) test is a correlation or regression in which the variables are themselves distance matrices summarizing pairwise dissimilarities among sample locations. We have already encountered Mantel's tests when we used them to correlate species compositional dissimilarity with environmental dissimilarity, to assess the explanatory power of ordinations based on dissimilarities (NMS in Chap. 4, Sect. 4.3.2), and again when we used the test to find an appropriate number of groups in classification (Chap. 5, Sect. 5.5.1).

Here we extend the family of Mantel's tests to include the case where one of the distance matrices is "space" itself, measured as geographic distance between samples. The inclusion of a geographic distance matrix allows Mantel's test to address spatial structure explicitly. In practice, the power and versatility of Mantel's test stem from the various ways that the test can be constructed.

6.3.2 *The Analysis*

Mantel's statistic is based on a simple cross-product term:

$$z = \sum_{i=1}^{n} \sum_{j=1}^{n} x_{ij} y_{ij} \tag{6.6}$$

where x and y are variables measured at locations i and j ($i \neq j$) and n is the number of samples. Mechanically, this is done by writing the elements of the lower triangle of the distance matrix as a vector, which will have $n(n - 1)/2$ elements for n sample locations. The Mantel correlation is normalized, analogous to a Pearson product-moment correlation:

$$r = \frac{1}{(n-1)} \sum_{i=1}^{n} \sum_{j=1}^{n} \frac{(x_{ij} - \bar{x})}{s_x} \cdot \frac{(y_{ij} - \bar{y})}{s_y} \tag{6.7}$$

and the s_x and s_y are standard deviations for variables (distance matrices) x and y. This standardized equation allows one to consider variables of different measurement units within the same framework, rescaling the statistic to the range of a conventional correlation coefficient bounded on $[-1,1]$. In practice, a negative Mantel correlation is rare. The magnitude of correlation is often comparatively small even when highly significant statistically (Dutilleul et al. 2000; Goslee 2010).

Note that Mantel's test is based on linear correlation and hence is subject to the same assumptions that apply to a common Pearson correlation (i.e., nonlinear relationships between variables will be degraded or lost in the linear correlation). Moreover, the test of spatial dependence is averaged over all distances in the simple Mantel's test, and so this test cannot discover changes in the pattern of correlation at different distances (scales). The Mantel correlogram (below) overcomes this problem.

Because the elements of a distance matrix are not independent, Mantel's test of significance is evaluated via permutation procedures. In this, the rows and columns of the distance matrices are randomly rearranged while maintaining the symmetry of the matrices. Mantel statistics are recomputed for these permuted matrices, and the distribution of values for the statistic is generated via many iterations (Manly 1986, 1991; Legendre 2000; Goslee and Urban 2007).

In the discussion that follows, we will be concerned with three distance matrices:

1. Species compositional distances, denoted SPP, e.g., as Bray-Curtis or other measures.
2. Environmental distances, using an appropriate measure. This matrix will be coded ENV. Recall Chap. 3 (Sect. 3.2.4) for more on these distance measures.
3. Geographic distances (GEO). These distances might be in two dimensions (e.g., latitude, longitude) or three (adding altitude), and they might be computed as Euclidean or as more nuanced ecologically weighted distances (e.g., as least-cost paths between samples, based on a surface representing resistance to dispersal (Urban et al. 2009; Dale and Fortin 2014; and see Urban (2023), Chapter 6)).

6.3.2.1 Mantel's Tests: Cases

Because Mantel's test is simply a correlation between distance matrices and the distance matrices can be variously defined, the test can assume a variety of forms as special cases. These are, in fact, variants of the same case but are interpreted somewhat differently. There are several variants of interest here (Legendre and Fortin 1989):

Simple Mantel's Test A simple Mantel's test involves two distance matrices. There are several of interest here. If one distance matrix is species similarity and the other matrix is geographic distance, the research question is "Are samples that are close together also compositionally similar?". This is equivalent to testing for overall autocorrelation in the dependent matrix (i.e., averaged over all distances).

If one matrix is environmental similarity and the other matrix is geographic distance, then the simple test asks "Are samples that are close together also environmentally similar?" This is the multivariate case of environmental autocorrelation.

Often, a comparison of interest is between a species compositional matrix and one of environmental distances. That is, *are samples that are environmentally similar also similar in species composition?* This is the question at the crux of the environmental control model of autocorrelation, i.e., as a result of environmental dependencies.

We have previously considered a simple Mantel's test using a *contrast matrix* to assess group differences. In this, the matrix is coded 0 if two samples are in the same group, else 1. The test assesses the extent to which samples in the same group are more similar than samples in different groups, the Mantel equivalent of a multivariate *F*-test. We used this to choose an appropriate number of groups (levels) in classification (Chap. 5).

The Mantel Correlogram An extension of a simple test with a geographic distance matrix (above) is to partition or subset the analysis into a series of discrete distance classes. That is, a first distance matrix is evaluated for all pairs of points within the first distance class; then a second matrix is scored for all pairs of points within the second distance interval, and so on. This is actually a series of contrast matrices, with each distance class in turn contrasted against all other distances. In this, pairs of samples are coded 0 if they are in the same distance class, else 1. The result of this analysis is a Mantel correlogram, analogous to an autocorrelation function but performed on a (typically multivariate) distance matrix.

Partial Mantel's Test on Three Distance Matrices The Mantel's test of interest here is a partial regression on three distance matrices: species dissimilarity, environmental dissimilarity, and geographic distance (Legendre and Fortin 1989). The analysis in this case is partial regression (Smouse et al. 1986), and two partial correlation (or regression) coefficients are of interest: $r_{SPP\text{-}ENV|GEO}$ and $r_{SPP\text{-}GEO|ENV}$. The partial $r_{SPP\text{-}ENV|GEO}$ asks whether there is an effect of the environment on species composition, after controlling for spatial structure in species composition. The partial $r_{SPP\text{-}GEO|ENV}$ asks whether there is residual spatial structure in composition after removing the effect of environment. This latter partial would include the effects of a pure spatial process (e.g., local dispersal) as well as the effects of any unmeasured spatial constraint or legacy.

These partials can be subtle to interpret and it can help to phrase them explicitly. The partial $r_{SPP\text{-}ENV|GEO}$ asks: *Are samples that are environmentally similar more compositionally similar than their spatial proximity would suggest?* Conversely, the partial $r_{SPP\text{-}GEO|ENV}$ asks: *Are samples that are in close proximity more similar in composition than their environmental similarity would suggest?* The relative magnitudes of these partials, relative to the simple Mantel's tests above, are what allow the inference about the relationship between environment and geographic distance.

Partial Mantel's on Multiple Predictor Variables Often, knowing that the environment has some relationship with the dependent variable of interest is not sufficiently satisfying: we wish to know *which* variables are actually related to the dependent variable. The logical extension of Mantel's test is multiple regression in which the predictor variables are entered into the analysis as individual distance matrices (Smouse et al. 1986; Manly 1986). In this, for example, dissimilarity on elevation would be coded as the absolute difference in elevation between a pair of samples (i.e., as univariate Euclidean distance). As a partial regression technique, Mantel's test not only provides an overall test for the relationships among distance matrices but also tests the contribution of each predictor variable for its pure partial effect on the dependent variable. If geographic location is included as one of the predictor matrices, then the test returns the pure spatial residuals (the effect of "space itself") as well as the partials for each of the predictor variables. Lichstein (2007) has elaborated this approach to multiple regression on ecological distance matrices.

Thus, the flexibility of Mantel's test provides for a wide range of reasonably explicit hypothesis tests. The onus is on the investigator to pose hypotheses and interpret the analysis in a meaningful way. Goslee (2010) has illustrated the relationships among Mantel's tests on univariate distance matrices (i.e., on a single variable), compound distance matrices (i.e., distances on multiple variables), and multiple regressions on distance matrices.

6.3.3 Interpretation and Presentation

By convention, Mantel's test is presented in the framework of path analysis (Leduc et al. 1992). In this, the underlying conceptual hypothesis is made explicit: space "causes" environmental variation, environmental variables might "cause" species distribution, and there might be residual spatial variation in the species that is not "caused" by the measured environmental variables ("pure spatial" residuals). In fact, these spatial residuals are unaccountable and as such are thus fodder for further study.

This same information is often presented in tabular form, in which the tabled matrix is split at the diagonal into simple and partial correlations (Legendre and

Table 6.1 Tabular form for a partial Mantel's test with three matrices **Y** (e.g., species), **X** (environment), and geographic space **G**

	Y	**X**	**G**
Y	–	r_{YX}	r_{YG}
X	$\rho_{YX\mid G}$	–	r_{XG}
G	$\rho_{YG\mid X}$	$\rho_{XG\mid Y}$	–

Elements in the upper diagonal are simple Mantel correlations, the lower, partial correlations

Fortin 1989). The matrix representation of a Mantel's test for three environmental variables would take the form illustrated in Table 6.1. In this, the upper-diagonal elements are simple correlations and the lower-diagonal elements are partials. In practice, one would table the coefficients as well as their significance levels (*P*-values). A predictor might have a high simple correlation but a much lower (even nonsignificant) partial if it was itself correlated with another predictor variable.

In the case of multiple predictor matrices, the path diagrams and corresponding tables can get a bit more cumbersome, but the idea is the same. An analysis with multiple independent predictor matrices plus space itself would include a large set of partial correlations, but in practice we are mostly concerned with the simple correlations between species and each of the predictors as well as the partials for each predictor by itself.

Path analysis is, of course, a matter of interpretation and it has been used and abused in ecological applications (Petraitis et al. 1996; Legendre and Legendre 2012). Because path analysis is based on correlation, finding a significant correlation between two variables actually cannot prove cause; yet the converse *is* true: failing to find a correlation between two variables certainly argues against a causal relationship. Thus, conservatively interpreted, path analysis provides a useful framework for the interpretation of partial regression such as in Mantel's test. (We will return to path analysis in more depth subsequently, as it provides the logical basis for structural equation modeling in Chap. 7.)

6.3.3.1 Reporting

Presentation of Mantel's test should describe the input data sets (what the variables are, sample sizes, etc.) as well as any pre-processing pertinent to the analysis (data screening, any transformations). Beyond this, the analysis depends on:

☑ The choice of distance measures for all input data sets and the motivation for these (and note that data transformations can have a large impact on these)
☑ Mantel correlograms, if appropriate (these are useful in EDA)
☑ Reporting of the simple Mantel's test results
☑ Reporting of any partial test(s)
☑ A narrative explanation of how the correlations (especially partials) should be interpreted ecologically

These results can be presented in a table (as in Table 6.1) and/or a path diagram.

6.3.4 Illustration: Sierran Forests

As an illustration of Mantel's test, we can revisit the Sierran forest data that we have explored in previous chapters. For this, we will use a compositional distance matrix computed as extended Bray-Curtis dissimilarities (i.e., the same data used for ordination in Chap. 4 and classification in Chap. 5). The environmental distance matrix is computed as Mahalanobis distances, because of the correlations among the variables (Goslee and Urban 2007) (recall Chap. 3, Sect. 3.2.4). The geographic distances are Euclidean, based on sample locations recorded as UTM coordinates (easting and northing, in m).

The Sierran forest data show spatial autocorrelation in tree species abundances and environmental factors to distances of approximately 2000–2500 m. Species composition autocorrelated to somewhat longer distances (~3000 m) than the environmental factors (Fig. 6.4).

Urban et al. (2002) illustrated a set of simple and partial Mantel's tests using the same data featured here. They found strong spatial structure in species composition (as Bray-Curtis distances but not extended distances) and in the environmental variables (as Euclidean distances; this was before Mahalanobis distances were readily available in software). The partial for environment, controlling for geographic distance, was small but significant, underscoring the spatial structure in the environmental variables. Because Mantel's tests are no longer the preferred method for this accounting (see below), we will not delve further into this analysis here.

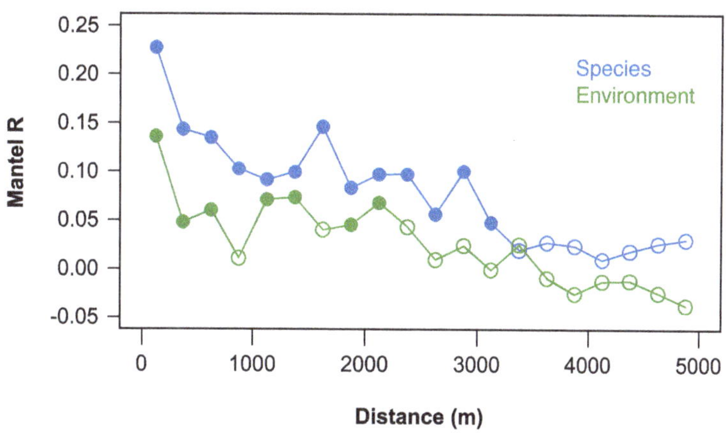

Fig. 6.4 Mantel correlograms for species (extended Bray-Curtis distances) and environment (Mahalanobis distances) for the forest data from Sequoia National Park. Solid symbols are significant at $P < 0.05$; open symbols, nonsignificant

6.3.5 Problems with Mantel's Test

Recently Mantel's test has been criticized (Guillot and Rousset 2013). One critique applies to the partial test in which "pure spatial" effects are distinguished from the effects of environmental variables. The issue is that the permutation test used in this case removes (randomizes) the environmental variable and also any spatial structure associated with that variable. That is, "environment" cannot be separated from "space" cleanly because they are confounded.

It might be argued, however, that the permutation test *should* remove the spatial structure associated with an environmental variable: an environmental variable and its spatial structure are essentially the same thing. Removing that variation should, in principle, allow other spatial influences to be revealed at that same or similar scales. For example, consider a hypothetical case where tree species composition is influenced by topographic moisture as measured by a topographic convergence index (TCI), which might scale on the order of tens to hundreds of meters (Urban et al. 2000; see TCI in Fig. 6.2). Forest composition might also be influenced by local seed dispersal, which acts at similar scales. Removing the effect of TCI might reveal residual spatial structure at that scale, which could be attributable to seed dispersal (or, again, some other spatial process or unmeasured constraint at that scale).

The main issue with Mantel's test in this application is its sensitivity or error rate (Legendre 2000; Legendre and Fortin 2010; but see Somers and Jackson (2022) and Quilodrán et al. (2023)) and that it does not seem to have the power that newer, alternative tests can provide (Legendre et al. 2005, 2008; Legendre and Fortin 2010; Dale and Fortin 2014). This newer approach is developed below. Legendre and colleagues suggest reserving Mantel's tests for applications that are best framed in terms of distance matrices and using the approach described next for cases that can be analyzed in terms of the primary data matrices. Distance-based cases would naturally include, for example, applications in landscape genetics (Wagner and Fortin 2013) where between-sample genetic similarities are the focus; this would include the fundamental issue of genetic *isolation by distance* (Wright 1943).

It is worth emphasizing here the practical value of Mantel's tests for landscape management. A Mantel correlogram (Fig. 6.4) indexes the spatial scaling of the *distance decay* of ecological similarity (Nekola and White 1999). This is a measure of *beta*-diversity, the turnover of species composition along geographic or environmental gradients. This information can inform conservation planning. For example, Wiersma and Urban (2005) used empirical estimates of rates of species turnover to recommend appropriate spacing of nature reserves in the Yukon, Canada. The logic is that reserves that are very close together (i.e., within the range of autocorrelation or self-similarity) would tend to support the same species and thus be redundant. Reserves spaced farther apart would tend to support different species and in this provide more conservation value. This is the issue of *complementarity* in reserve design, an issue that frames our initial approach to site prioritization in Chap. 8. This example also raises the question that sites that are closer together might be better connected via dispersal as compared to sites farther apart. This implies a possible trade-off between local conservation value and connectivity—a trade-off we address explicitly in Chap. 8.

6.4 Multivariate Spatial Regressions

In this section, we revisit ordination analysis (Chap. 4) and extend this to include spatial structure. To do this, we will need to merge two lineages and perspectives: ordinations constrained by ancillary predictors and spatial predictors in regression.

6.4.1 Indirect Versus Constrained Ordinations

We have already touched on the constrained ordinations that are counterparts to the three lineages or underlying models for indirect ordination (Chap. 4). A constrained analysis forces the ordination axes to be expressed in terms of (typically, environmental) predictors provided as an ancillary data set—a multivariate regression. In the case of the linear model of principal components analysis (PCA), the constrained approach is redundancy analysis (RDA). With weighted averaging, canonical correspondence analysis is the constrained version of correspondence analysis (reciprocal averaging). With distance-based ordinations, nonmetric multidimensional scaling (NMS) has its (approximate) constrained counterpart in distance-based RDA (dbRDA), which is a constrained principal coordinates analysis (PCoA). It will be timely to review how these techniques work, as injecting spatial structure into the analysis essentially means constraining an ordination with predictors that are explicitly spatial.

With the linear model, the constrained approach (RDA) is a fairly straightforward extension of PCA. In this, each species in the dependent data set is modeled as a linear function (by multiple regression) of environmental predictors chosen by the researcher. Thus, for m species and p environmental predictors, there are m regressions, each of which is constructed from the p predictors. These regressions are then used to predict the abundances for each species on each sample plot. In turn, the set of *predicted* species abundances is assembled into a new data set and ordinated via PCA. The result is a set of new axes (PCs) that summarize the amount of variability in species composition that can be accounted by the environmental predictors.

In the case of the weighted-averaging ordination, the algorithm of reciprocal averaging provides an easy heuristic on the constrained version (Fig. S4.6 in Supplement S4). In the indirect approach, sample scores are used to compute species scores by weighted averaging, and these species scores are then used to compute sample scores as weighted averages of the species scores. This process iterates until both sets of scores (samples and species) converge. To constrain this algorithm, an additional step is inserted. After computing sample scores from the species scores, new sample scores are estimated by regressing the weighted-average scores on the environmental predictors. These scores are typically called LC scores because they are linear combinations of the environmental predictors. These scores are then used to compute species scores as weighted averages (WAs) of the LC scores; the species WA scores are used to recompute sample WA scores; the WA sample scores are regressed on the environmental predictors to yield new sample LC scores; and so on, until all scores converge. There are eigenanalysis solutions to the indirect and constrained ordinations, but the reciprocal averaging algorithm helps make it clear how the scores are related to each other.

In distance-based ordinations, NMS is the popular indirect technique. The constrained alternative is a bit less straightforward, as it borrows from the PCA lineage. In this, note first that a computationally expedient alternative to the NMS algorithm is to instead perform PCA on a distance matrix. This is principal coordinates analysis (PCoA). As compared to PCA, PCoA is a *Q*-mode eigenanalysis of an *n* x *n* matrix of ecological distances between pairs of samples (e.g., as Bray-Curtis dissimilarities); recall that PCA is an *R*-mode eigenanalysis of an *m* x *m* (for *m* species) or *p* x *p* (for *p* environmental predictors) correlation or covariance matrix. The result of PCoA is *n* new axes—principal coordinates—that summarize the main trends in dissimilarities among samples. As with PCA, we expect most of the variation to be expressed on the first few axes. To constrain this analysis, the principal coordinates of compositional dissimilarity are then analyzed with RDA, yielding compositional ordination trends (axes) that are predictable from the environmental variables. This approach is distance-based RDA, or dbRDA (Legendre and Anderson 1999; McArdle and Anderson 2001). There are subtleties to all of these techniques, to be sure! Legendre and Legendre (2012) describe these approaches in depth.

In each constrained ordination, we can inject spatial structure by using environmental predictors that have spatial information. This might be via locational variables (latitude, longitude, distance to the coast, etc.) or more sophisticated methods. To that end, we turn to spatial regressions more generally and consider how to account for spatial structure in such models.

6.4.2 Multivariate Regression with Spatial Predictors

The emerging approach for dealing with environmental constraints and spatial structure is multivariate regression with environmental as well as explicitly spatial predictors. Multivariate regression is the generalization of multiple regression to include multiple response variables as well as multiple predictors. In the cases of interest here, these variables are species abundances and constrained ordinations (RDA, CCA, or dbRDA) provide the regression framework. Explicitly spatial predictors are generated from a geographic distance matrix; these spatial components are called Moran eigenvector maps (MEMs) and are related to autocorrelation functions (recall Moran's *I*, sect. 6.2.1).

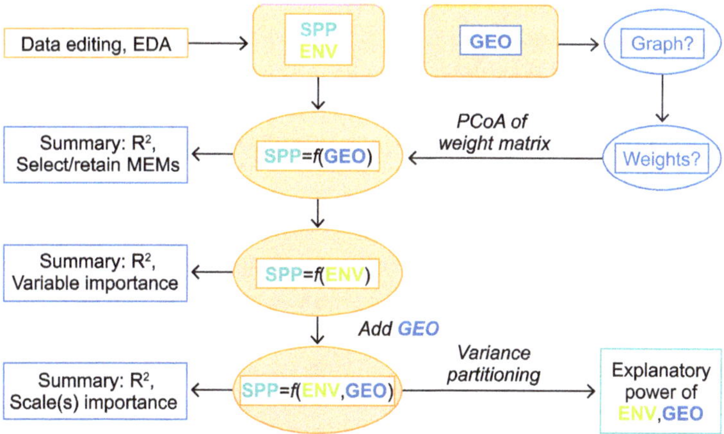

Fig. 6.5 Workflow for a constrained ordination using environmental as well as spatial predictors. Preparation of the species and environmental data sets are much as with other ordinations. The geographic data are used to construct a network model (a graph), the distances in which are then used to construct spatial weights. A PCoA of the weight matrix generates Moran eigenvector maps (MEMs), which are pure spatial predictors. The rest of the analysis is a partial regression

The workflow for a spatial, constrained ordination follows the general workflow shown in Fig. 6.3 but includes enough extra details that it is worth expanding that workflow to highlight the extra steps (Fig. 6.5). The analyses illustrated here were done with redundancy analysis, using R packages *ade4* and its dependencies (Dray and Dufour 2007) and *vegan* (Oksanen et al. 2021; R Core Team 2021). The analyses followed recommendations of Borcard et al. (2011), Dray et al. (2012), Legendre and Legendre (2012), Bauman et al. (2018), and especially Dray (2020).

6.4.2.1 Moran Eigenvector Maps

Moran eigenvector maps are created from a distance matrix comprised of between-sample geographic distances (typically Euclidean). To begin, a network model (a graph) is constructed for the samples. There are several options for the form of the graph. These typically are complete graphs (all samples are connected) but rather sparse representations of the network, emphasizing short-distance connections (and see below).

The between-sample distances extracted from the graph are then weighted to emphasize closer distances. Again, the aim here is to emphasize direct short-distance connections and local spatial structure.

Principal coordinates analysis of the (sparse, weighted) distance matrix generates new variables—principal coordinates, which are the MEMs—that decompose the trends in the distance matrix into independent elements. For a data set with

n samples, there are n new variables (MEMs). These MEMs are an empirical estimate of the "pure spatial" predictors implied by Eq. (6.4).

It is not easy to describe MEMs in simple language: they are the main trends in "distances apart" in the spatial network matrix. The MEMs are explicitly spatial constructs and mutually independent. In the simplest case of a data set consisting of samples evenly spaced along a linear transect, the MEMs obtained are a series of sinusoidal waves of decreasing wavelength or scale: coarse scales first (long waves), followed by successively finer-scale patterns. These MEMs are relatively tidy in terms of their order, scales, and interpretation. Perhaps this is because all of the "distances apart" along a transect are equivalent (e.g., any two samples can be 1, 2, 3, and so on units apart and all of these separation distances are essentially the same no matter where along the transect they occur).

In more realistic cases of two-dimensional sampling designs with irregularly spaced sample locations, the MEMs are not as orderly but still represent sinusoidal patterns of various scales. In these cases, the MEMs unavoidably also contain locational information beyond "distance apart," because the unique distances between sample pairs depend on the precise locations of each of the samples. Legendre and Legendre (2012, Chapter 14) provide helpful illustrations of MEMs derived from a series of sampling arrangements including a transect, a regular grid, and irregularly spaced samples.

6.4.2.2 Constrained Ordinations with MEM Predictors

The approach for modeling with environmental as well as spatial predictors is an iterative process (Fig. 6.5). In each step, the analysis is a multivariate regression in which species composition is modeled in terms of multiple predictor variables. Again, this regression could be conducted with redundancy analysis (RDA), canonical correspondence analysis (CCA), or distance-based redundancy analysis (dbRDA) (see Supplement 4S, Borcard et al. (2011), and Legendre and Legendre (2012) for more details on these tools).

As a first model, the constrained ordination is fitted using the environmental predictors only. This accounts for the explanatory power of the environmental variables, and the analysis includes a summary of this explanatory power. The *residuals* of this regression represent variability in species composition that is unrelated to the measured environmental variables.

In one sense, a second regression completes our task. Modeling the residuals of the first regression (SPP~ENV) on the MEMs (SPP~GEO|ENV or residSPP~GEO) captures that spatial structure in species composition that is unrelated to the environmental variables. The significant MEMs in this regression suggest the scales of this structure.

Before going further, it might be worth underscoring the ordering of this analysis. Because the MEMs account collectively for all of the spatial information in the data set, they can predict essentially anything measured on the samples—including the environmental variables. We reduce this effect by beginning with a sparse network of between-sample connections and then pre-screening the MEMs to retain only those MEMs related to species composition. Still, to get a clearer picture of the relative explanatory power of environment as compared to geographic distances, we will need several models to parse the environmental and spatial predictors. The models provide different inferences:

SPP~ENV The first model reveals the environmental effects (including any spatial effects inherent to the environmental factors). By contrast, a model of SPP~GEO would capture spatial structure in species (including environmental effects that are spatially structured). The overlap in spatial signals in these two models is what forces partial regression models (next).

SPP~GEO|ENV This partial regression reveals the residual spatial structure in species composition after accounting for environmental factors. By contrast, the model SPP~ENV|GEO would show the pure environmental effects after controlling for spatial structure.

SPP~ENV + GEO The full model provides the total explanatory power from the environmental as well as spatial predictors. This includes all of the effects and is used to compute the residuals (see below).

This accounting is common to multiple regression in general, when using correlated predictors (recall our efforts to understand the relative importance or correlated predictors in species distribution models in Chap. 2).

6.4.2.3 Variance Partitioning

The final result of this analytic process is a set of models with which we can partition the amount of species compositional variability due to (a) the measured environmental variables, (b) the amount of compositional variability that is spatial and related to measured environment, (c) the variability explained by the MEMs but that is unrelated to the environmental variables, and (d) the amount of residual compositional variability that is unrelated to environment and also unstructured spatially (i.e., the residuals of the full model).

This partitioning requires a set of models, because only some of the components can be estimated directly (Table 6.2, Fig. 6.6, after Legendre and Legendre 2012). (Note that Fig. 6.6 is *not* how these fractions are typically illustrated as a Venn diagram; this is perhaps mostly a matter of software concessions. Here the figure is designed to be consistent with Venn diagrams used in other chapters in this book.)

In practice, multiple models are used and the fractions are computed by differencing (Legendre and Legendre 2012).

Table 6.2 Variance components for a set of models using two sets of predictors, set 1 and 2, with dependent variable Y

Model	Fitted	Residuals
Y~s1	a + b	c + d
Y~s2	b + c	a + d
Y~s1 + s2	a + b + c	d
Y~s1|s2	a	d
Y~ 2|s1	c	d

See Fig. 6.6 for component labels

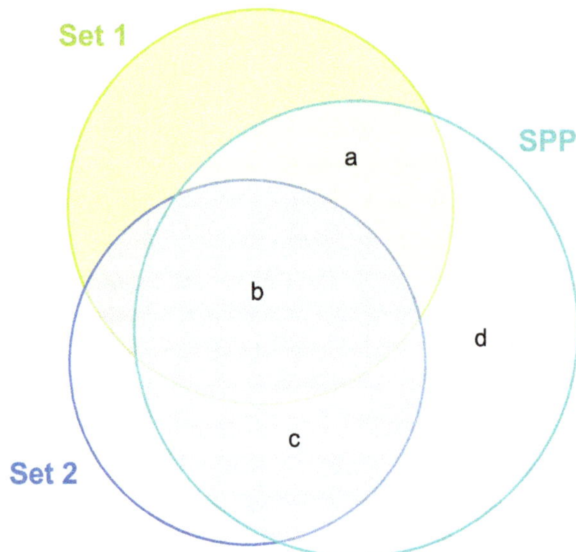

Fig. 6.6 Schematic of variance partitioning for models with two sets of predictor variables. Components are defined in Table 6.2 (after Legendre and Legendre 2012)

6.4.3 Interpretation, Presentation, and Reporting

These analyses generate an impressive volume of information and it takes some effort to parse the many results. The interpretation proceeds as with any regression: evaluating the model's statistical significance, explanatory power, and the relative importance of the predictors.

As these models are based on ordinations, they also are interpreted from this perspective (and recall Chap. 4): What *are* the ordination axes? How do species sort along the axes? Which environmental variables are correlated with which axes? And for the spatial model, which MEMs are correlated with the ordination axes derived from the residual variation in species composition not related to the environmental variables? At what scale(s) are these spatial residuals expressed?

These evaluations will generate a series of tables and figures as with any ordinations, as well as tables that summarize the regressions:

☑ Sample sizes and descriptions of the SPP, ENV, and geographic (GEO) data sets, along with details about data screening, transformations, and any exploratory data analysis that influenced subsequent analyses

For the spatial processing:

☑ Choice of graph model for the geographic coordinates of the samples
☑ The function and effective distances used to generate graph weights
☑ Preselection of MEMs for further analysis (e.g., by selection of MEMs significantly related to species composition)

For the multivariate regression analyses:

☑ Choice of multivariate regression model (RDA, CCA, dbRDA) and why
☑ Overall R^2 (adjusted) and significance of the models
☑ Interpretation of the axes: variance per axis (singly and cumulative; table)
☑ Species loadings on the axes (weighted-average positions or correlations) (table or joint plot)
☑ For the environmental regression, correlations between the axes and environmental variables (table or biplot)
☑ For the partial regression, correlations between the axes and the spatial MEMs
☑ Joint biplots as appropriate: sample scores, species positions, and biplot vectors for the environmental variables and/or spatial MEMs
☑ Maps of the partial axes on MEMs controlling for space (i.e., the axis values plotted into geographic space)
☑ Variance partitioning to distinguish environmental and spatial explanatory power (table and/or Venn diagram)

These are illustrated with examples in the following section.

6.4.4 *Illustration: Sierran Forests*

We will illustrate spatial multivariate regression using the Sierran forest data that we have explored previously. In this case, the samples are located in clusters of three to four samples with the clusters themselves distributed over a longer elevation gradient (recall Fig. 3.3).

6.4.4.1 Pre-processing of the Spatial Data

Preparation of the MEMs was conducted using the *ade4* package (Dray and Dufour 2007) and its dependencies in R, following the protocol outlined by Dray (2020). This entails constructing a graph model of the sample locations, choosing a

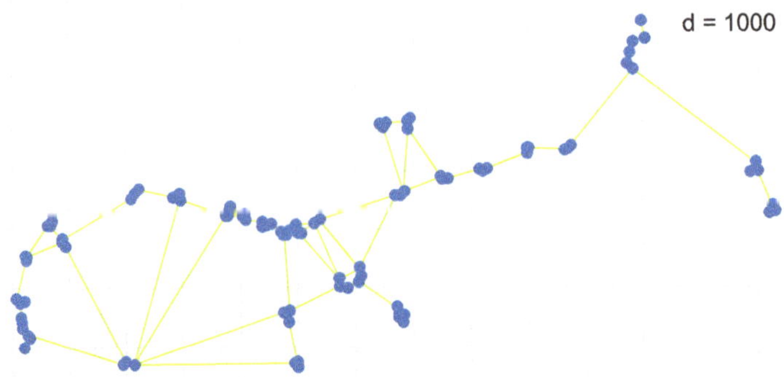

Fig. 6.7 A Gabriel graph connecting the sample locations shown in Fig. 3.3. The underlying grid is of 1000-m partitions. See text for explanation of a Gabriel graph

distance-weighting function, creating the MEMS, and screening MEMs for their relevance to species composition.

A variety of graphs were assessed from these points, with a Gabriel graph (Fig. 6.7) ultimately proving more effective (marginally) than other alternatives such as a minimum spanning tree or various distance-thresholded graphs. A Gabriel graph is constructed by drawing a circle with a diameter corresponding to the line between any two samples. If there is no other sample located within that circle, the link that connects those two samples is included in the graph. The result is a completely connected but rather sparse graph that emphasizes relatively short-distance connections.

From the distances represented in the graph, a weighting matrix was estimated by various functions of distance. In this instance, the final weights were based on a negative-exponential function of distance computed to tail off at a maximum distance of 3000 m. This distance was, in turn, chosen based on the Mantel correlograms for species composition and environmental factors (Fig. 6.4). This choice was corroborated by comparing it to other distance functions.

It should be noted there that the selection of the graph—and the form of the weighting function—was done by completing the first regression analysis in Fig. 6.6 (i.e., SPP on GEO) using alternative graphs and alternative weighting functions and then choosing the combination that provided the best result. That is, this is a process of iterative model selection. It should also be noted that several alternative graphs and weighting functions provided rather similar results to those shown here.

A few MEMs constructed from this stage of the Sierran analysis illustrate the range of spatial scales represented. Scale is inferred from the distances between strong positive versus negative values (Fig. 6.8).

After selecting a graph and weighting function, a constrained ordination was conducted with the spatial predictors only. This model was post-processed using a forward stepwise analysis, and those MEMs that contributed significantly to the solution were retained (Dray 2020). This step typically omits many of the MEMs as being unrelated to species composition. Recall, from an $n \times n$ distance matrix

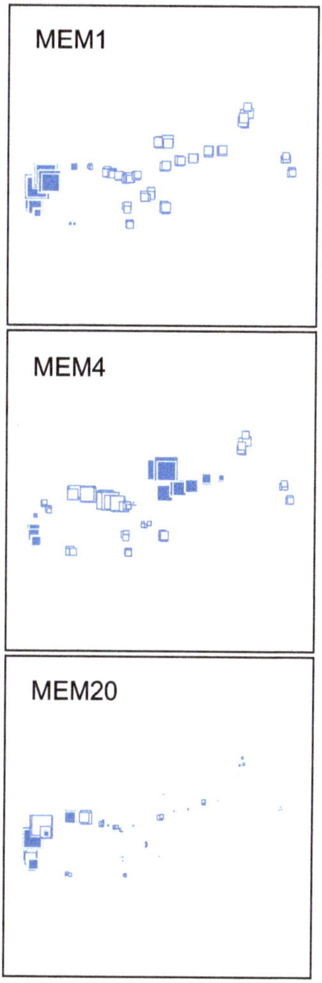

Fig. 6.8 A selection of MEMs constructed by principal coordinates analysis of the weighted Gabriel graph in Fig. 6.7. Symbols sized proportional to magnitude; filled symbols, positive; open, negative. The distances between strong positive and negative values indicate scale. The top panel shows a large-scale MEM; the middle panel, intermediate; and bottom, finer-scale pattern

representing the graph, there will be n MEMs: a *lot* of spatial predictors (here, 99). In this case, half of the MEMs have negative eigenvalues, which correspond to negative autocorrelation. Because we are interested in environmental control and do not have any measurements that might imply negative autocorrelation (e.g., due to competition among tree species), we discarded the MEMs with negative eigenvalues (see discussion in Legendre and Legendre (2012, Chapter 14)). In this example, 22 MEMs with positive eigenvalues were retained for further analysis.

6.4.4.2 Multivariate Regressions

In this illustration, we used RDA so that the analysis was a constrained linear model on species abundances. This decision diverges from exploratory data analysis (Chap. 3) and NMS ordinations considered previously (Chap. 4), which embraced the nonlinear species response to the dominant elevation gradient. Here, we used RDA because it allows the analysis to track individual species as compared to the aggregate species response in NMS (Chap. 4) and dbRDA, which are based on compositional dissimilarities. In this, we followed the recommendation of Legendre et al. (2005, 2008) and Legendre and Fortin (2010) to use actual data matrices, when available, rather than distance matrices.

The species data were modified using the Hellinger transformation (Legendre and Gallagher 2001), which improves the geometry of the projection of the multidimensional species data onto a few (often, two) Euclidean axes. (We might note here that we did not choose to use canonical correspondence analysis, because it has its own projection issues that are not as readily addressed via data transformations.)

These transformed data were then analyzed using RDA with the 13 environmental factors and 22 MEMs retained in screening. Three analyses are summarized here. These include the base-case model of species in terms of environmental predictors (SPP~ENV), the partial regression of the residuals of the environmental model on the MEMs (SPP~GEO|ENV), and the full model (SPP~ENV + GEO). Analyses shown here were conducted using function *rda* in the *vegan* package (Oksanen et al. 2021) in R (R Development Team 2021).

SPP~ENV In the Sierran case, the RDA using only environmental variables was highly significant ($P < 0.001$) and accounted for 35.3% (adjusted R^2) of the variation in species composition, indicating a substantial amount of variability related to the environmental variables. In this implementation, the significance tests are generated by permutation (999 trials), and the adjusted R^2 reflects sample sizes and the number of variables in the model. Three RDA axes were highly significant ($P < 0.001$) and a fourth marginally so ($P < 0.10$).

Elevation was highly significant in the RDA model ($P < 0.001$), while slope angle, pH, transformed aspect, litter depth, and mean soil depth were significant at $P < 0.01$; C-N ratio was significant at $P < 0.05$. The first axis was clearly identifiable as an elevation gradient, with subsequent axes identified in terms of other environmental factors (Table 6.3).

Species sort on the first axis to reveal the elevation gradient, as indicated by *Pinus contorta* (PIco) and *Pinus monticola* (PImo) with positive scores and mid- to lower elevation species such as *Abies concolor* (ABco), *Pinus lambertiana* (PIla), *Pinus ponderosa* (PIpo), *Quercus kelloggii* (QUke), and *Calocedrus decurrens* (CAde) with low (negative) scores. Subsequent axes sort more subtle differences among species. A joint biplot of this solution emphasizes the elevation gradient in terms of species sorting (as their weighted-average positions on each axis) and correlations

Table 6.3 Correlations between the environmental variables and the first three axes of an RDA with environmental variables only

Variable	RDA1	RDA2	RDA3
Elevation	0.94	−0.22	
Slope		0.25	0.62
TAspect		−0.20	−0.47
TSI	0.19	0.21	
xLitter	−0.18	−0.17	
xDepth	−0.66	−0.43	0.19
sDepth			
pH	−0.70	−0.46	0.28
C	0.35		
C.N	0.55	−0.23	0.47
P	−0.25	−0.40	−0.28
ECEC	−0.38		0.55
Clay		0.28	0.44

Blank entries are for correlations with $P > 0.05$; environmental variables described in Table 3.6

with the environmental variables (Fig. 6.9). Relative to the indirect NMS ordinations of Chap. 4, RDA also emphasizes the dominant elevation gradient.

SPP~GEO|ENV A second model was constructed as a partial RDA, modeling SPP on GEO after controlling for ENV. This is a model of GEO with the residuals of the first model (above). Note that this is a new multivariate regression: the residuals from the environmental analysis comprise a new data set, and this second analysis aims to summarize patterns in the residuals collectively. That is, the axes of this solution need not correspond to the axes of the first model. This second model was highly significant ($P = 0.001$) and explained 28.4% (adjusted R^2) of the residual variation in SPP after controlling for ENV.

The first three partial axes are statistically significant ($P < 0.001$) with a fourth less strongly significant ($P < 0.05$). Many of the MEMs were significant: eight at $P < 0.001$, two at $P < 0.01$, and four at $P < 0.05$.

As with other ordinations, the partial dbRDA can be post-processed to explore correlations between the MEMs and each axis and to calculate species scores (as weighted averages) on each axis. But it can be easy to overinterpret these summaries; in particular, it can be tempting to try to relate individual MEMs to species composition. This is not a good practice (Legendre and Legendre 2012), as the risk is to interpret idiosyncrasies in the MEMs that depend on the sampling design and the (selected) between-sample distances.

Instead, it can be useful to plot selected partial axes in geographic space, to explore geographic trends in the axes without looking too closely at any individual MEMs contributing to that axes except in terms of their relative scaling. For example, a plot of the first partial axis (Fig. 6.10) shows substantial variation at mid-elevations (center of the figure) where Jeffrey pine (PIje) occurs separately from

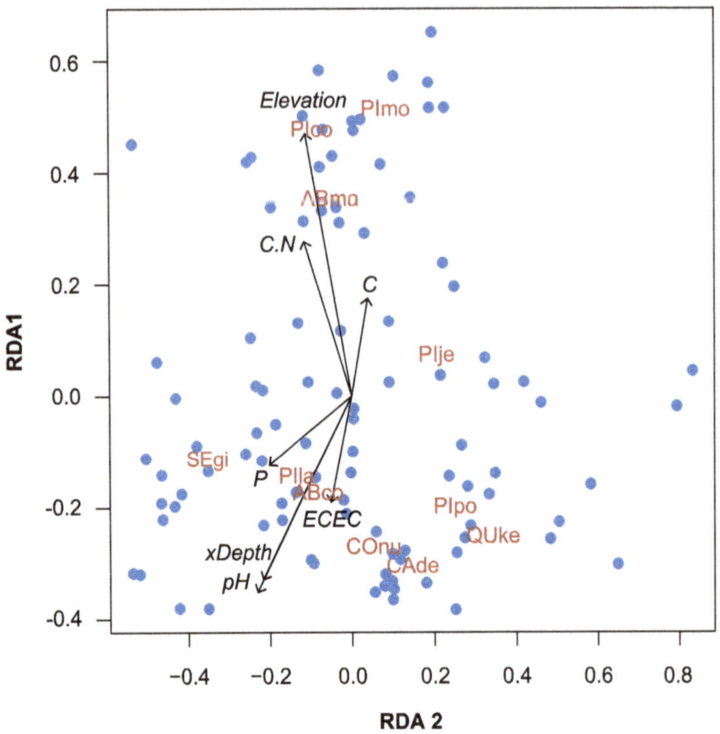

Fig. 6.9 Joint biplot of the RDA of species composition in terms of the environmental variables. Dots are sample points. Species are identified by codes (see Table 3.5) at their weighted-average position on each axis. Correlation vectors for the environmental variables (described in Table 3.6) are shown only for variables significant at $P < 0.01$

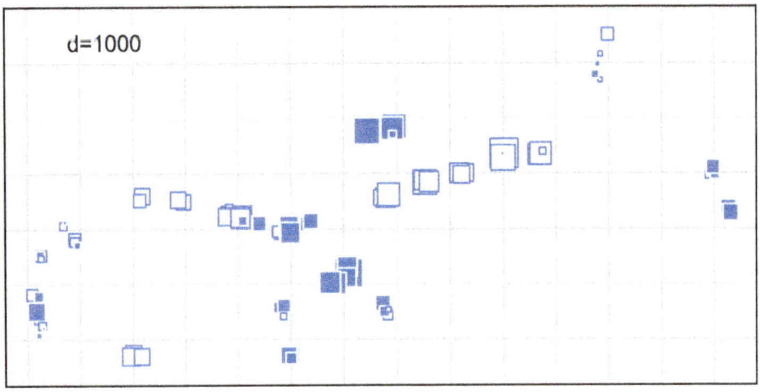

Fig. 6.10 Bubble plot of sample scores on the first axis of the partial RDA on MEMs after controlling for environmental variables (see text for explanation). Symbols sized relative to magnitude of sample scores (solid, positive; open, negative)

samples characterized by red fir (ABma) and white fir (ABco) (compare to Fig. 3.7). The environmental ordination separates these forest types but with only modest explanatory power from the environmental variables (and recall the NMS ordinations in Chap. 4, Figs. 4.9, 4.10, and 4.11).

SPP~GEO + ENV A full model was estimated using the environmental factors as well as the MEMs. This analysis was also highly significant ($P = 0.001$), capturing a total of 63.7% (adjusted) of the variance.

The particulars of this analysis are not really interesting here. What we need from this model is its overall explanatory power, which we will use in variance partitioning (below).

6.4.4.3 Variance Partitioning

Partitioning the explanatory power of environmental as compared to spatial predictors requires multiple models, which allows the estimation of the unique components of explanatory power by differencing (Table 6.2). Here, this analysis was conducted using function *varpart* in R package *vegan* (Oksanen et al. 2021), which itself uses function *rda* as used above.

For this data set, variance partitioning shows a strong spatial effect that is shared between the environmental variables and the MEMs (Fig. 6.11). A total of 33.1% of the variation is shared between the environmental and spatial predictors. The MEMs explain 28.4% variance in species composition after controlling for the environmental variables, while the environmental variables contribute only 2.2% after controlling for their spatial structure. Collectively, the environmental and spatial predictors explain 63.7% of the variation in species composition (the full model, above), leaving 36.3% as residual variation (which is unstructured spatially).

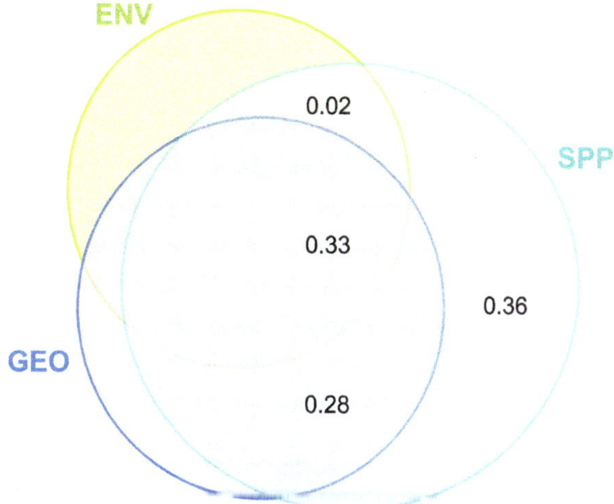

Fig. 6.11 Variance partitioning for the Sierran forest data, using RDA. See Table 6.3 and Fig. 6.7 for definitions of the components. Ellipses are only approximately to scale or proportion

6.4.4.4 Interpretation

Collectively, the results of RDA and variance partition underscore the importance of the environmental variables and that these variables are strongly structured spatially. This is corroborated by supporting analyses (scaling of selected environmental variables, Fig. 6.2; Mantel correlograms, Fig. 6.4). We can further corroborate this general conclusion by correlating the MEMs with the NMS ordination featured in Chap. 4: the indirect ordination axes are correlated with several MEMs, especially the second axis (results not shown). It might also be noted here that this same general result was provided by partial Mantel's tests and by a variance partitioning based on distance-based RDA (neither result shown here). This implies that the ecological result is strong and relatively robust to the details of alternative analytic approaches.

It is worth reiterating what the variance partitioning suggests ecologically. The nonspatial environmental influence (component *a* in Fig. 6.6, here 2.2%) represents environmental influences that are not also represented in the MEMs. This typically implies spatial variability that is finer-scale than the MEMs (which themselves depend on sample locations), or random variation. In this system, soil factors might be one candidate (Urban et al. 2000). This component is quite negligible here.

Component *b* (33.1%) is variation shared by environmental variables and the MEMs, which is to say, spatial structure in the environmental variables. In this instance, the environmental variables are strongly structured spatially, as already noted. To underscore this meaning of this component relative to component *a* (above), the low partial (*a*) does not mean that the environmental variables are not important; it means they are spatially structured.

Spatial variation unrelated to the environmental variables is harder to identify. This component (*c*, 28.4%) might include unmeasured environmental variables that are spatially structured (here, cold air drainage is one candidate), localized (and unobserved) biotic interactions such as competition, spatial processes such as seed dispersal, the legacies of contagious disturbances (here, especially fire), and so on. We do not know what these factors might be, and so knowing the spatial scaling of these (as revealed in mapping the partial RDA axes; Fig. 6.10) can suggest candidates for a follow-up study. The aim here is to transfer, iteratively, the unknown spatial variation in component *c* into the identifiable component ($a + b$). This is the general recipe for untangling spatial structure in landscape-scale data (Levin 1992; McIntire and Fajardo 2009; Legendre and Legendre 2012; Dale and Fortin 2014).

6.5 Further Reading

Spatial statisticians are heavily invested in spatial models and most texts cover the subject in some depth. Haining (1993) and Cressie (1993) are standard references but are somewhat difficult as these models appear sprinkled throughout the texts.

Haining's (2003) text seems to package these issues in a more accessible format (Chap. 9). These texts are written by statisticians, for statisticians. Legendre's (1993) paper on autocorrelation provides a solid foundation for the approaches described here. Wagner and Fortin (2005) provide an overview of spatial issues for ecologists.

While Mantel (1967) is the original authority on the test, the Legendre group is responsible to a large extent for the popularization of Mantel's tests (Legendre and Fortin 1989; Borcard et al. 1992; Leduc et al. 1992; Legendre and Troussellier 1988; Fortin and Gurevitch 1993; Dutilleul et al. 2000; see especially Legendre and Legendre (2012)). Manly (1986, 1991, 1997) has also played a strong role in introducing the approach to ecologists.

Legendre et al. (2005) presented a new approach for partitioning spatial structure in community data, based on *principal coordinates* analysis of truncated *neighbor matrices* (PCNM). MEMs are the general case of PCNM, and while the emerging approach of constrained ordinations using MEMs has been championed primarily by Legendre and colleagues (Legendre et al. 2005, 2008; Legendre and Fortin 2010; Borcard et al. 2011; Dray et al. 2012; Legendre and Legendre 2012; Dale and Fortin 2014), it is now being used and refined by others as well (e.g., Wagner 2013).

Dale and Fortin (2014) cover a wide range of topics on spatial ecology, in a format aimed at ecologists. Relative to this chapter, their coverage includes sampling design, autocorrelation and regressions, Mantel's tests, graph (network) models, and constrained ordinations using MEMs. Fletcher and Fortin (2018) cover spatial analysis of communities in some depth, in a format aimed at ecologists and emphasizing analyses in the R environment.

6.6 Summary and Prospectus

Interpreting spatial structure in community-level landscape data is a natural evolution of analytic approaches long used by ecologists. The decision to "embrace space" invites a new level of analysis and supports a new realm of ecological insights.

Constrained ordinations, as multivariate regressions with environmental as well as spatial predictors, are emerging as the technique of choice for many ecological applications concerned with species composition, environment, and spatial structure. It is fair to say that this approach is a major step beyond the descriptive ordinations we considered in Chap. 4. In particular, it is difficult to attach an intuitive interpretation to an MEM variable other than its being a spatial structure at a particular scale; locational effects also are a part of this but also invite overinterpretation. As complicated as this approach might be, it offers substantial insights into the spatial structure of data sets and a regression-based perspective on the relative importance of explanatory variables. Beyond this, comparative testing using simulated and real data sets suggests that the MEM-based approach with constrained ordinations has more power and better sensitivity as compared to Mantel's tests.

There is still work to be done in evaluating this approach and arriving at a consensus on "best practices." As it took decades for ecologists to come to a consensus on how to use ordinations and classification techniques (if we are actually there!), the recent evolution of constrained ordinations with MEMs is rather remarkable. The next few years should be interesting indeed.

References

Augustin NH, Mugglestone MA, Buckland ST (1996) An autologistic model for the spatial distribution of wildlife. J Appl Ecol 33:339–347

Bauman D, Drouet T, Fortin M-J, Dray S (2018) Optimizing the choice of a spatial weighting matrix in eigenvector-based methods. Stat Rep 99:2159–2166

Borcard D, Legendre P, Drapeau P (1992) Partialling out the spatial component of ecological variation. Ecology 73:1045–1055

Borcard D, Gillet F, Legendre P (2011) Numerical ecology with R. Springer, New York

Bradshaw GA, Spies TA (1992) Characterizing canopy gap structure in forests using wavelet analysis. J Ecol 80:205–215

Cressie NAC (1993) Statistics for spatial data. Wiley, New York

Dale MRT, Fortin M-J (2014) Spatial analysis: a guide for ecologists, 2nd edn. Cambridge University Press, Cambridge

Dormann CE, McPherson JM, Araujo MB, Bivand R, Bollinger J, Carl G, Davies RG, Hirzel A, Jetz W, Kissling WD, Kuhn I, Ohlemuller R, Peres-Neto PR, Reineking B, Schroder B, Shurr FM, Wilson R (2007) Methods to account for spatial autocorrelation in the analysis of species distributional data: a review. Ecography 30:609–628

Dray S (2020) Moran's eigenvector maps and related methods for the spatial multi-scale analysis of ecological data. Vignette for R package *adespatial*. Available at https://cran.r-project.org/web/packages/adespatial/vignettes/tutorial.html. Accessed 2020

Dray S, Dufour A (2007) The *ade4* package: implementing the duality diagram for ecologists. J Stat Softw 22:1–20

Dray S, Pelissier R, Couteron P, Fortin M-J, Legendre P, Peres-Neto PR, Bel-lier E, Bivand R, Blanchet FG, De Caceres M, Dufour A-B, Heegaard E, Jom-bart T, Munoz F, Oksanen J, Thiroulouse J, Wagner HH (2012) Community ecology in the age of multivariate multiscale spatial analysis. Ecol Monogr 82:257–275

Dutilleul P, Stockwell JD, Frigon D, Legendre P (2000) The Mantel-Pearson paradox: statistical considerations and ecological implications. J Agric Biol Environ Stat 5:131–150

Fletcher R, Fortin M-J (2018) Spatial ecology and conservation modeling: applications with R. Springer Nature Switzerland AG

Fortin M-J, Gurevitch J (1993) Mantel tests: spatial structure in field experiments. In: Scheiner SM, Gurevitch J (eds) Design and analysis of ecological experiments. Chapman and Hall, New York, pp 342–359

Goslee SC (2010) Correlation analysis of dissimilarity matrices. Plant Ecol 206:279–286

Goslee SC, Urban DL (2007) The *ecodist* package for dissimilarity-based analyses of ecological data. J Stat Softw 22:7

Guillot G, Rousset F (2013) Dismantling the Mantel tests. Methods Ecol Evol 4:336–344

Haining R (1993) Spatial data analysis in the social and environmental sciences. Cambridge University Press, Cambridge

Haining R (2003) Spatial data analysis: theory and practice. Cambridge University Press, Cambridge

Keitt TH, Urban DL (2005) Scale-specific inferences on vegetation-environment relationships using wavelet regression. Ecology 86:2497–2504

Keitt TH, Bjornstad ON, Dixon PM, Citron-Pousty S (2002) Accounting for spatial pattern when modeling organism-environment interactions. Ecography 25:616–625

Leduc A, Drapeau P, Bergeron Y, Legendre P (1992) Study of spatial components of forest cover using partial Mantel's tests and path analysis. J Veg Sci 3:69–78

Legendre P (1993) Spatial autocorrelation: trouble or new paradigm? Ecology 74:1659–1673

Legendre P (2000) Comparison of permutation methods for the partial correlation and partial Mantel tests. J Stat Comput Simul 67:37–73

Legendre P, Anderson MJ (1999) Distance-based redundancy analysis: testing multispecies responses in multifactorial ecological experiments. Ecol Monogr 69:1–24

Legendre P, Fortin M-J (1989) Spatial pattern and ecological analysis. Vegetatio 80:107–138

Legendre P, Fortin M-J (2010) Comparison of the Mantel test and alternative approaches for detecting complex multivariate relationships in the spatial analysis of genetic data. Mol Ecol Resour 10:831–844

Legendre P, Gallagher ED (2001) Ecologically meaningful transformations for ordination of species data. Oecologia 129:271–280

Legendre L, Legendre P (2012) Numerical ecology, 3rd English edn. Elsevier, New York

Legendre P, Troussellier M (1988) Aquatic heterotrophic bacteria: modeling in the presence of spatial autocorrelation. Limnol Oceanogr 33:1055–1067

Legendre P, Borcard D, Peres-Neto PR (2005) Analyzing beta diversity: partitioning the spatial variation of community composition data. Ecol Monogr 75:435–450

Legendre P, Borcard D, Peres-Neto P (2008) Analyzing or explaining beta diversity? Comment. Ecology 89:3238–3244

Levin SA (1992) The problem of pattern and scale in ecology. Ecology 73:1943–1967

Lichstein JW (2007) Multiple regression on distance matrices: a multivariate spatial analysis tool. Plant Ecol 188:117–131

Lichstein JW, Simons TR, Shriner SA, Franzreb KE (2002) Spatial autocorrelation and autoregressive models in ecology. Ecol Monogr 72:445–463

Manly BFJ (1986) Randomization and regression methods for testing associations with geographical, environmental, and biological distances between populations. Res Popul Ecol (Kyoto) 28: 201–218

Manly BFJ (1991) Randomization and Monte Carlo methods in biology. Chapman and Hall, London

Manly BFJ (1997) Randomization, bootstrap and Monte Carlo methods in biology, 2nd edn. Chapman and Hall, London

Mantel N (1967) The detection of disease clustering and a generalized regression approach. Cancer Res 27:209–220

McArdle BH, Anderson MJ (2001) Fitting multivariate models to community data: a comment on distance-based redundancy analysis. Ecology 82:290–297

McIntire EJB, Fajardo A (2009) Beyond description: the active and effective way to infer processes from spatial patterns. Ecology 90:46–56

Moran PAP (1950) Notes on continuous stochastic phenomena. Biometrika 37:17–23

Nekola JC, White PS (1999) The distance decay of similarity in biogeography and ecology. J Biogeogr 26:867–878

Oksanen J, Blanchet FG, Kindt R, Legendre P, O'Hara RB, Simpson GL, Solymos P, Stevens MHH, Wagner H (2021) *vegan*: community ecology package. R package version 1.17-11. http://CRAN.R-project.org/package=vegan

Petraitis PS, Dunham AE, Niewiarowski PH (1996) Inferring multiple causality: the limits of path analysis. Funct Ecol 10:421–431

Plant RE (2012) Spatial data analysis in ecology and agriculture using R. CRC Press, Boca Raton

Quilodrán CS, Currat M, Montoya-Burgos JI (2023) Benchmarking the Mantel test and derived methods for testing association between distance matrices. Mol Ecol Resour. https://doi.org/10.1111/1755-0998.13898

R Development Core Team (2021) R: A language and environment for statistical computing. R Foundation for Statistical Computing, Vienna. ISBN 3-90005107-0. http://www.R-project.org/

Renshaw F, Ford ED (1984) The description of spatial pattern using two-dimensional spectral analysis. Vegetatio 56:75–85

Smouse PE, Long JC, Sokal RR (1986) Multiple regression and correlation extensions of the Mantel test of matrix correspondence. System Zool 35:627–632

Somers KM, Jackson DA (2022) Putting the Mantel test back together again. Ecology 103(10): e3780

Urban DL (2023) Agents and implications of landscape pattern: working models for landscape ecology. Springer Nature AG, Cham

Urban DL, Miller C, Stephenson NL, Halpin PN (2000) Forest pattern in Sierran landscapes: the physical template. Landsc Ecol 15:603–620

Urban DL, Goslee S, Pierce KB, Lookingbill TR (2002) Extending community ecology to landscapes. Ecoscience 9:200–212

Urban D, Minor E, Treml E, Schick R (2009) Graph models of habitat mosaics. Ecol Lett 12:260–273

Wagner HH (2013) Rethinking the linear regression model for spatial ecological data. Ecology 94: 2381–2391

Wagner HH, Fortin M-J (2005) Spatial analysis of landscapes: concepts and statistics. Ecology 86: 1975–1987

Wagner HH, Fortin M-J (2013) A conceptual framework for the spatial analysis of landscape genetic data. Conserv Genet 14:253–261

Wiersma YF, Urban DL (2005) Beta-diversity and nature reserve system design: a case study from the Yukon, Canada. Conserv Biol 19:1262–1272

Wright S (1943) Isolation by distance. Genetics 28:114–138

Chapter 7
Structural Equation Models

Abstract Ecological data sets tend to present a tangled web of correlations among variables. Structural equation modeling (SEM) is a method for explicitly analyzing this web of correlations, by declaring why the variables should be correlated. A structural equation model is based on a path diagram that poses causal interactions among variables, which might include indirect effects mediated by several variables. A SEM also poses a measurement model, which might include latent variables representing concepts that can be measured only indirectly via indicator variables (e.g., the concept of water quality as indicated by sediment or contaminant loadings). The path model implies a pattern of covariation among variables; this implied covariance provides the basis for testing both the path and measurement model. The process of SEM is an iterative workflow from an initial conceptual model, though estimating and evaluating a formal SE model (or perhaps several of these), to post-processing that summarizes all of the direct and indirect effects in the model. As ecologists increasingly deal with targets such as water quality, environmental health, and other latent concepts, SEM should be a valuable part of the ecological toolkit.

7.1 Introduction

We began with the task of species distribution modeling (Chap. 2), in part, because it introduces most of the issues that plague landscape analysis: ecological data are messy, intercorrelated, and spatially structured. In Chap. 2 we dealt with the correlations by removing redundant predictors, to improve the model fit and ease of interpretation, and we engaged in some modest gymnastics to infer the relative importance of predictor variables. In Chaps. 4 and 5, we dealt with redundancy and noise by aggregating it away—using ordinations (Chap. 4) or classification tools (Chap. 5). Then we added spatial structure, still in aggregate form (Chap. 6).

Here we return to multivariate webs of correlations, addressing these through the lens of path analysis and structural equation models. This follows a thread that is largely separate from the analytic traditions in ecology, and it borrows rather heavily from social sciences. The approach developed here will build on concepts and tools from Chap. 4; in turn, this material will provide a conceptual bridge to Chap. 8.

D. L. Urban, *Landscape Ecology: A Task-Oriented Perspective*,
https://doi.org/10.1007/978-3-031-72251-6_7

Structural equation modeling (SEM) has been described as the marriage of path analysis and factor analysis. But SEM is a rather large topic and a full consideration would broach a number of related issues in inferential design. Indeed, Grace (2006) argues that the logic of SEM should invite us to think carefully about how we practice statistics, even how we *think* about ecological data (see also Shipley 2000). The essence of this argument is that nature is a multivariate, interrelated web—and we should embrace this rather than trying to avoid it. SEM is a tool that embraces this multivariate web of interactions.

SEM is a framework for interpreting correlations among variables in terms of hypothesized causal relationships. SEM consists of two models. The first is a *causal* (*structural*) model that poses a pattern of interactions among variables diagrammed as causal *paths*. For example, "A causes B and B causes C" is a causal path from *A* to *C*. *Path analysis* is a method for structuring and interpreting a regression analysis to emphasize such relationships. The second model is a *measurement* model, which admits that many of the "things" we find interesting in ecology are difficult to measure directly; and so, we devise empirical *indicator variables* for these elusive *latent factors*. For example, we use a term such as "water quality" but this is a concept; what we measure are indicator variables such as dissolved oxygen, nitrogen and phosphorus concentrations, turbidity, and so on. The link between concepts and indicators evokes factor analysis (Sect. 4.3.2). In SEM, the causal model implies a pattern of correlation (or covariance) among the measured variables, linking the causal and measurement models and providing a means to test both.

While SEM might seem a rather substantial departure from the general workflow of this book, we have previously taken the first few steps in this direction. In setting up species distribution models (SDMs), we began with a conceptual model that used ecological theory and natural history to suggest factors that might explain species distributions. We then devised empirical measurements to capture this ecology: the ecological and data models underpinning the statistical SDM (Chap. 2, Sects. 2.2.1 and 2.2.2). This same process will guide us in defining latent variables and their indicators in SEM. Similarly, both SDM and constrained ordinations (Chap. 6) are regressions and use the same set of tools in estimating and interpreting models (estimation, variable importance, variance partitioning).

The Task at Hand

Most generally, SEM is a framework for the analysis of patterns of covariance in data sets. From this perspective, many familiar tools (multiple regression, factor analysis, ANOVA, MANOVA) are special cases of SEM. In many particular cases, we will explore structural models in which we expect variables to interact (i.e., be correlated) via indirect paths involving multiple variables (e.g., the "A causes B causes C" case posed above). These instances are addressed directly and explicitly in SEM. In short, *SEM is a crucial tool for analyses in which we expect variables to be correlated because of the way they interact along paths.*

It might be helpful here to address an issue that often arises with an introduction to path analysis and SEM. We are taught from an early stage of our training that *correlation does not imply causation*. That is true. But the converse is not true;

indeed, *causation does imply correlation*. The pattern of covariation implied by the path model is precisely what provides the leverage to use SEM to understand complicated natural systems.

We will delve into SEM by starting with the logic of path analysis, which is a special (simple) case of SEM. We will then enrich this model by layering on the concept of latent variables and indicators. We will explore a few hypothetical cases as a way to illustrate the process of building a SE model. Finally, we will consider the potential for a variety of multivariate applications in ecology, emphasizing modern implementations of SEM.

SEM, at its core, is a procedure for dissecting the pattern of covariances implied by a conceptual model of how a system works. From this perspective, the declaration of a path model for a given system might be considered a prerequisite[1] to any efforts in research or management of that system: How can we aim to work with a system without also declaring some notion of how it works? This conceptual model—even if never implemented as a SEM—will provide a foundation for site prioritization and structured decision-making (Chap. 8) as well as the interpretation of system change over time (Chap. 10).

7.2 Conceptual Foundations

Path analysis and SEM have an episodic history (reviewed by Shipley 2000; Grace 2006) that might partially explain why it is not as well-known among ecologists as it might be or should be. Path analysis was devised by Sewall Wright (1921), the hugely influential geneticist and evolutionary biologist. But his method was somewhat at odds with prevailing approaches at that time, as espoused by statistical powers including Fisher (who championed randomized trials as the primary basis for causal inference) and Pearson (who argued that correlations or "associations" were all that we have access to—not causation). And so path analysis fell by the wayside...until it was picked by social scientists decades later and extended in ways (see below) that established "modern" path analysis and SEM.

Here, we begin with the initial form of path analysis, but most of this chapter is focused on modern SEM.

[1] I often argue to students (and anyone else who will listen) that they should not embark on any research or management applications for their system until they can produce a plausible conceptual model of how it works. While we all might believe we have such a model in mind, the act of specifying the model, in substantial detail, is harder than it might seem. It is worth the effort.

7.2.1 Path Analysis

A common problem in ecological analysis is that many of our analytic techniques are purely correlational, that is, they imply nothing of causality. This can be a particular problem in landscape ecology because the logistics of conducting experiments at very large spatial scales frequently forces us to rely on inferential methods rather than direct experimental manipulations that could demonstrate causes more directly.

Path analysis is a way to organize a correlational analysis to emphasize posited causal relationships (Petraitis et al. 1996). In its original form, this was not so much an analysis itself, but rather a way of post-processing and interpreting an analysis.

Path analysis recognizes that there are a few ways that two variables might be correlated:

1. There is a *direct* causal relationship (a *simple path*, either $A \rightarrow B$ or $B \rightarrow A$).
2. There is an *indirect* causal relationship via causal chains (a *compound path*, $A \rightarrow M \rightarrow B$, so A and B are correlated).
3. There is a noncausal correlation because both variables are caused by a third measured variable ($C \rightarrow A$, $C \rightarrow B$, so A and B are correlated but spuriously so).
4. There is a noncausal correlation due to (unspecified, unmeasured) correlated causes, which is unanalyzable.

The way to sort through these relationships is to specify explicitly a model—a *path diagram*—that poses a hypothetical relationship among variables, specifying *why* variables should be correlated. Again, this model is interpretative; labeling a path as "causal" does not make it so.

Path Coefficients and Effects

The goal of the original version of path analysis was to decompose correlations into their causal and noncausal components. The basic approach is linear regression, and the focus is on partial regressions. By convention, the analysis is framed in terms of standardized partial regression coefficients, which express the pure effect of one variable on another if all other variables are held constant. The standardization yields predictions in terms of z-scores, to avoid complications due to different measurement units on the variables. Also by convention, r's are used to denote correlations and p's to denote causal *path effects* (which are standardized partial regression coefficients).

The results of path analysis are typically presented by attaching the path coefficients (p's) to the arrows in a path diagram. Additionally, total effects may be collated for all variables in the analysis. Total effects include direct causal paths as well as indirect effects via causal chains through intermediate variables (Fig. 7.1).

Path effects are standardized correlations for simple paths, while for compound paths the effects are multiplied over the path. Thus, for the left-hand path diagram in Fig. 7.1, in which the correlation between A and B is coincidental:

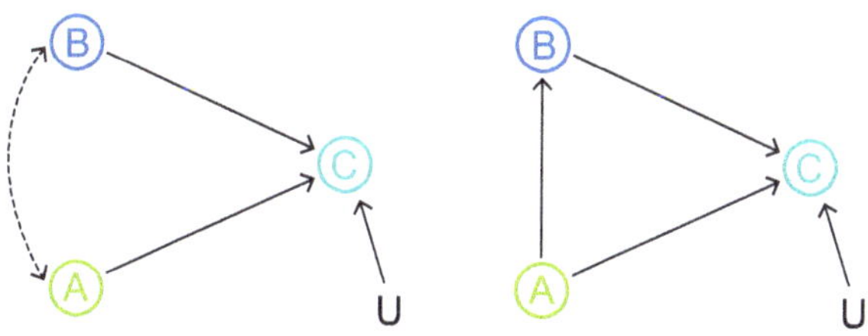

Fig. 7.1 Examples of path diagrams illustrating relationships among three measured variables A, B, and C. The U represents all unmeasured (unknown) variables influencing variable C. Causal relationships are denoted by solid arrows and noncausal correlations by dashed double-headed arrows

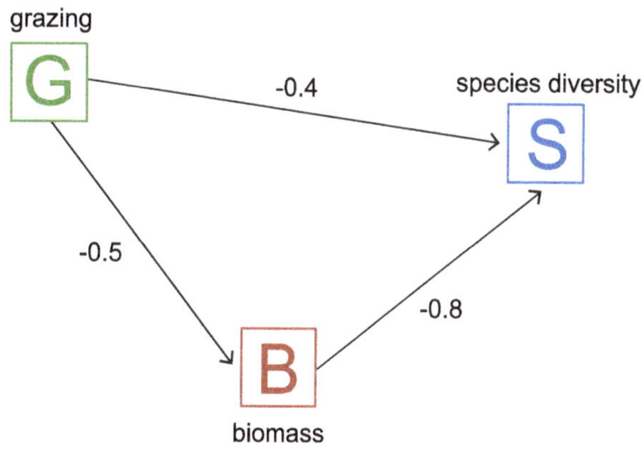

Fig. 7.2 A grazing system in which grazing (G) affects species richness (S) as well as plant biomass (B), which also affects species richness. Coefficients are standardized regression coefficients. (Redrawn from McCune and Grace (2002), with permission of the authors)

$$r_{AC} = p_{AC} + r_{AB}p_{BC}$$
$$r_{BC} = p_{BC} + r_{AB}p_{AC}$$

Here, A's total *effect* on C, e_{AC}, is the sole direct path $e_{AC} = p_{AC}$; likewise, B's effect on C is $e_{BC} = p_{BC}$. The associations mediated by the correlation r_{AB} are unanalyzable. For the right-hand path diagram, the situation is different:

$$r_{AC} = p_{AC} + p_{AB}p_{BC}$$
$$r_{BC} = p_{BC} + r_{AB}p_{AC}$$

In this case, B's total effect on C is simply p_{BC}, its direct effect, while A's effect on C is the sum of its direct and indirect effects, $e_{AC} = p_{AC} + p_{AB}p_{BC}$. Note that the correlation structure in these two cases is identical; what differs is the interpretation of causal paths (the direction of the arrows). (To be strictly correct, variable B in the right-side model should also have an error term U connected to it, as we do not know whether B is completely caused by A.)

Grace (2002) presents a nice illustration of alternative ways that paths might play out in ecology. A grassland system provides the setting (Fig. 7.2). In this example, grazing has a negative effect on plant biomass (i.e., herbivores eat plants), and biomass has a negative effect on plant species richness (i.e., through competitive exclusion). So the indirect effect of grazing on richness through biomass is positive (i.e., negative x negative). But grazing also reduces richness directly, perhaps through selective foraging.

In this case, the effect of grazing (G) on richness (S) is the sum of the direct and indirect paths, which is $-0.4 + (-0.5 \times -0.8) = 0.0$, which is also the simple correlation between G and S. The simple correlation between B and S is -0.6 ($-0.8 + -0.5 \times -0.4$), which yields a regression R^2 of $p_{GS}\, r_{GS} + p_{BS}\, r_{BS} = (-0.4 \times 0.0) + (-0.8 \times -0.6) = 0.48$. Thus, grazing is uncorrelated with richness but explains 48% of the variance in richness when analyzed in terms of causal paths. An instance like this, perfectly offsetting effects, is probably quite rare in natural systems; but the more general case of offsetting causes that confound observed correlations is probably not uncommon.

It might be worth underscoring the difference between this partitioning of the variance and more conventional regressions. In the corresponding multiple regression, $S \sim B + G$, the analysis would have the same R^2 (0.48) but could not reveal the indirect path. In the paired univariate regressions, $S \sim G$ and $S \sim B$, grazing would not emerge as significant because of its (lack of) correlation, while biomass would yield an r^2 of 0.36 and not include the indirect effect of grazing.

The original version of path analysis amounts to an after-the-fact partitioning of pure and partial coefficients from a conventional regression analysis. The analysis makes the usual assumptions that apply to regression and adds a few more that are specific to the paths:

1. Relationships are linear, additive, and causal.
2. Residuals are uncorrelated with variables previously entered into the model.
3. There is no reciprocal causation.
4. All variables are interval scale (no ranks or categories).
5. All variables are measured without error.

Of course, these assumptions are unlikely to be met perfectly by ecological data, but as with regression the results seem robust to minor departures from these assumptions. Modern methods of analyzing path relations in advanced applications of structural equation modeling ("modern" SEM) can relax all of these assumptions except the assumption of causality.

7.2.2 *Factors and Structural Equation Modeling*

Factor analysis enters into SEM because the entities that are posed in causal paths are often not measure directly. Instead, they are represented by specific indicators. In the language of SEM, these factors are termed *latent variables* if they are considered to be unmeasurable in principle. In the context of this discussion, latent variables often are *common factors* represented by indicator variables (see Supplement 4S.2.3 for more on factor analysis). The measured indicators are termed *observed* or *manifest*

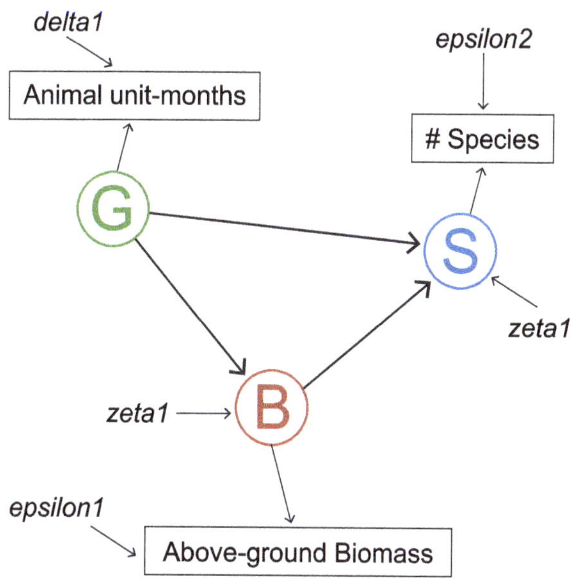

Fig. 7.3 A hypothetical grazing system in which grazing (G) influences species diversity (S) as well as plant biomass (B)—which also influences species diversity. Here, the ovals indicate latent variables associated with indicators in rectangles (e.g., animal unit months for grazing pressure, measured aboveground biomass for total biomass, richness for species diversity). Each indicator is associated with an error term (*deltas* for exogenous indicators, *epsilons* for endogenous indicators), as are the endogenous latent variables (which are not completely specified by the paths; errors denoted with *zetas*)

variables—the former term used to remind us that these are actual data, and the latter term used to connote the hypothesis that the underlying latent variable *caused* the indicator (i.e., it is a measurable manifestation of the latent variable). From this perspective, the path analysis example above is a simple case of SEM in which all of the variables are observed.

We can revisit the grassland example above and expand the model (Fig. 7.2) to more fully embrace the conventions of modern SEM (Fig. 7.3). This is essentially the same model as above, but we have now made it explicit that we are measuring three indicators (animal unit months, aboveground biomass, and species richness; drawn in boxes in the diagram) to represent three latent variables (ovals in the diagram). We have also admitted that the indicators are not without error; by convention, exogenous indicators have errors (or *disturbances*) denoted with deltas (δ), while errors on endogenous indicators are denoted by epsilons (ε). An *exogenous* variable is one that takes on values due to forcings outside the system; the only arrows on these variables point *from* the variable. *Endogenous* variables are affected by other variables in the system as modeled; that is, they have arrows pointing *to* them (as well as perhaps *from* them). Latent variables also have prediction errors, because the causal paths might not determine them completely. These errors are denoted with zetas (ζ).

7.2.3 Modern SEM

The addition of latent variables to path analysis represents a major innovation and one that reflects a disciplinary investment in common factors by social scientists. Social scientists routinely are interested in concepts such as "intelligence" (including verbal or other aspects of this). These factors are typically quantified with indicator variables derived from surveys; various aptitude or personality quizzes are familiar examples. In these, the questions are indicators and the research task is to devise indicators that are unambiguous, unbiased, and repeatable. In applications, the investigator is interested in the common factors—the indicators are not of interest.

A second and perhaps more important extension to path analysis was the generation of a means to estimate and evaluate the model globally. By contrast, the initial method for path analysis tested each "arrow" in the system but there was no way to assess the overall model. Modern path analysis was framed in terms of covariances and fit using maximum likelihood methods, which provided the basis for an overall test of the model. This capability also provides the means to compare alternative models, a compelling feature of SEM.

Conditional Independence and Model Fitting

The path model in SEM represents two models. One is the *causal model*, a hypothesis about causal relationships in the system. The second model is the *measurement model*, posed in terms of indicator variables for the latent variables. These are related, in that the causal model implies, for example, how indicators should covary.

The goal of the SEM analysis is to estimate and interpret coefficients that populate the path model. From this, the structural model is evaluated in terms of the posited causal paths (the arrows among latent variables), while the measurement model is evaluated to assess how well the indicators capture the latent variables and to estimate the relative explanatory power of each of the indicators in the model. The entire model comprises a set of regression coefficients along with a set of variances, covariances, and error variances. These are estimated simultaneously.

The key to the solution of a SEM is that the causal model implies a particular pattern of covariation among the variables. For example, if *X1*, *X2*, and *Y* are linked by a compound path ($X1 \rightarrow X2 \rightarrow Y$), they *must* covary in a way consistent with that path. This is the so-called test of mediation in path analysis—that the effect of *X1* on *Y* is mediated by *X2*. Another way to say this is to note that *unless X2* is available empirically, *X1* and *Y* will be uncorrelated. Again, the correlation between *X1* and *Y* is the correlation between *X1* and *X2* multiplied by the correlation between *X2* and *Y*. The correlation between *X1* and *Y* is conditional on *X2*; *X1* and *Y* are *conditionally independent*. Thus, if we can fix *X2* (experimentally or statistically), we do not need to know *X1* to know *Y*: *X2* is all we need. That is, $r_{X1,Y|X2} = 0$, or in SEM terms, $X1 \perp Y|X2 = 0$ (read "*X1* is conditionally independent of *Y*, given *X2*") (Fig. 7.4).

The key to the path model is that collectively it implies a set of covariances and partial covariances expected from the model. In modern SEM, a key innovation was the derivation of a solution, estimated using maximum likelihood methods, that is expressed in terms of the covariances and partial covariances. If the data do not

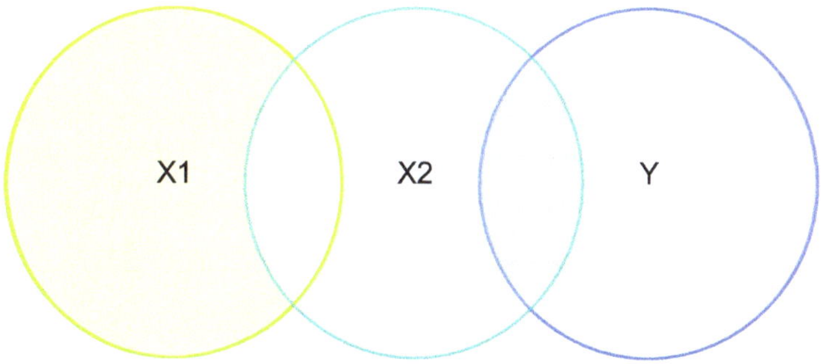

Fig. 7.4 A Venn diagram to illustrate one case of *d*-separation or conditional independence. Circles represent variance in each variable, and overlap is covariance. Here, *X1* is conditionally independent of *Y* because its influence on *Y* is mediated by *X2*. The partial correlation between *X1* and *Y*, given *X2*, is 0

match this pattern, the model is wrong. Importantly, this fit assesses both the causal and the measurement model. Inspection of the fitted parameters, their errors, and significance values can suggest where the model might be revised to better fit the data (and see below).

7.3 The Procedure of SEM

The workflow for SEM consists of the usual process of data preparation, model fitting, and model evaluation. But fitting a SEM often comprises its own workflow (Fig. 7.5), as the fitting and evaluation often is done in an iterative way.

In what follows, we focus on "modern" SEM and its maximum likelihood estimation. We extend this approach to include newer innovations subsequently. The workflow is adopted primarily from Grace et al. (2010, 2012).

7.3.1 Data and Data Preparation

Because it is based on regression, SEM is subject to the data screening and assumptions common to regressions. Beyond this, there is an emphasis on screening variables in terms of what they represent (e.g., if indicating for latent variables) and the role they will play in the analysis.

While original path analysis was expressed in terms of standardized coefficients, modern SEM often uses unstandardized as well as standardized solutions. As a reminder, standardization is particular to a data set, so presenting results in standardized forms means that they cannot be compared directly to studies based on other datasets. By contrast, unstandardized estimates reflect the measurement units

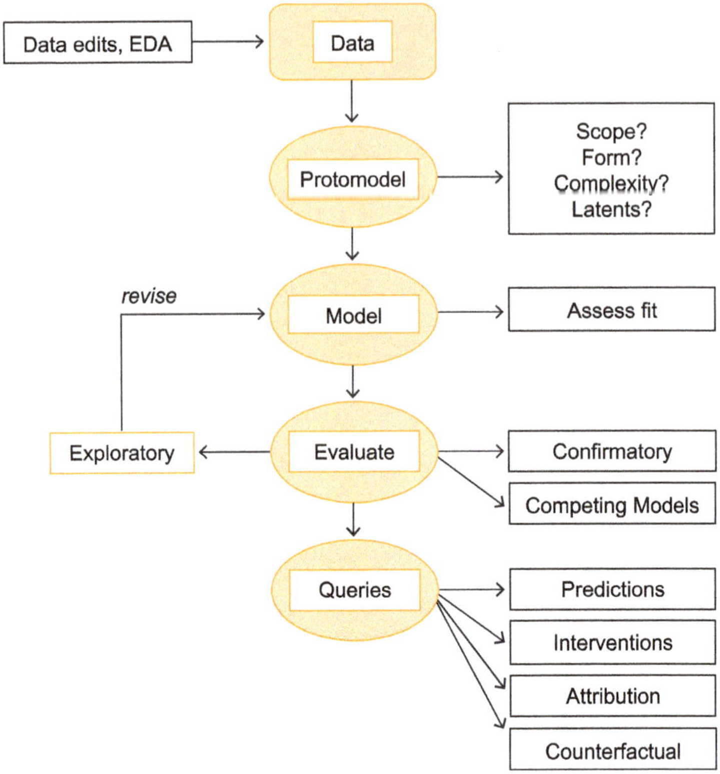

Fig. 7.5 Workflow for structural equation modeling (adapted from Grace 2006; Grace et al. 2012). The step from prototype (conceptual model) to a final fitted model can be rather involved (see text). The evaluation stage is separated based on how the model is being used: with confirmatory SEM or competing models, there is a single evaluation; in exploratory modeling, the process might iterate through multiple revisions

of the variables, and so terms cannot be compared easily within the same analysis. Grace (2006) and colleagues suggest using a compromise approach and scaling variables so that they vary on a similar scale (e.g., 1–100) so that the results can be interpreted readily but without standardization.

While SEMs could, in principle, be modeled with spatial predictors (recall Sect. 6.4.2.), in practice this is rarely done; the models are typically nonspatial. . .although they might invoke explanatory or response variables that are, in reality, spatially structured.

7.3.2 *Observed, Latent, and Composite Variables*

The initial example considered here to illustrate path analysis (Fig. 7.2) amounts to a SEM using only observed variables. In the social sciences, SEM is heavily invested in latent variables indicated by observed variables. It is useful here to revisit these approaches.

In the simplest case, an SE model can be framed in terms of observed variables only. This simplifies the analysis somewhat. But if we have several possible observed variables that are loosely related, we expect these to be correlated to some degree. At this point, three alternatives emerge in modern SEM.

Observed Variable One approach is to choose a single indicator variable for each term in the conceptual model, resulting in an *observed-variable* or *manifest* SEM. This model would have no latent variables. This is a common approach in the natural sciences (and see below).

Latent Variable A second approach is to frame these correlated variables as multiple indicators for the same latent factor. For example, we might measure impervious surfaces, road density, and any number of other manifestations of urban development. In this perspective, impervious surface area and road density are indicators for the latent variable *urban development*. In SEM, the estimated model is concerned primarily with the effect of development on a response variable (say, water quality, itself also a latent variable). The indicators are considered to be rather uninteresting in themselves; they are redundant and substitutable.

Composite Variable By contrast, we might actually be interested in assessing the relative importance of impervious surfaces as compared to roads in affecting water quality. In this case, we would use both observed variables in the model, and not specify a latent variable. In estimating the model, the path coefficients would allow us to determine which (if either) of these variables had the stronger influence on water quality. This would be especially appropriate if we expect the mode of action to differ among variables. For example, impervious surfaces affect water quality via overland flow and washing of sediments and contaminants into streams. Roads, by contrast, might contribute heavy metals to streams via emissions from car exhausts. We expect these effects to be correlated because most roads are impervious while not all impervious surfaces are roads.

This is to say, development has a number of interrelated manifestations, many of which might affect water quality. From this perspective, the effect of development on water quality is the *sum* of the influences from the separate observed variables. These cumulative influences can be aggregated after the fact from an SEM simply by adding up their individual effects. In this example, the effects of development on water quality would be the sum of the influence of impervious surfaces and roads. This aggregate construct is a *composite* variable in SEM (Grace 2006; Grace and Bollen 2008; Grace et al. 2010). (As a matter of detail, a composite variable is diagrammed as a hexagon, and the arrows point from the indicators to the composite, signaling that the indicators cause the composite.) This approach of aggregating composite variables seems likely to be especially important in ecological applications (see below).

One final consideration in the choice to use latent variables is the possible aim to generalize the model results to other systems. If this is the aim, then latent variables can be used to express the general model and its expectations. Others could apply the same model—perhaps using different indicator variables—to test its generality. By contrast, an observed-variable approach restricts the conversation to the selected indicator variables (or composites based on these).

7.3.3 The Process of SEM

Building and evaluating SE models is a process that evolves in several stages. The first step is to specify a *construct model*, which is a prototype[2] of the path diagram. This is essentially a conceptual model that represents what theory (or working hypothesis) suggests about relationships among the ecological concepts of interest. That is, the construct model illustrates how we *believe* the system works. This model might be diagrammed with ovals and arrows only. This model is *formalized* by posing the details of the SEM, in terms of latent variables (if used), indicators, covariances, and error terms.

The details of this can become rather complicated, and so for any given application, the full model might be reduced in scope; in this, we might ignore certain complexities or otherwise narrow the scope of the model. This amounts to removing some of the boxes and arrows before fitting the model. These decisions depend on the explicit goals of the application, of course, but often also reflect available data, spatial and temporal scale, and other considerations.

The model to be fitted might still include a lot of terms! These will include regression coefficients for causal paths, variances, or covariances among pairs of variables and error terms. It is often useful to constrain the model with empirical estimates of terms to the extent possible. The parameters of the specified model are then *estimated*. Again, this amounts to solving for all of the terms simultaneously, subject to the stated constraints.

In models estimated using maximum likelihood methods, the fit is evaluated in terms of a Chi-square test on model deviance (recall this test from species distribution models in Chap. 2). Note also that the goal of the analysis is to find that the model as posed is *not inconsistent with the data*. That is, the aim is to find a *non*significant *P*-value—to *not* reject the hypothesis that the data match the model. In practice, multiple models might be evaluated to choose the most likely model from among several plausible alternatives. In confirmatory mode, this approach explicitly tests the posed model using the measured data (and see Sect. 7.3.4). This approach is in contrast of the conventional approach in regression, which assesses the model (and data) against the null model of randomness. This capacity to evaluate the overall causal model explicitly is an important feature that

[2]Grace and colleagues refer to this prototype model as a metamodel. But there is another tradition in ecology in which metamodels are after-the-fact constructions of models of models. So I will use construct or prototype model here.

distinguished modern SEM from the original formulation of path analysis. In particular, modern SEM is often used to compete alternative models against each other, using the data to arbitrate the best fit.

In model estimation, it is typical to evaluate alternative models successively, perhaps by modifying hypothesized paths or covariances in order to find the best-fitting model. In model revision, this often consists of deleting arrows that were posed in the model but were not significant in the fitted model, or adding new paths that were not included in the original model.

In software implementations of SEM, a powerful aid to model refinement is the provision of *modification indices*, which are suggestions of new arrows that might be considered to improve the overall fit; these suggestions are based on observed correlations within the data set. As with many modeling exercises, this stage of model evaluation might continue indefinitely, as one tries to revise the model to be as general and robust as possible. In practice, one would retain the first (reduced) model that is consistent with the data. Again, a strength of SEM is the capacity to compete models of similar complexity, letting the data choose the better model.

Ideally, validating a SE model involves tests using data independent of the model-building exercise—as with any model. To do this, the model must work in terms of unstandardized covariances rather than standardized correlations (recall any standardization is particular to a data set and cannot be extended to new data sets). To provide for broader generalization and richer insights, modern SEM typically estimates standardized as well as unstandardized parameters.

7.3.4 Model Evaluation and Interpretation

A SEM is a regression, and like any regression, there are several items to evaluate: (1) whether the model is significant, (2) its explanatory power, (3) the relative importance of predictor variables, and (3) whether it can provide useful predictions about unobserved cases or scenarios. With SEM, each of these can be a richer exploration than with a simple regression.

The test of significance, a Chi-square test based on maximum likelihood estimation, is unusual in that we aim for nonsignificant results: that the model is not inconsistent with the data. As with species distribution models fitted as GLMs or GAMs, the maximum likelihood criterion provides a means for comparing (or competing) alternative models. An ANOVA using the Chi-square test explicitly compares one model to an alternative (provided these are nested models using the same data). For two models that are both nonsignificant, there is no clear rule on choosing the better model (except, perhaps, in the case of a marginally nonsignificant versus a clearly nonsignificant test, e.g., $P = 0.06$ versus $P = 0.5$). For any given system, there might be alternative models that are equivalent in terms of their fit and explanatory power; only further model tests or manipulative experiments can resolve these.

As a regression, an SEM provides an estimate of its explanatory power in terms of R^2 value. This is an aggregate estimate, perhaps reflecting multiple paths to the response variable(s). In many SEMs, there also might be intermediate R^2 values for endogenous variables that are generated along causal paths.

Interpreting variable importance in a SEM is simpler than in a multiple regression, in the sense that the role of each variable is specified in the model. Each variable might be involved in more than one path or noncausal correlation, and each term will have its associated *P*-value. Predictor variables might play multiple roles, and it is interesting to decompose the overall effect into its component parts: the direct effects of a predictor on the dependent variable via a simple path, as well as the indirect effects via compound paths. In SEM, it is conventional to collect and report these components explicitly: direct, indirect, and total effects.

The process of fitting an SEM is typically iterative. At each iteration, the model might be revised by removing nonsignificant paths or adding new paths or correlations based on the preliminary fit. This iteration can lead to model over-fitting to the training data set. Given this tendency, Grace et al. (2012, Grace 2006) suggest three alternative cases in model evaluation:

1. *Strictly confirmatory:* If the model was constructed based on prevailing theory, then there should be little model revision. In evaluating the model, the decision is straightforward: the model either fits the data or it does not. This case is probably more typical in the social sciences than in ecology.
2. *Competing models:* In some applications, there might be alternative and equally plausible conceptual models or hypotheses. In such cases, the competing models would have different topologies (different path models), and again, the evaluation is a single decision: whichever model better fits the data is supported.
3. *Exploratory modeling:* In perhaps most applications, the construct model is a working hypothesis and model fitting involves some iteration and revision. This process should lead to a more robust working hypothesis. Exploratory modeling can lead to substantial learning—the discovery of new paths or effects not known previously. But the resulting model is best interpreted as a new hypothesis, with provisional acceptance pending independent validation.

7.3.5 Model Queries

Once validated, a SEM can be used in applications—just as with any other regression. But the range of applications is rather broader with SEMs, precisely because of the logic of folding *why* into the relationships among variables and the nuance provided by direct and indirect paths. Exploring an SEM might involve different types of queries (the bottom tier in Fig. 7.5) (Grace et al. 2012).

Some model explorations are prospective and entail exploring the consequences of a predictor variable taking on a new value in the future. These are model *predictions* in the sense of most regression models. SE models can also be used to conduct virtual experiments, in which the investigator intervenes in the model to set one of more values (i.e., of predictors or mediating variables) to assess the implied consequences; these are forms of model-based *scenarios*. SEMs can also be used for retrospective explorations. *Attribution* studies are one example, in which interventions are used to assess the relative plausibility of various paths in the model as a way to account for observed outcomes (i.e., might *this* have been the cause of what we have observed?). Finally, retrospective queries can be used to evaluate counterfactuals (e.g., to assess the implications of alternative starting points or baseline conditions).

7.3.6 Reporting and Communication

Reporting the results of a SEM analysis would typically detail a stepwise process from prototype model, through any intermediate models, to a final model. These models might be represented as a sequence of path diagrams, with the final model diagrammed in terms of significant effects (arrows) describing the topology of the model. By convention, significant arrows are retained and drawn with line thicknesses proportional to effect sizes; nonsignificant relationships often are removed from the path diagram. In cases where a nonsignificant relationship is obtained but the conceptual model or theory demands that relationship, it would be reported but drawn with a dashed line.

The final model would be presented in terms of its path diagram(s), test of significance (Chi-square test), direct and indirect effects (as partial R^2 values), and total explanatory power (R^2). These latter elements would be collated into a table similar to a typical regression summary but with the added information on partial (indirect) effects. Collectively:

☑ Description of the data (what they are, what they represent, samples sizes, etc.)
☑ Data screening and editing, including standardization or relativizations
☑ Bivariate summaries of correlations (matrix)
☑ Conceptual model (proto- or construct model diagram, with explanation and justification of claims)
☑ Final model (diagram, with proportional arrows, etc.)
☑ Table of model coefficients, errors, and *P*-values, probably standardized, along with unstandardized results (perhaps in an appendix)
☑ Overall test of model significance (Chi-square value, *df*, *P*-value)
☑ Narrative summary of the main direct and indirect effects and discussion of any model revisions (e.g., paths removed or added)

7.3.7 Extensions

The discussion thus far has illustrated the form of SEM that is probably in widest use currently. This is typically a hybrid of observed and latent variables (natural scientists tend to use latent variables less than social scientists), with relationships presumed linear and estimated using maximum likelihood methods. But SEM is a rapidly evolving practice and there are several developments that are promising to make it even more applicable to natural systems (Grace 2006; Grace et al. 2012). Collectively, Grace et al. (2012) suggest that these extensions represent an emerging third-generation SEM.

Categorical Response Variables While original path analysis was restricted to continuous variables, newer implementations can also admit categorical response variables (note that categorical exogenous variables are not a complication in SEM). This is especially germane for models that include experimental contrasts such as categorical treatment levels.

Multigroup Models In many instances, we expect different groups to behave differently in the same model. For example, sexual dimorphism might lead to different responses for males as compared to females. In a multigroup model, the topology of the model is the same for both groups but the coefficients can vary.

Hierarchical Models In some systems, the causal effects in the model might inherit from or be influenced by higher-level effects. For example, individual responses might be nested within a species-level response, or species-level responses might vary within habitat guilds (e.g., ground-nesting versus canopy-nesting birds). Multilevel models permit these hierarchical relationships.

Piecewise Estimation While modern SEM is usually estimated globally, an emerging approach uses local estimation (Grace et al. 2012; Lefcheck 2016). By contrast to the global fit, which uses all the available information, a local fit uses only part of that information.

The local fit is applied to subsets of variables, subject to a test of *d*-separation (think *d* for dependency; Shipley 2000; Grace et al. 2012). These cases are candidates for conditional independences in the model, and are identified for all pairs of variables that are not directly connected by a path, controlling for common ancestors (but not common descendants) and inclusion in compound paths. There might be several instances or claims of *d*-separation in any given path model, many of which are redundant because they are implied by related claims. The *basis set* of claims is the minimum set of claims that imply all the rest. The piecewise solution fits the basis set, resulting in a *P*-value for each case. From these, an overall test of significance can be estimated in terms of Fisher's *C* (Lefcheck 2016):

$$C = -2\sum_{k} \ln (P_k) \tag{7.1}$$

for the *k* cases. *C* is distributed approximately as Chi-square and can be evaluated accordingly.

Piecewise estimation has two important consequences: (1) It allows more complex functional forms (e.g., nonlinear fits), because the estimates are based on the data points themselves (recall that the maximum likelihood estimate uses only the covariance matrix, not the samples that generated that matrix); and (2) it reduces the chance that local misspecification of a model will propagate errors to other parts of the model.

Bayesian SEM Bayesian methods have been in practice as a solution technique for SEM for some time, but there is more potential here in terms of incorporating existing information as priors—including estimates derived from other studies.

Bayesian methods tend to outperform maximum likelihood methods for smaller sample sizes while providing similar estimates for large sample sizes (maximum likelihood estimators are asymptotic, thus more accurate for larger sample sizes). As a rule of thumb, Grace et al. (2012) recommend a ratio of samples to estimated coefficients, $d = n/a$ (for n samples and a coefficients), of 5 or more; a ratio of <5 would suggest using a Bayesian estimate.

As a trade-off, Bayesian methods do not provide a measure of overall model fit, as is provided by maximum likelihood solutions. Some practitioners use maximum likelihood methods and corroborate this solution with a Bayesian version. Bayesian methods are not as widely available yet in software but are emerging.

7.4 Applications

It will be useful to explore a few examples in some detail, as applications typically evolve over a characteristic logical sequence of model conceptualization, refinement, estimation, and summary.

7.4.1 Human Disturbance and Biotic Integrity of Wetlands

Grace et al. (2010, 2012) have provided two in-depth illustrations on the iterative process of applying SEM in ecology. Here we follow the illustration of Grace et al. (2012), in which their concern was a relationship between human disturbance and biotic integrity in wetlands of Acadia National Park in Maine, USA. An index of biotic integrity was strongly and negatively correlated with an index of human development pressure; the question is "why?" Because they stepped through the entire process of generating, evaluating, and applying a SEM, we will work through this example in sequence. They also provide a much deeper discussion of the steps highlighted briefly here.

They began their study by first posing a rather general path model and then refining this in terms of relationships they wished to evaluate as the objectives of their study. In this, the initial model is a prototype which is fleshed out in subsequent refinements (Fig. 7.6). This effort to come to an initial model can be quite challenging, as it forces the investigator to address issues of spatial and temporal scale,

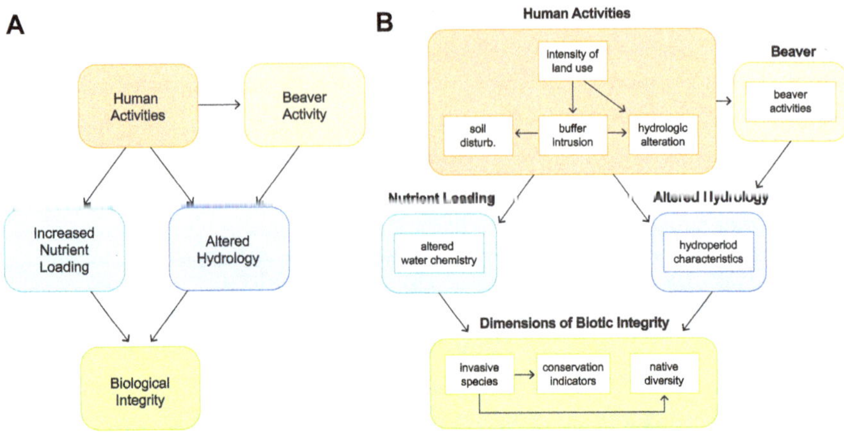

Fig. 7.6 Prototype path model for the Acadian wetland study (**a**) and (**b**) a more refined version of the model that poses relationships more explicitly in terms of measurable variables. (Figures 7.6–7.10 redrawn from Grace et al. 2012, permission licensed via Creative Commons)

available data, and what is known (or knowable) about the system—all subject to the needs and objectives of the modeling exercise.

This model was simplified somewhat to arrive at an initial (full) model framed in terms of measured variables only (a manifest model, without latent variables) (Fig. 7.7). In this, the effect of beavers was *absorbed* into human land use effects, because of the complicated way that beaver activity is tied to historical and current land use practice. Because the sample size was relatively small compared to the number of parameters to be estimated, they used Bayesian methods to fit the model.

On revision, a final path model was accepted that included many but not all of these relationships, as well as a few new ones discovered during model exploration (Fig. 7.8). Again, beaver activity was included in the path model but only indirectly so in the SEM. Three new paths emerged in model revision, and five paths that were postulated in the initial causal diagram were found to be nonsignificant for this data set.

The final model is presented with paths labeled with their coefficients and R^2 values for intermediate and the final endogenous variables (Fig. 7.9). In this, Grace et al. adopted a compromise approach to presenting the coefficients: instead of standardizing (which is particular to the data set) or showing unstandardized terms (which are difficult to interpret at a glance), they present results quasi-standardized as the results produced from the minimum and maximum values observed for each variable, divided by that variable's range.

This final model was used to explore management alternatives, in an application that explicitly incorporated direct and indirect effects. In this instance, the application compares the status quo (no action) to the outcomes predicted from two hypothetical management alternatives posed as model interventions: to manage invasive vegetation directly, by removing it from riparian zones or by treating

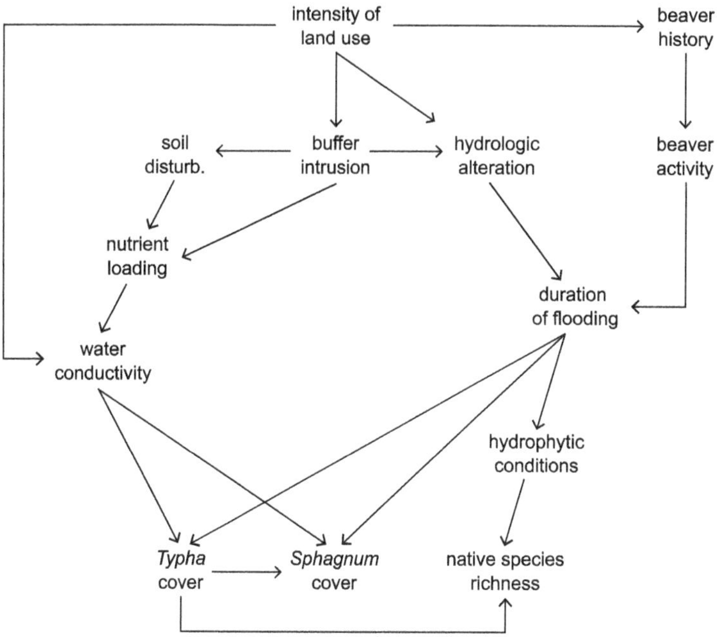

Fig. 7.7 Initial path model of the Acadian wetland system, framed in terms of measured (manifest) variables only. (From Grace et al. 2012)

water quality directly. While the regressions implicit in the path model would provide an average prediction, in this case the predictions were simulated by generating a large number of predictions—each based on a set of 200 model coefficients drawn randomly from the mean and standard deviation of the estimated model coefficients. In sum, the simulations provide an average expectation as well as an estimate of the uncertainty of the model predictions (Fig. 7.10).

This example of using a path model to explore management alternatives will serve as a bridge to the next chapter, in which we develop this path model as one component of the structured decision-making process that often underlies site prioritization.

7.4.2 *The Urban Stream Syndrome*

The urban stream syndrome is a tangled web of correlations stemming from development infrastructure and its impacts on hydrology (Paul and Meyer 2001; Allan 2004; Meyer et al. 2005; Walsh et al. 2005; Hassett et al. 2018; and see Urban 2023, Chapter 9). This web can be organized by emphasizing a few mediating pathways, which include stream hydraulics affecting channel morphometry, energy levels (heat and light), nutrient loadings, and the delivery of contaminants to the stream. These

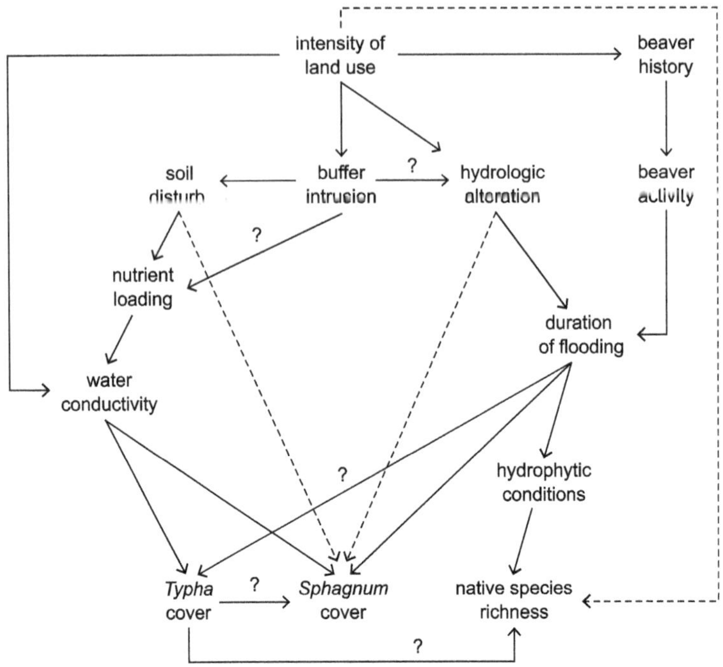

Fig. 7.8 Revised path of the Acadian wetland system, reduced from the initial path model shown in Fig. 7.6. (From Grace et al. 2012). Dotted lines are new relationships discovered in model revision; question marks denote paths found to be not important in this data set

mediating factors, in turn, influence in-stream biota and ecosystem processes (Fig. 7.11).

Bernhardt and colleagues collected a synoptic sample of low-order urban streams along a gradient of development intensity, in an effort to disentangle this web of interactions. Perhaps most tellingly, they were unsuccessful in fitting a SE model of the entire system—there are nearly as many arrows as observations in their data set! But such a complicated prototype model need not be estimated in its entirety in order to be useful. This conceptual model has been explored more narrowly to model thermal pulses in urban streams (Somers et al. 2013) and chemical flashiness of urban streams (Blaszczak et al. 2019).

For example, Somers et al. (2013) used SEM to reveal that stream temperature at baseflow is largely explained by reach-level predictors (e.g., canopy closure, channel incision) while stormflow temperature (thermal pulses) are explained by watershed-scale predictors (e.g., impervious surface area).

In urban streams in Melbourne, Australia, Caressa and Parris (2013) found indirect effects of impervious surface area mediated by aquatic vegetation to influence amphibian communities. Lu et al. (2024) used SEMs based on a construct model similar to Fig. 7.11 to highlight indirect effects of landscape development on aquatic macroinvertebrates in streams of the southeastern USA; their results underscored the importance of altered flow regime and pesticides in urban streams.

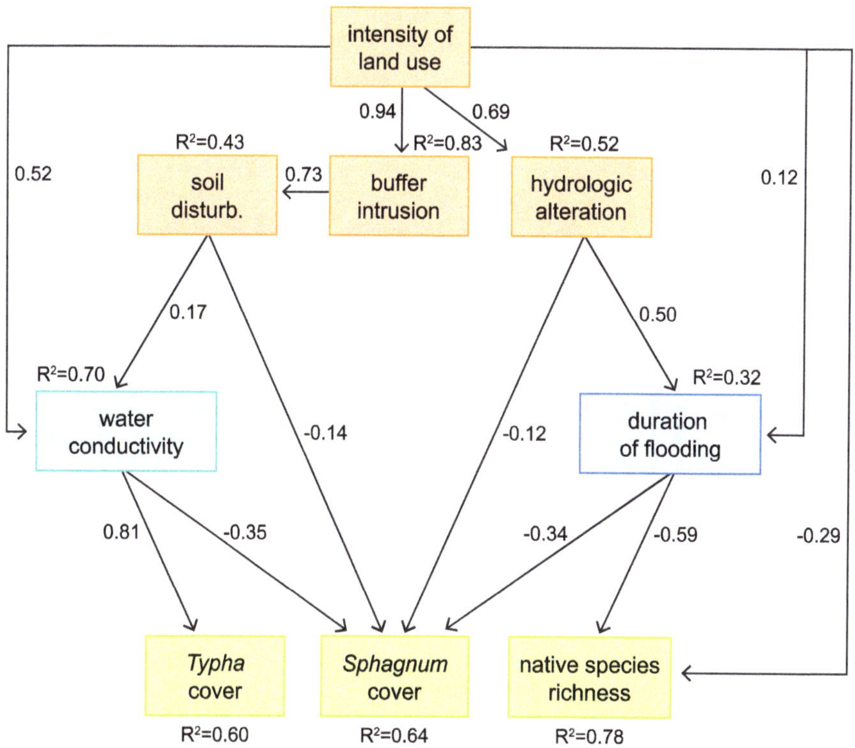

Fig. 7.9 Final model reporting, with R^2 for predicted endogenous variables and quasi-standardized path coefficients for all dependencies (see text, from Grace et al. 2012). Note that beavers were removed as a separate term and are absorbed into land use effects

The larger importance of using a construct such as Fig. 7.11 is that it can serve as a comparative framework for other studies of urban streams. In general, we expect that urban streams function according to this model, yet we also expect that the various pathways should be expressed in different ways in different systems (Booth et al. 2016; Urban 2023, chapter 9). These differences would appear as different patterns in the relative importance of the mediating paths. To be the most general, the model should be implemented in terms of latent variables, which would permit the use of different empirical indicators as appropriate in different systems.

7.4.3 *Urbanization and Diversity: The Four Filters Hypothesis*

Urbanization also produces a web of interactions that influence biodiversity in developed landscapes. Williams et al. (2009) isolated these in their "four filters"

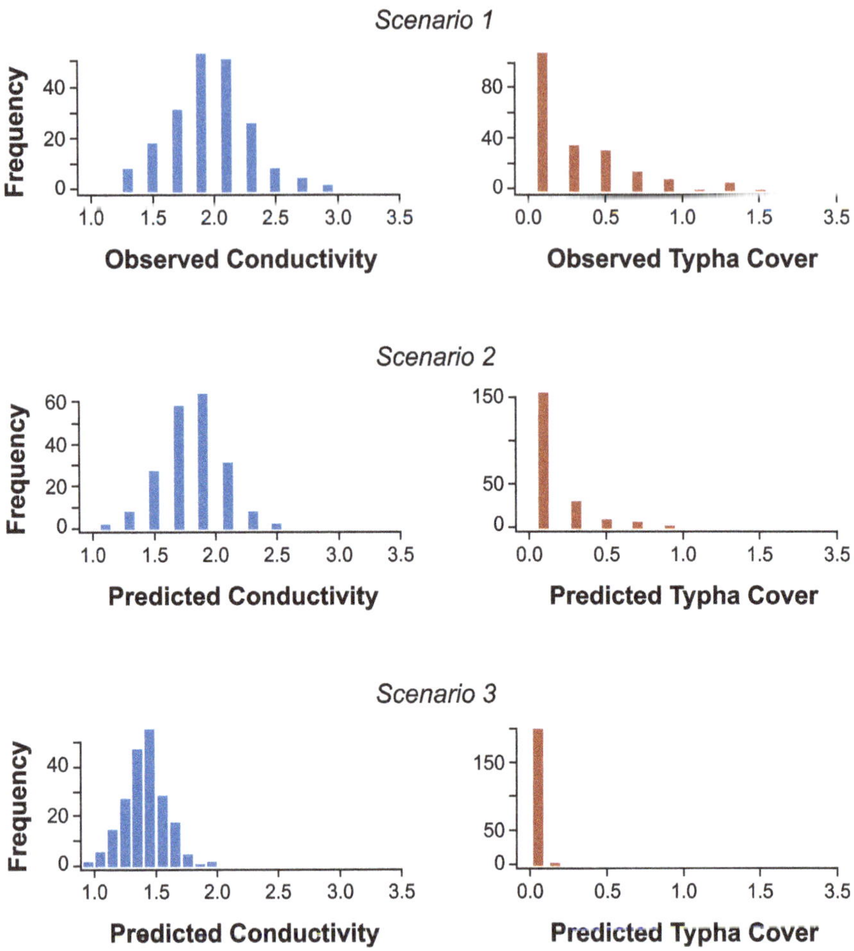

Fig. 7.10 Model predictions used to evaluate alternative management options for Acadian wetlands (from Grace et al. 2012). Scenario 1 is the "do nothing" baseline; scenario 2 entails physical removal of invasive plants from riparian zones, while scenario 3 uses direct treatment of water quality. Model predictions were generated from 200 replicates, stochastic realizations of model parameters based on their errors or estimation

hypothesis: that urbanization acted via habitat conversion, fragmentation, changes in environmental variables (e.g., temperature), and the actions of people (fertilization, planting ornamentals).

Lopez et al. (2018) evaluated the four filters hypothesis as a structural equation model (Fig. 7.12). They then estimated the model several times, focusing in turn on the response variable taxonomic diversity, as well as phylogenetic and functional type diversity, and also fitting the model for native as compared to exotic plant species. In all this, the topology of SEM serves as a sort of experimental control,

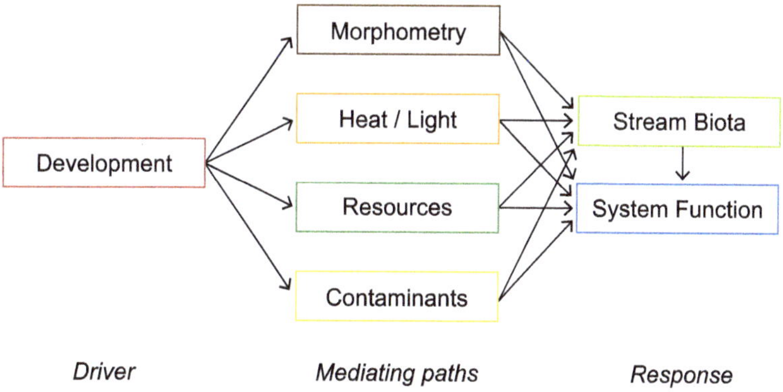

Fig. 7.11 The urban stream syndrome, emphasizing four mediating paths via which urbanization influences the biota and ecosystem function of urban streams. (Bernhardt et al., unpublished, after Poff et al. 1997; figure reproduced from Urban 2023, permission conveyed via Copyright Clearance Center, Inc.)

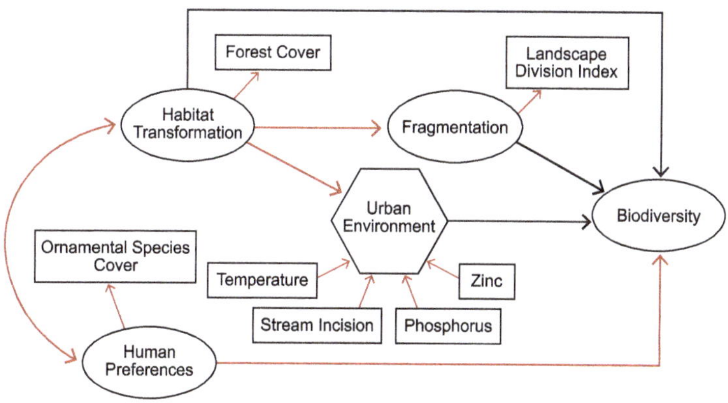

Fig. 7.12 The four filters hypothesis on urbanization effects on plant biodiversity (Williams et al. 2009), posed as a SEM (Lopez et al. 2018). Black arrows are positive; red arrows, negative effects. The ovals are posed as latent factors, and rectangles are measured variables; the hexagonal factor is a composite variable. (Reproduced from Urban (2023) and Lopez et al. (2018); permission conveyed via Copyright Clearance Center, Inc.)

allowing the several models to reveal the relative importance of the paths for the different aspects of biodiversity.

In their model, they opted to represent the urban environment as a composite variable (hexagon in Fig. 7.12), as they were interested in the relative importance of temperature, physical alteration of streams (incision), nutrients (phosphorus), and contaminants (zinc) as influences on plant biodiversity.

7.4.4 Experiments as SEMs

Finally, it should be underscored that SEM offers a powerful framework for evaluating ecological experiments (Shipley 2000; Grace 2006). In this, the path model poses the logic of the experimental design and the hypothesized mode of action. Fitting the model provides an explicit test of the path model, as well as the opportunity to discover alternative modes of action (or outcomes!) beyond those originally posed.

For example, Whalen et al. (2013) used SEM as a framework in which to evaluate the role of mesograzers in a coastal marine seagrass system. The SEM itself was a food web diagram, and the authors' interest was in the relative importance of top-down as compared to bottom-up control of the system. Here, eutrophication contributed a bottom-up influence, while experimental chemical treatment of the grazers perturbed the potential top-down control of epiphytic algae. The dynamics of the system were complicated by seasonal changes in grazer abundances, which emerge as a shifting pattern of control in the food web. SEM provided a useful framework in which to explore all of these issues.

7.5 Further Reading

Sokal and Rohlf (1995) and Legendre and Legendre (2012) cover original path analysis in some depth. There is a massive literature on structural equations, including very many books written by statisticians or social scientists and at least one dedicated journal (*Structural Equation Modeling*). Shipley (2000) and Gotelli and Ellison (2004) offer more up-to-date coverage of modern SEM for ecologists. Pearl (2009) provides a deep dive into causality. Grace and his colleagues have been quite active in introducing SEM to ecologists. Grace's chapter in McCune and Grace (2002, Chapter 30) offers an easy introduction for ecologists, including an emphasis on the role of factor analysis in SEM and how SEM relates to more familiar regressions. Grace's (2006) book similarly focuses on SEM applications for ecologists and evolutionary biologists. Grace and Bollen (2008) provide a useful introduction to latent and composite variables, while Grace et al. (2010) provide an in-depth discussion of SEM for ecologists. Grace et al. (2012) have provided a comprehensive guide to SEM that focuses on the path models. Grace also maintains a website (www.structuralequations.org, now hosted at https://www.usgs.gov/centers/wetland-and-aquatic-research-center/sci-ence/quantitative-analysis-using-structural-equation) on the topic, which features a wealth of tutorials and guides to software packages. Fan et al. (2016) provide a review of recent ecological applications of SEM.

7.6 Summary and Prospectus

Structural equation modeling is extremely well developed in some disciplines, especially in the social sciences, but it is a relative new approach for ecological models. SEM incorporates a logical path model as well as a measurement model and so is especially attractive in that it admits the multivariate nature of nature and embraces a multifaceted model of how nature works. This approach seems to be a natural fit for ecological applications that invoke concepts inviting latent variables (e.g., water quality, ecosystem health). It seems likely that SEM will become increasingly important as an analytic framework for ecological applications.

The path model that provides structure to SEM is especially appropriate for applications that pose a specific mode of action and expectation of management interventions framed as experiments. Such path models—even as conceptual models—are increasingly playing a role in adaptive management and structured decision-making (e.g., Olander et al. 2018; Qiu et al. 2018). We explore these in the next chapter.

References

Allan JD (2004) Landscapes and riverscapes: the influence of land use on stream ecosystems. Annu Rev Ecol Evol Syst 35:257–284

Blaszczak JR, Delesantro JM, Zhong Y, Urban DL, Bernhardt ES (2019) Watershed urban development controls on urban streamwater chemistry variability. Biogeochemistry 144:61–84

Booth DB, Roy AH, Smith B, Capps KA (2016) Global perspectives on the urban stream syndrome. Freshw Sci 35:412–420

Caressa S, Parris KM (2013) Multi-scale, direct and indirect effects of the urban stream syndrome on amphibian communities in streams. PLoS One 8(7):e70262

Fan Y, Chen J, Shirkey G, John R, Wu SR, Park H, Shao C (2016) Applications of structural equation modeling (SEM) in ecological studies: a review. Ecol Process 5:19

Gotelli NJ, Ellison AM (2004) A primer of ecological statistics. Sinauer, Sunderland, MA

Grace JB (2002) Structural equation modeling. In: McCune B, Grace JB (eds) Analysis of ecological communities. MjM Software, Gleneden Beach, Oregon, pp 233–256

Grace JB (2006) Structural equation modeling and natural systems. Cambridge University Press, Cambridge

Grace JB, Bollen KA (2008) Representing general theoretical concepts in structural equation models: the role of composite variables. Environ Ecol Stat 15:191–213

Grace JB, Anderson TM, Olff H, Scheiner SM (2010) On the specification of structural equation models for ecological systems. Ecol Monogr 80:67–87

Grace JB, Schoolmaster DR, Guntenspergen GR, Little AM, Mitchell BR, Miller KM, Schweiger EW (2012) Guidelines for a graph-theoretic implementation of structural equation modeling. Ecosphere 3:73

Hassett BA, Sudduth EB, Somers KA, Urban DL, Violin CR, Wang S, Wright JP, Cory RM, Bernhardt ES (2018) Pulling apart the urbanization axis: patterns of biophysical degradation and biological response across urban streams. Freshw Sci 37:653–672

Lefcheck JS (2016) Piecewise SEM: piecewise structural equation modeling in R for ecology, evolution, and systematics. Methods Ecol Evol 7:573–579

Legendre L, Legendre P (2012) Numerical ecology (3rd English edn). Elsevier, New York

Lopez BE, Urban D, White PS (2018) Testing the effects of four urbanization "filters" on forest plant taxonomic, functional, and phylogenetic diversity. Ecol Appl 28:2197–2205
Lu J, Calabria J, Hawkins GL (2024) Exploring potential mechanisms linking urban landscapes to instream stressors and macroinvertebrates in the Southeast Piedmont. Ecol Indic 163:112114
McCune B, Grace JB (2002) Analysis of ecological communities. MjM Software Design, Gleneden Beach, Oregon
Meyer JL, Paul MJ, Taulbee WK (2005) Stream ecosystem function in urbanizing landscapes. J N Am Benthol Soc 24:602–612
Olander L, Mason S, Warnell K, Tallis H (2018) Building ecosystem services conceptual models, National Ecosystem Services Partnership Conceptual Model Series No. 1. Nicholas Institute for Environmental Policy Solutions, Duke University, Durham, NC. https://nicholasinstitute.duke.edu/conceptual-model-series
Paul MJ, Meyer JL (2001) Streams in the urban landscape. Annu Rev Ecol Syst 32:333–365
Pearl J (2009) Causality, 2nd edn. Cambridge University Press, Cambridge
Petraitis PS, Dunham AE, Niewiarowski PH (1996) Inferring multiple causality: the limits of path analysis. Funct Ecol 10:421–431
Poff NL, Allan JD, Bain MB, Karr JR, Prestegaard KL, Richter BR, Sparks RE, Stromberg JC (1997) The natural flow regime. Bioscience 47:769–784
Qiu J, Game ET, Tallis H, Olander LP, Glew L, Kagan J, Kalies EL, Micha-nowicz D, Phelan J, Polasky S, Reed J, Sills EO, Urban D, Weaver SK (2018) Evidence-based causal chains for linking health, development, and conservation actions. Bioscience 68:182–193
Shipley B (2000) Cause and correlation in biology: a user's guide to path analysis, structural equations and causal inference. Cambridge Univ. Press, Cambridge
Sokal RR, Rohlf FJ (1995) Biometry—the principles and practice of statistics in biological research, 3rd edn. Freeman, New York
Somers KA, Bernhardt ES, Grace JB, Hassett BA, Sudduth EB, Wang S, Urban DL (2013) Streams in the urban Island: spatial and temporal variability in temperature. Freshw Sci 32:309–326
Urban DL (2023) Agents and implications of landscape pattern: working models for landscape ecology. Springer Nature AG, Cham
Walsh CJ, Roy AH, Feminella JW, Cottingham PD, Groffman PM, Morgan RP II. (2005) The urban stream syndrome: current knowledge and the search for a cure. J N Am Benthol Soc 24: 706–723
Whalen MA, Duffy JE, Grace JB (2013) Temporal shifts in top-down vs. bottom-up control of epiphytic algae in a seagrass ecosystem. Ecology 94:510–520
Williams NSG, Schwartz MW, Vesk PA, McCarthy MA, Hahs AK, Clements SE, Corlett RT, Duncan RP, Norton BA, Thompson K, McDonnell MJ (2009) A conceptual framework for predicting the effects of urban environments on floras. J Ecol 97:4–9
Wright S (1921) Correlation and causation. J Agric Res 10:557–585

Chapter 8
Site Prioritization

Abstract Prioritizing sites for protection or special management is a fundamental task in conservation practice. But even simple prioritization schemes can be complicated by other (sometimes competing) objectives. Here we begin with the task of assembling a system or reserves to represent the full set of targets as parsimoniously as possible (the minimum representation problem), as a way to illustrate various approaches to this task. The focus in this is on the greedy heuristic algorithm. We then extend this approach to consider additional objectives: species rarity, habitat quality or patch geometry, and connectivity. These alternative objectives invite structured decision-making in the form of multi-attribute decision analysis. Conservation practice often targets ecosystem services, which can be complicated by the engagement of a heterogeneous and spatially distributed set of stakeholders. The workflow for structured decision-making for these applications begins with goal-setting and works through the definition of objectives, empirical indicators for these objectives, a model of how the objectives might be attained, and a decision process that is as inclusive, deliberate, and transparent as possible.

8.1 Introduction

Site prioritization is the ranking of candidate sites in terms of some explicit criterion or criteria. That criterion might be biodiversity, contribution to watershed protection, scenic view, recreational value, or whatever. We will return to definitions shortly, but for now, a *criterion*, as used here, is a measure, a quantity (a noun). An *objective* relates to a criterion but is aspirational and directional: we might want to increase (a verb) biodiversity by a set amount over a given time span. Protecting land is an aspiration; various "30 by 30" initiatives (protecting 30% of the area by 2030) are objectives.

Prioritizing is assigning a ranking to various alternatives, with that ranking representing relative importance or urgency. The ranking might be to identify the best possible sites, sites to be conserved or protected; or we might be interested in identifying the worst sites, perhaps as candidates for restoration; or we might wish to balance these: the best sites on some criteria, if only they were restored on some

D. L. Urban, *Landscape Ecology: A Task-Oriented Perspective*,
https://doi.org/10.1007/978-3-031-72251-6_8

other criteria. A ranking is often implicit in or easily derived from a basic inventory (Chap. 1). For example, if we inventory plant or animal species, species richness provides for a simple and useful prioritization for conservation value. Similarly, habitat classification techniques that estimate habitat suitability (Chap. 2) also provide a simple prioritizing of conservation value for the focal species.

But site prioritization can address a heterogeneous set of objectives, and the task of prioritizing for multiple purposes often can lead to competing objectives that force decisions about which objective(s) we should favor and how to trade these off against other objectives. In this chapter we begin with the relative straightforward task of ranking sites in terms of their relative conservation value, with this value defined solely in terms of species richness. We then extend this task to consider a range of other criteria that might contribute to conservation value, again including diversity but also attending habitat geometry (patch size, amount of edge) and spatial context (especially connectivity).

Similarly, we can expand this discussion beyond the consideration of sites as the units of prioritization. We might also consider, more generally, management options such as alternative management practices that might help us meet our objectives (i.e., alternative *means* to these *ends*; see below). Or we might consider alternative policy instruments intended to help us reach our objectives.

This broadening of scope invites the guidance of structured decision-making, in this case, a multi-criteria decision framework (Gregory et al. 2012). This framework can be extended readily to multiuse objectives typical of most landscape management contexts (Huang et al. 2011; Hemming et al. 2022).

Recently, conservation planning has extended itself to a much broader set of targets beyond focal species or community types (*targets* are the units by which criteria are tallied). Increasingly, targets are framed in terms of ecosystem services. Ecosystem services are the benefits that ecosystems provide to humans (MEA 2005). Targeting ecosystem services is a seemingly simple extension of the same multi-criteria framework but is complicated by the explicit translation of ecological criteria to measures of stakeholder preferences and social impact. We close this chapter with a consideration of prioritizing sites for the provision of ecosystem services.

The emphasis here is on objective methods for estimating the relative value of management alternatives, in terms of explicit objectives and empirical indicators of these. This framework, in turn, will be a key element in the task of ecological assessment, the final chapter in this book.

In terms of the overall workflow of this book (Fig. 2 in Preface), site prioritization is a natural follow-up to inventory and, for focal species, species distribution modeling. In turn, prioritization invites continued monitoring and assessment over time. We turn to these in the next two chapters.

8.2 Context: Site Selection Logic

The task of site selection is a part of systematic conservation planning, which took root in the late 1990s and following decade (e.g., Pressey et al. 1993; Csuti et al. 1997; Ferrier et al. 2000; Poiani et al. 2000; Groves et al. 2002; Possingham et al. 2002; Moore et al. 2003). There has been enormous growth in this area, but the basics remain much the same even as the tools and applications continue to evolve (Moilanen et al. 2009; Game et al. 2013; Kukkala and Moilanen 2013; Tallis et al. 2017; Qiu et al. 2018). Here we are focused on site prioritization, an application that is narrower in scope than conservation planning but a key task in that larger process.

To begin, we might consider the aims of any analyses conducted to support site prioritization. There are three options. One is to actually *solve* the problem: to use an algorithm that considers the options and returns the identification of the best (optimal) solution. A second approach offers an approximation of the optimal solution (or several of these), which suggestion(s) might not actually be optimal (but will be nearly so). An alternative approach is to do analyses that will *inform* any decisions about priorities. In this, the aim is to consider the alternatives and provide some information about the relative merits of alternative choices—but without actually making a choice (i.e., the choice is left to the practitioner).

We will touch on all three approaches but focus on the last approach, *decision support*: providing objective information in support of decision-making. The logic in this is that many of the steps are the same, whether informing or optimizing. But many management decisions are not *optimal* but rather *defensible* given the particulars of the decision-making process. In land protection, for example, which properties are conserved depends on the conservation value of the sites...but this also depends on many other realities including landowner willingness to participate and available funding. The logic of decision support, in this context, is to inform the decision while retaining flexibility given all of the other variables that influence decisions.

Let us begin with the relatively straightforward task of selecting a set of candidate sites to constitute a nature reserve system. The goal of the reserve system is to include all of the targets (e.g., one occurrence of each species), in a minimum set of reserve sites (e.g., number of sites or total area or overall cost). This is known as a "covering" or "minimum representation" problem (Margules and Pressey 2000; Possingham et al. 2002; Moore et al. 2003). There are many ways to solve this problem, ranging from simple algorithms that can work by inspection of small systems to formal optimization procedures that require specialized expertise and software. Here we focus on an intuitive approach that will lead naturally to extensions we will consider later in this chapter.

8.2.1 Optimization (and Approximations Thereof)

There are various solutions to the minimum representation problem (Possingham et al. 2002). These include linear programming methods for optimization and a simulated annealing technique that works by iteratively refining an approximate solution.

Optimization: Linear Programming Linear programming methods (see overviews by Moore et al. (2003) and Önal and Briers (2006)) find the exact (optimal) solution to the minimum coverage problem, by posing this as an optimization that maximizes a single objective (e.g., representing all targets) while minimizing a constraint (e.g., total cost or area). This approach suffers two drawbacks: (1) it is computationally demanding and so is feasible for only smaller problem sets (but see below); and (2) the method is essentially a "black box" to most end users and so is difficult to interpret and communicate. Some conservation organizations explored optimization decades ago but largely abandoned the approach in favor of less demanding and more intuitive methods.

Computational complexity is becoming less problematic given modern computing power. Newer implementations of the approach (Hansen et al. 2022) are much more promising, as clever coding allows the approach to be extended to multiple objectives (e.g., by coding a second objective as a function of the first).

Approximately Optimal: Simulated Annealing An alternative to strict optimization is an approximate solution to the optimal solution. One popular implementation of this approach is MARXAN (Possingham et al. 2002; Ball et al. 2009). In this, the program attempts to optimize a set of objectives subject to a set of constraints. In MARXAN, the aim is to minimize an objective function defined as the sum of:

- Cost per site (sites are termed "planning units")
- A user-provided penalty for not meeting conservation targets (e.g., not capturing all species; priorities for species or targets may be defined via a "species protection factor" that favors some targets over others)
- An optional user-provided penalty for not meeting a total cost threshold
- Optionally, a user-provided premium on boundary lengths shared among selected sites
- Optionally, other considerations (minimum parcel size, distances among sites)

The boundary length option was originally intended as a means of increasing connectivity, but in effect it tends to generate compact clusters of adjacent sites—efficient for stewardship and management but not really connectivity in the common use of that term. Newer approaches with MARXAN contribute more directly to connectivity (Daigle et al. 2019).

The penalties and other modifiers are specified in a common currency for programming convenience, and there is some craft to specifying these. Fortunately, the software is very well supported (https://marxansolutions.org) and there is a large and international users group who also helps support applications.

The MARXAN algorithm, termed *simulated annealing* (a reference to metal-working, in which iterative heating and cooling temper or harden the metal), is a successive approximation method. In this, sites are randomly entered into or removed from the solution set (reserve system), with changes evaluated in terms of the objective function. Rather large and random changes are allowed initially, but later in the process, the thresholds for allowable changes and the retention of these become more restrictive (this is the "annealing" bit). At the end of the process, MARXAN uses a greedy heuristic (see below) to "finish" the solution. The end result is a solution that is nearly optimal; but the approximation algorithm (like all such methods) cannot guarantee an exact solution. By contrast to strict optimization, the approximation is very fast and can handle very large data sets.

In practice, the model typically is run many times, and a count of how often each site is selected as part of the solution is tallied. This tally, converted into a proportion, is termed "summed irreplaceability." Sites with very high value are nearly always selected (a site with a score of 1.0 would be in every solution, hence "irreplaceable"), while sites with very low scores would be, by consensus, of low conservation value. The advantage of the collection of solutions is that the end user is provided with a set of highly ranked sites, with some flexibility in terms of protection priorities.

8.2.2 The Greedy Heuristic Algorithm

A greedy algorithm is one that finds a solution as rapidly as possible (most numerical approximation algorithms are greedy). A heuristic algorithm "learns" as it progresses, so that later decisions in the process depend on decisions at previous steps. A greedy heuristic algorithm finds a solution quickly, by making at each step a decision that is optimal (although perhaps not uniquely so) given the previous steps.

The algorithm can be explained using a simple example. Consider a set of six candidate sites that collectively support seven species (Table 8.1). Inspection of the data matrix indicates that site 6 should be selected first, as it has the most species (four). For the second step, the algorithm now looks at the remaining sites and tallies, for each site, the number of *new* species it holds—species that are not already

Table 8.1 Hypothetical example of site selection via greedy heuristic algorithm

Site	SppA	B	C	D	E	F	G	Sum
1	1		**1**				1	3
2	1	1	1					3
3		1	1					2
4			1			1		2
5		**1**		**1**				3
6	**1**		**1**		**1**	**1**		4

Sites would be selected in order 6, 5, or 1 to capture all species on as few sites as possible. Bold entries are the novel species added at each step

Table 8.2 Species by site for the Columbia Plateau ecoregion (excerpted from Table 17.1 from Possingham et al. (2002))

Species	Site 2	3	5
Loggerhead shrike	1	1	1
Western burrowing owl	1	1	0
Grasshopper sparrow	1	0	1
Ferruginous hawk	1	1	0
Sage thrasher	1	1	0
Western sage grouse	1	0	1
Sage sparrow	1	1	0
American white pelican	1	1	0
Bald eagle	0	0	1
Forster's tern	0	1	0
Total species richness	8	7	4

Only three of the ten sites are shown from the original. (Copyright Springer-Verlag, permission conveyed via Copyright Clearance Center, Inc.)

represented in the reserve system (i.e., on site 6). While there are three sites with three species each, some of these are already on site 6. Site 5 has two new species and is selected next. This updating then repeats, and we search for the site that will provide the most new species that are not already represented in the reserve system (i.e., on either site 6 or 5). The third site to be selected is site 1, which provides a single new species. At this point, the reserve system is complete in that it has all seven species. This example illustrates both aspects of the greedy heuristic algorithm: at each step, the relative merits of other sites depend on which sites (and, so, which species) have already been made (the heuristic part), and because the biggest improvement is selected at each step, the algorithm finds the solution as quickly as possible (the greedy part).

To be clear, in this example all species are considered equivalent in terms of conservation value. We will revise this assumption shortly. The criterion is simple species richness, and the objective is to maximize this in the fewest sites.

The key to a greedy heuristic is *complementarity* (Kukkala and Moilanen 2013). The best site, at any iteration, is the site with the highest value on the criterion (here, species richness) while also being as different as possible from sites already in the reserve system. These are the sites that confer the most novel contribution of value.

The greedy heuristic approach is appealing in its simplicity, and indeed, for small systems the correct solution can often be identified easily by inspection. A drawback to the approach is that it is not guaranteed to find the correct answer—even for small systems. Possingham et al. (2002) illustrated this failing with a simple example of ten species on eight censused sites (Table 8.2). The problem arises, in this case, because the logical first choice—the site with the most species, site 2—has species that are completely redundant with other sites. In this example, two other sites (3 and 5) would support all the species, making the logical first site unnecessary. This failure is not peculiar to this case but typical of all sequential solutions to the minimum representation problem (Possingham et al. 2002).

8.2.3 Extensions to Greedy Heuristics

While the greedy heuristic algorithm is an imperfect solution to site prioritization, its simplicity makes it a useful as a means to illustrate more general issues in site prioritization. These extensions include a variety of other criteria invoked as objectives for conservation. A partial list (adopted loosely from The Nature Conservancy 1996; Groves et al. 2002) might include the following:

- *Ecological uniqueness*: Species rarity, endemism, or vulnerability to natural or human-caused stressors; here, the logic is that uniqueness confers a higher conservation value.
- *Habitat geometry*: Patch size, core area, edge or perimeter length, or other aspects of patch shape or configuration; these concerns are often motivated by edge effects that confer a higher conservation value to larger or more compact (less edgy) sites (reviewed by Urban 2023, Chaps. 7 and 8).
- *Connectivity*: This might refer to large-scale connectivity, near the scale of the geographic range of a species; more locally, *permeability* might be the focus, as it affects local movements (Anderson et al. 2016, 2023). In either case, higher connectivity confers a higher conservation value.
- *Threats*: Regional pressures such as encroaching human development, or, more recently, climate change; the relative conservation value of sites under high threat might be equivocal: practitioners might react to threat with greater urgency, or they might be reluctant to invest in sites viewed as lost causes.
- *Feasibility*: Issues related to the logistics of acquiring (cost) and protecting the site (including long-term stewardship). Beyond costs, sites that are adjacent to other protected sites are often easier to manage and steward over time—a simple matter of travel time and efficiency.

Each of these can be addressed using a greedy heuristic algorithm, with some minor modifications. For example, species rarity can be targeted, for small systems, by inspection; this amounts to targeting species that occur on only a single site first and then complementing these sites with other sites that offer new species that are less rare (Possingham et al. 2002). More generally, if each species is coded in the data table with a "rarity score" that takes on larger values for rarer species (e.g., a score of "10" meaning 10 times as rare as a common species with a score of 1), then the simple greedy heuristic outlined above would capture rare species first and common species only after the rare species had been accounted. Any scoring scheme that scales the same way could work similarly for endemism, vulnerability, or other attributes of concern.

Habitat geometry or condition can be used in a greedy algorithm easily, if the conditions are coded as relative or effective area. In such cases, a greedy pursuit of total area will accomplish one such aim. Beyond this, if core (interior) habitats are more valued than edges, then tallying core area instead of total area will do what is desired. For habitat condition, habitat suitability as modeled on a scale of [0,1] (as with many species distribution models; Chap. 2) can be used as a proxy for

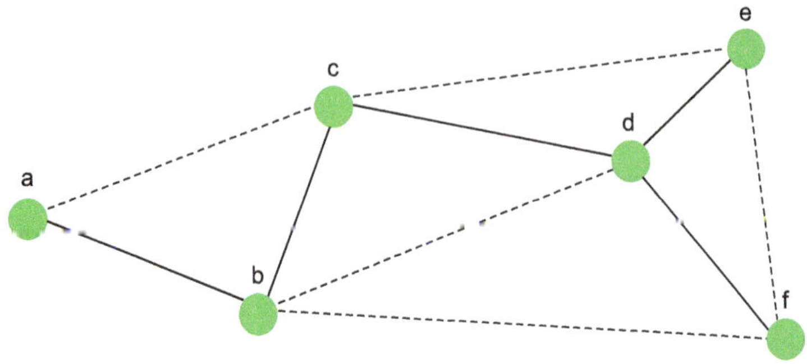

Fig. 8.1 Greedy algorithm for connectivity, as illustrated by the construction of a minimum spanning tree for a set of sites. The selection is for the shortest connecting distance from any site to any site already in the reserve system. Here, that sequence is DE, DF, DC, CB, BA. Dotted lines are candidate connections that are not part of the minimum spanning tree

effective area. That is, if each cell in a (raster) habitat patch is coded on [0,1], then the sum of those suitabilities is essentially an effective area: a 1-ha patch of cells with suitability 1.0 has effective area 1.0, while lower suitabilities result in a lower effective area. Any estimate of suitability can be used this way, so long as it scales appropriately as effective area.

Habitat connectivity is an intriguing illustration of this general algorithm. For example, consider a set of candidate reserves that are at various locations within the study area. One way to maximize connectivity of the reserve system would be to minimize the collective distances between pairs of sites in the system. How to find this connected set? If we tally the distances between pairs of candidate sites, then a greedy heuristic algorithm that minimizes the distance from a new site to any other site already in the system will accomplish the objective. That is, begin with the pair of sites that are closest together. Then add to the system the site that is closest to either of the sites already in the system and so on. This process assembles a *minimum spanning tree* for a graph of the reserve system (Urban and Keitt 2001), and indeed, this is the classic and optimal algorithm for finding a minimum spanning tree (Prim 1957; Fig. 8.1). (As with any sequential solution, it is also true that this algorithm will find *a* minimum spanning tree but not necessarily a unique one.)

The emphasis here on greedy heuristics is not because the approach is infallible, but rather because it is intuitive. Importantly, the heuristic aspect of the algorithm also emphasizes that it is the relative change or *marginal value* provided by a site that confers its value to the reserve system being assembled.

Urban (2002) devised a decision support tool, PORTFOLIO,[1] for conservation planning that is based on a multi-criteria implementation of a greedy heuristic algorithm. The program includes estimates of several components of conservation value:

---

[1]The name is deliberate, to connote a reserve system as a collection of holdings with various risks and rewards. The aim is to maintain a balanced portfolio. The software is available from the author.

- *Species richness*, as a simple tally of the number of species. In a greedy algorithm, only *novel* species not already in the reserve have incremental value. That is, the focus is on *complementary* richness.
- *Species rarity*, based on a user-defined estimate of relative rarity (higher scores for rarer species). Rarity can be scored simply (any *supplementary* rarity is good) or as complementary rarity (only novel rarity is scored).
- *Habitat patch geometry or condition:* as discussed above, any measure of total area, core/edge geometry, or habitat condition, expressed in terms of relative or effective area.
- *Connectivity*: from graph-based models, the algorithm attempts to maximize:

$$C = \sum_i \sum_j A_i A_j P_{ij} \tag{8.1}$$

for all sites i and j in the system, based on habitat areas A and dispersal likelihoods

$$P_{ij} = e^{kd_{ij}} \tag{8.2}$$

for between-distance d_{ij}. Here, k is a distance-decay coefficient ($k < 0.0$) that describes how steeply dispersal likelihood decreases with increasing distance. This approach is similar to the approach for a minimum spanning tree (Fig. 8.1) but includes terms for a "donor" as well as a "target area" effect (Hanski and Ovaskainen 2000). The aim is to maximize total (probabilistically) connected area.

The program also accepts optional indices of site-level habitat heterogeneity or environmental buffering capacity, threat, and acquisition cost. At each iteration, the program displays the incremental *change* in each metric that would result from the addition (or removal) of a candidate site in the reserve system. The user is prompted to choose a next site, and the process updates.

Backward and Stepwise Greedy

One useful embellishment of the greedy algorithm is incorporated into PORTFOLIO. This is the capacity to run *backward*. That is, once a few sites are included in the reserve system, the program also tallies the relative (incremental) values that would obtain if any site already in the reserve system were to be removed. Thus, the common result with a greedy algorithm—that a site chosen early in the process might become redundant later—is solved by allowing redundant sites to be discarded or swapped for another site.

The backward approach to a greedy algorithm is important in another instance. With connectivity, for example, a small stepping-stone patch might not contribute substantially to overall connectivity if considered as an incremental addition to the reserve. By contrast, that stepping-stone—if it connects sets of patches that are themselves locally connected—can show an enormous impact on overall connectivity if deleted (red patch in Fig. 8.2). This result reflects a subtle but important shift in focus in this algorithm; the focus can be on site-level criteria (i.e., incremental values

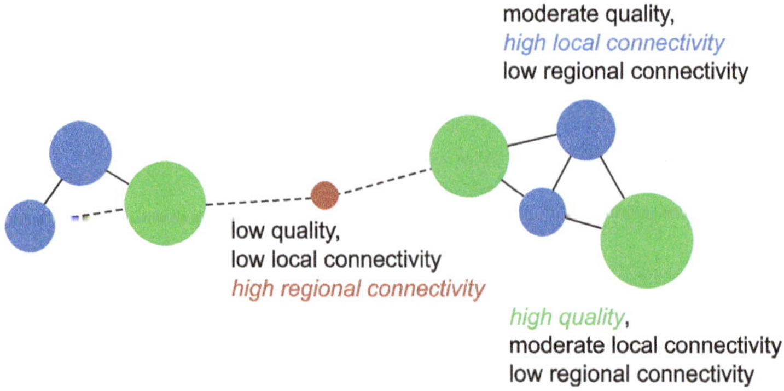

Fig. 8.2 Illustration of the relative importance of local as compared to regional connectivity, in the context of variation in patch size and habitat quality. Patches might be valued due to habitat quality (green patches), local connectivity (blue), or long-distance connectivity (red). In particular, the red patch's value can be seen most readily if it is removed from the system, disconnecting the patches on the left side from those on the right

of sites); or the focus can also reflect the role of individual sites in the larger context of the reserve system (Eq. 8.1). Stepwise processing is not crucial to discovering sites with network-level importance—but it often makes it much easier.

This approach is equivalent to a "full stepwise" approach to multiple regression, which greatly expands the capacity of the algorithm to explore the relative value of sites in the larger context of the reserve system (Urban and Keitt 2001). For example, this approach might be used to assess the contribution of sites to erosion potential or watershed protection (both of which are context-dependent), or to explore viewsheds (again, context matters).

PORTFOLIO was intended primarily as a teaching tool, and it nicely illustrates many of the issues that arise when attempting to assemble a system of reserves. At the simplest level, the selection protocol emphasizes the incremental value represented by any single site. But these incremental values can be hard to isolate in the aggregate. For example, sites that are valuable for any given reason (e.g., richness, rarity, or connectivity) might also be valuable for other reasons. These instances of *leverage* or *co-benefits* are not part of the greedy algorithm (although a decision support tool should reveal them). In PORTFOLIO, these instances often arise when there are sites that are tied in terms of their value on a given attribute and so invite further consideration of "tiebreaker" rules. For example, we might target species richness but invoke rarity or connectivity in cases where candidate sites offer the same incremental value on richness. Reciprocally, the greedy exercise often reveals that focusing on one criterion rather than another (e.g., richness versus rarity, rarity versus connectivity) leads to competing objectives that imply a need to make trade-offs among the objectives.

These two key concepts—marginal values and multiple, competing objectives—provide an easy entry into the realm of structured decision-making and multi-criteria decision analysis (Huang et al. 2011, Gregory et al. 2012; Hemming et al. 2022).

This framework underlies much of site prioritization, although the tools might not be invoked explicitly in decision-making.

8.3 Structured Decision-Making

Structured decision-making, as the term implies, is a process for injecting rigorous logic and transparency into decision-making (Clemen 2001; Gregory et al. 2012; Johnston et al. 2015; Hemming et al. 2022). The aim is to contrast a set of management (decision) alternatives in terms of their relative (marginal) values on each of several criteria or objectives. This is a large topic with a variety of approaches, but it can be summarized in a few key steps. These steps include (1) defining the problem clearly, (2) identifying alternative actions and the likely outcomes of these actions, and (3) defining how these alternatives will be evaluated (Johnston et al. 2015).

The framework invites a large set of processes and supporting tools (Hemming et al. 2022), but we will focus on a few key pieces:

1. The objectives and how these will be measured (an *objectives hierarchy*)
2. A model that suggests how the alternatives will effect changes in the objectives (a *means-ends* model)
3. A summary table or *decision matrix* that captures this information to inform the decision

The overall workflow implied by these steps underscores the discrete components of the analysis while also emphasizing the iterative and adaptive nature of the process (Fig. 8.3). This is also a simplified version of the process, in that in real

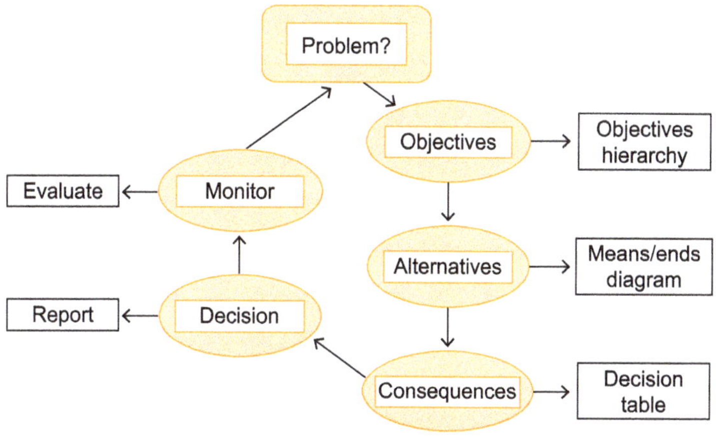

Fig. 8.3 Workflow for site prioritization, as a structured decision-making process. This process maps neatly onto the adaptive management framework, in that it begins and ends with an explicit consideration of overall goals and whether these are being met

applications, there are often iterative revisions that would be represented more honestly (if messily!) by additional arrows connecting most steps of the overall process. And while this workflow applies to structured decision-making, it might also apply to site prioritization more generally.

8.3.1 Objectives Hierarchies

An objectives hierarchy outlines the objectives in terms of a larger, overarching goal while also indicating how each objective will be measured. For example, an objectives hierarchy for a conservation program might identify biodiversity preservation as its goal. Specific objectives might include biodiversity itself, as measured in a variety of ways: species richness, rarity, endemism, or (lack of) exotic or invasive species; alternatively, one might index biodiversity potential in terms of biophysical proxies or habitat variety. Similarly, one might consider habitat geometry in terms of size, shape complexity, or amount of edge. Conservation planning also addresses spatial context, such as encroaching developing threats or habitat connectivity (Fig. 8.4). Again—though for simplicity the diagram does not show this—objectives are directional: the aim is to increase or decrease the values that measure the criteria (attributes). That is, we want to increase richness or rarity, increase the variety of proxies, increase patch area, decrease edge, increase connectivity, and decrease threat. Objectives also are specific in terms of desired change (how much) and timescale (how soon).

The specific indices used to measure objectives are *indicators* or indicator variables. Good indicators in general, and ecological indicators in particular, share a number of attributes: they are quantifiable with low measurement error, are discretely scaled and repeatable over time and space, have low observer bias, and—

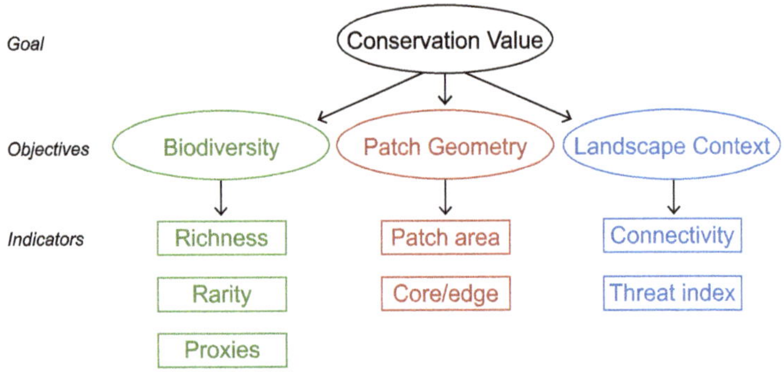

Fig. 8.4 An objectives hierarchy for conservation planning, targeting three objectives, each with two or more empirical indicators. This scheme corresponds to the decision support tool PORTFOLIO (Urban 2002). Here, ovals represent concepts, while rectangles are empirical indicators

most importantly!—capture the information they are intended to capture, unambiguously.

In an objectives hierarchy used for decision-making, it is also important that the indicators (and the objectives they serve) are logically separable in that they represent qualitatively different things of interest (i.e., not redundant measures of the same thing). For example, in Fig. 8.4, the indicators for biodiversity, patch quality, and landscape context can be separated logically and empirically.

Devising an objectives hierarchy can get subjective, in cases where it is not clear whether two indicators both apply to the same objective or if they are sufficiently distinct as to warrant being separate objectives or sub-objectives. For example, in Fig. 8.4 are connectivity and threat both indicators for a common objective, landscape context? Or might these be elevated to the level of objectives? The means-ends model (below) can help guide this choice about representation. Keeney and Gregory (2005) offer additional guidance on choosing how to measure objectives. In practice, the act of declaring the hierarchy—especially if done in open collaboration with other stakeholders or decision-makers—can help guide these decisions.

The identification of objectives and indicators recalls factor analysis (Chap. 4), although in an objectives hierarchy these indicator relationships might not be developed statistically in the formal way that factor analysis is conducted. The relationship between objectives and their indicators also evokes the relationship between latent variables and their indicators in structural equation models (Chap. 7).

An objectives hierarchy is a simple, graphic declaration of what is *valued* in the decision context: it summarizes what we want to accomplish and *why*. The means-ends model (below) describes *how* we will accomplish these objectives.

8.3.2 *Means-Ends Models*

Path models are diagrams that show how management actions (or policy alternatives) effect a change in the criteria identified in the objectives hierarchy. These models go by many names (e.g., path models, causal chains; see Qiu et al. (2018)), but *means-ends models* nicely convey their purpose in illustrating how management options (the *means*) might achieve the objectives (the *ends*). While the objectives hierarchy is a static *measurement* model of what we want, the means-ends model is a *process* model that focuses on actions and their consequences, of how we might get what we want.

It might be apparent here that the process of generating an objectives hierarchy and a means-ends diagram is exactly analogous to the prototype modeling stage in the workflow of structural equation modeling (Chap. 7, Fig. 7.5 and Sect. 7.3.3). What is different here is that the means-ends model might not ever be implemented and parameterized from data: it is often used simply as a way to organize the logic of the decision process (but see below).

Consider the specific instance of an objective to increase bird species diversity. We might envision a variety of paths (Fig. 8.5). In one case, increasing habitat patch

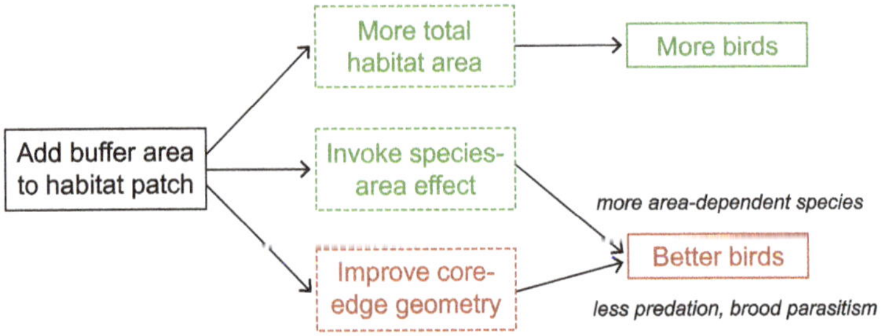

Fig. 8.5 Path diagram to show how adding habitat area might lead to higher bird species diversity. *Top (green) paths*: More total area supports more birds overall, which should lead to more species via passive sampling; this would also support more species that require large habitat patches. *Bottom (red) path*: Reducing edge would lead to reduced nest predation and brood parasitism (both edge effects) and so help support species vulnerable to these pressures. The diagram makes clear that the result of buffering patches is mediated by three distinct mechanisms

size or total habitat area would increase the number of individual birds, and, by chance, we would expect to find more species (a sampling effect implicit in the species-area effect; see Urban (2023), Chap. 5). We would also expect to support more species that have larger area requirements because they require large nesting territories or forage widely (e.g., raptors). By contrast, if we were to manage patch buffers to reduce forest edge, we might expect to reduce nest predation and brood parasitism by edge species and thus increase the likelihood of supporting vulnerable species (birds using low and open nests, or preferred hosts for brood parasites) (Fig. 8.5, bottom tier).

This example also illustrates two important points about how means-ends models and objectives hierarchies relate to each other. First, the objectives hierarchy contains only the final targets, not any intermediary conditions. That is, the hierarchy makes clear how the ends relate to each other, but not how they might be met. The path model makes explicit the way that we think managing habitat geometry might help us meet our objective in terms of bird diversity: we could increase patch size, or buffer edges, to improve patch geometry. There would also be path models for the objectives related to the other objectives in the hierarchy, i.e., diversity and landscape context (Fig. 8.4). These ultimately could be combined into a single (busy!) path model.

Second, the path diagram makes clear why we might choose to include two different indicators related to bird diversity: it is because different species are affected via different paths in these two examples, and so these are not redundant indicators. Likewise, a path model about landscape context would invoke different mechanisms for the effects of threats as compared to connectivity (although encroaching development might also influence connectivity).

It is worth underscoring that the path diagrams shown here are mostly conceptual. While based on reasonable expectations based on decades of research on birds in forest patches (Urban 2023, Chaps. 5, 6 and 7), it would be better to develop the model more fully so that it is empirical (e.g., as a regression) and so capable of predicting changes in species diversity given specific management actions. The end result of this process might be a structural equation model (Chap. 7) capable of predicting the consequences of management actions as these propagate through the system.

Current conservation applications tend to rely on conceptual diagrams more than fully parameterized structural equation models. But there is some movement toward increasing the rigor of the conceptual models, including efforts to quantify confidence in the component paths or submodels (e.g., Qiu et al. 2018). *Evidence grading* is one example of this trend (e.g., Tallis et al. 2017). In this, each component (path) of the conceptual model is assessed in terms the empirical evidence supporting it. This evidence might range from expert opinion (rigorously elicited) to formal meta-analysis from systematic reviews.

That said, practitioners do use, directly or indirectly, a wide range of models in support of conservation planning and decision-making. Zurrell et al. (2022) provide a useful review and synthesis, along with some guidance on applications.

8.3.3 The Decision Matrix

The aim in structured decision-making is to generate a table or *decision matrix* that contrasts alternative management options in terms of their expected effects on the specific indicators for the objectives. One way to arrange this is as a table with objectives (indicators) as its rows and management alternatives as the columns.

In the hypothetical example being developed here, we might want to contrast a management action that increases habitat patch area by making existing patches larger and more compact (e.g., by buffering them and reducing edge), compared to a plan to increase connectivity by connecting existing habitat patches via stepping-stone patches or corridors (e.g., wooded fencerows or corridors created by reforesting riparian zones). The other option would be to do neither, a business-as-usual default. A decision matrix would summarize these contrasts (Table 8.3). As neither management alternative concerns biophysical proxies or threats (as in Fig. 8.4), these criteria are not included in the table.

Note that we have included two indicators each for diversity and habitat geometry and one for connectivity (Eq. 8.1). The initial task is to fill in the cells of the table with estimates of the changes—in ecological units—expected from each of the alternatives. The "do nothing" scenario might be rather uninteresting unless we actually expect changes if we do nothing (e.g., if encroaching development implies a loss of species if we do not act soon). To fill in the table, we would need empirical estimates of the changes we would expect in each instance. This is, of course, why we would like to implement the means-ends diagram as an actual, working model! It is worth emphasizing that many forms of models might meet this need: expert systems, regressions, or simulation models might all provide the desired estimates.

Table 8.3 A decision matrix template for the hypothetic case of two management options to attempt to increase conservation value

Indicator	Do nothing	Buffer area	Corridors
Richness			
Rarity			
Habitat area			
% edge			
Connectivity			

Weighting Schemes: Multi-attribute Utility Theory

Decision analysis borrows heavily from multi-attribute utility theory (MAUT); indeed, it might be useful to think of multi-criteria decision analysis as an applied version of the more abstract MAUT (Gregory et al. 2012). An important step in this analysis is quantifying the relative importance or value of the alternatives itemized in the decision matrix (Table 8.3).

There are various ways to do this. Because we are developing examples that include multiple objectives as well as multiple indicators for each objective, we will adopt a two-stage approach to weighting. In the first stage, we will quantify the relative contributions of each indicator to a given objective. In the second stage, we will weight the objectives themselves. In practice, these stages are often combined but it will be helpful here (perhaps!) to separate them.

Typically, the first stage amounts to creating a weighting scheme that links the indicators to their objectives. Perhaps the most common approach is a simple weighted sum as an estimate of the score S for each site or option for an objective:

$$S = w_1 x_1 + w_2 x_2 + w_3 x_3 + \cdots + w_p x_p \tag{8.3}$$

where the w's are weights for each indicator x. The weights are usually normalized or relativized to yield scores on [0,1] or [0,100], to make it easier to compare scores across objectives.

In the second stage, the objectives are weighted to reflect the decision-maker's relative preferences across objectives. To be clear, the decision-maker or stakeholders prefer *all* of the objectives—that is why they were selected as objectives at the outset. But in instances where all objectives cannot be met simultaneously, one has to declare their relative importance. In practice, we might use another weighted sum (Eq. 8.3) to assign weights to each objective. The result of this second stage of weighting is a final score or ranking for each site, reflecting its performance on each of the objectives and on each of the indicators for each objective. Again, this can be done in a single pass but at some risk of loss of clarity.

In terms of decision support, the use of a weighting scheme is a substantial change from the starting point of ranking candidate sites on each criterion separately. To return to the example software PORTFOLIO, at each iteration of the decision process (choice of a site), the relative value of each site on each criterion (indicator) is displayed, and the user chooses a site. In this, the user decides how to weigh the different indicators and objectives and then makes a decision; then the process repeats. Using an overall weighting scheme provides a means of combining objectives and declaring a relative importance of these in the aggregate. But this combined weighting also might obscure important differences among sites. And again, the relative merits of different objectives themselves might change as a portfolio of sites is assembled: e.g., diversity might be the initial focus, with connectivity coming into play after the rarest species are accounted. We return to this issue below.

Weighting Schemes: Geospatial Models

A common implementation of multi-criteria site prioritization is done in a geographic information system (GIS). In this, each candidate site or parcel is characterized in terms of geospatial indicators for the various objectives of the application. In the GIS, the *attribute table* for the candidate sites (e.g., as a set of parcel polygons) comprises a spreadsheet of all of these indicator variables, for each site. It is a relatively simple matter to sort the spreadsheet on any selected column, to highlight the highest-ranking sites on each attribute. It is also straightforward to invoke "map algebra" to assign weights to each attribute (Eq. 8.3), to compute a weighted sum as an overall ranking among sites. This is a straightforward implementation of multi-attribute decision support.

As an illustration, the Triangle Land Conservancy in central North Carolina has four programmatic foci: protecting natural habitats, ensuring clean water, preserving working farms, and connecting people to nature (www.triangleland.org). They have prioritized sites on each of these criteria, using multiple geospatial data sources and weighting schemes based on staff's engagement with local stakeholders. These rankings can be used separately or in combination; for example, overlaying the habitat and water priorities reveals sites that rank highly on both objectives (and so provide co-benefits) as well as sites that rank highly on only one objective (and might invite trade-offs) (Fig. 8.6).

Implementing an objectives hierarchy via map algebra makes the weighted averaging within objectives clear, in that some weighted combination will generate a new GIS data layer that represents each objective. Similarly, a tool that assigns relative weights to the objectives themselves helps make this decision transparent. In practice, it is usually convenient to code the GIS model to make it easy to modify the weighting scheme both *within* and *across* objectives. This makes it easier to explore uncertainties about the weights, or to adjust the weights to particular decision contexts.

In land conservation the funding options might be restricted to particular programs so that, for example, water quality projects are eligible for program funding while species diversity projects are not. In a GIS model, it is straightforward to set the weights accordingly, to reveal those sites that would be highly prioritized under program guidelines.

Similarly, some funding sources might be restricted to particular jurisdictions (e.g., co-managed watersheds, counties, or other administrative units). In a GIS, the

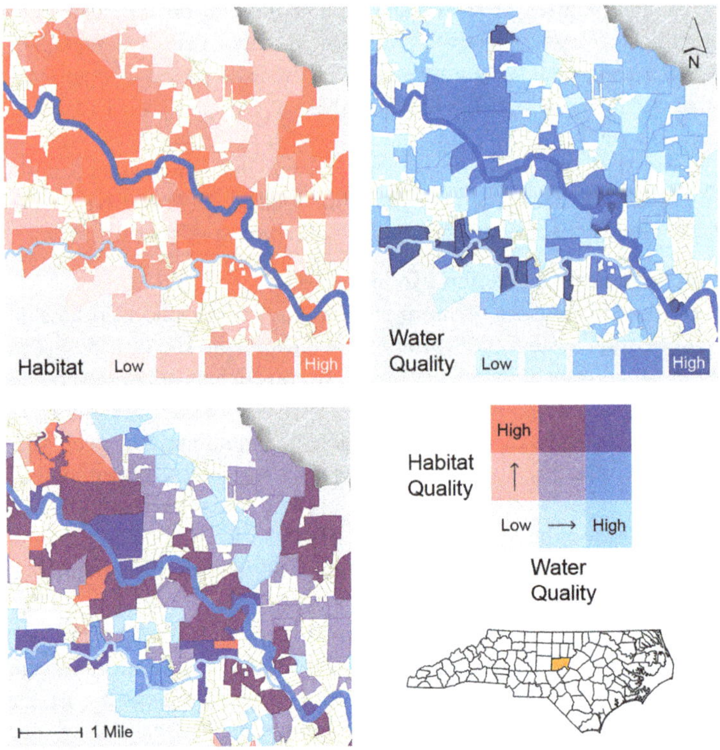

Fig. 8.6 Geospatial site prioritization for the Rocky and Deep River catchments in the Triangle region of the North Carolina Piedmont, USA. The color scheme highlights sites that are highly ranked for habitat quality (red scale) as compared to water quality (blue scale); sites highly ranked on both metrics, indicating co-benefits, are scaled to purple. Inset map shows watershed location in central North Carolina, USA. Image courtesy Chloe Ochocki, Triangle Land Conservancy

decision context can be clipped to that jurisdiction so that only local sites are put forward as options.

What is not as straightforward in this approach is the sequential nature of the selection algorithm and how the rankings of sites might change as sites are selected for protection for inclusion in the reserve system. To be clear, this approach of multi-criteria prioritization via map algebra is greedy in the sense that it reveals the relative ranking of alternatives in terms of the criteria. Often, this is sufficient for planning purposes: the practitioner merely needs to have a list of high-priority site to pursue over time. But a model implemented this way is not *heuristic*—the rankings are static, at the time of the assessment. To be heuristic, the rankings need to be re-evaluated at each iteration, to reflect changes in the decision context. This updating needs to be done deliberately (i.e., probably manually) in the GIS-based version of the approach. For example, to evaluate the value of a candidate site in terms of species richness, a multi-attribute model would identify the site's species richness, but it would not know which of those species are not already represented on other sites in the portfolio. Likewise, a site's importance to connectivity depends on which other sites are in the network. . .and this information needs to be updated as sites are accrued.

Algorithm: Recursive Site Deletion One way to make a simple multi-attribute ranking system more sensitive to system-level status is to use a recursive site-deletion algorithm (Keitt et al. 1997; Urban and Keitt 2001). In this, an index of conservation value is tallied over the entire collection of candidate sites, and this index is saved for reference. Then, each site is removed in turn, the index is recomputed with that site omitted, and the difference between that index and the (full) reference case is saved. Sites can then be ranked on the *change* in the index resulting from that site's removal. While Keitt et al. (1997) applied this to connectivity, the approach can be generalized to any system-level measure: richness, rarity, connectivity, watershed protection, viewsheds, and so on (Urban and Keitt 2001).

This approach could also be run in a forward-looking mode. Given a set of sites already protected, and a set of new candidate sites for protection, each of the candidates could be added to the system in turn and scored on the basis of its incremental improvement to the overall network.

This approach would make a simple multi-attribute ranking system behave more like a greedy heuristic algorithm, with each site's value assessed relative to the rest of the sites in the system.

In a simple sense, the weighted scores for the alternatives in the decision matrix—however derived—suggest the "correct" decision: the highest-ranking alternatives. But real applications are rarely so simple. In particular, applications can get complicated when stakeholders enter into the decision process.

8.3.4 Ecological Performance and Stakeholder Preferences

The decision matrix above is interesting and might even be useful. . .but it is not the desired endpoint here. What Table 8.3 summarizes is *ecological performance*, or, rather, changes in ecological performance expected from each of the management alternatives. The decision process might be driven simply by this ecological performance, with the actual decision-making often done by an analyst or manager who is essentially acting on behalf of some set of (presumed) stakeholders.

Here we expand the decision process to consider stakeholder preferences for the potential consequences of alternatives implied by the means-end model. These preferences are expressed at two levels. First, stakeholders might have different reactions to varying levels of ecological performance on a given objective (e.g., for a given change in bird species richness). Second, they might have different preferences for the objectives (e.g., richness or rarity as compared to connectivity). We consider these two valuations in turn.

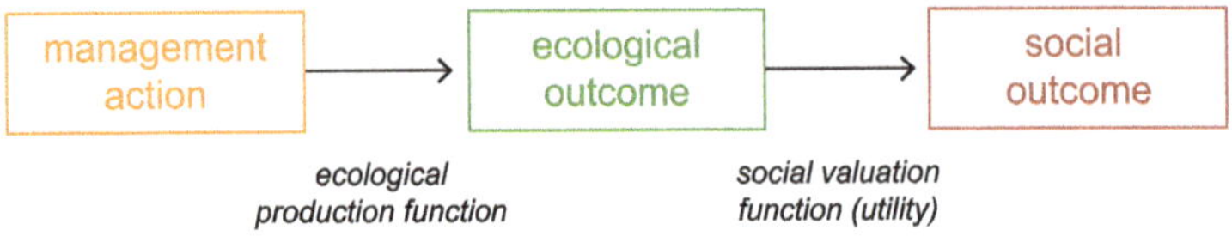

Fig. 8.7 Schematic of the translation from measures of ecological performance (i.e., based on ecological or biophysical indicators) to stakeholder preferences for those outcomes

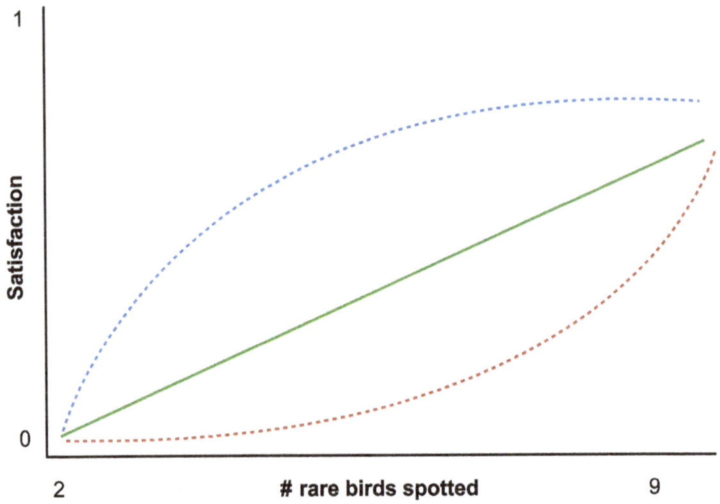

Fig. 8.8 Examples of utility functions. Concave downward (upper curve) implies "diminishing returns" in stakeholder preference as ecological performance increases; concave upward (lower curve) implies a greater marginal change is required to elicit the same level of satisfaction

In general, this next stage of decision analysis entails translating measures of ecological performance into estimates of *stakeholder preferences* for those levels of performance (Fig. 8.7).

Utility Functions

Relative stakeholder preferences for different levels of ecological performance are typically estimated by dimensionless scales on [0,1], using *utility functions* (Fig. 8.8). *Utility* is a measure of how much a person (or stakeholder group) favors a given level of performance, as compared to other levels of performance. Perhaps the simplest utility function would be linear, implying that utility increases in proportion to ecological performance—"more is better" regardless of the actual levels. In reality, utilities are often nonlinear and frequently concave downward to reflect a common "diminishing returns" response. For example, while a stakeholder might value an increase in species from 3 to 5 species, they might put less value on the same incremental increase from 23 to 25 species. In some cases, the function might be concave upward, which would imply that a bigger increase in ecological performance is required to elicit a change in stakeholder value (this might also reflect risk aversion).

Utility curves might be hypothetical, but the aim is to estimate these empirically from actual stakeholders or representatives of stakeholder groups. There are various ways to do this. One simple approach is the *ratio method*. In this, the stakeholder assesses different levels of ecological performance and assigns the lowest-ranked alternative to an arbitrary score of 1. They then rank the other levels relative to this baseline. For example, a level that is twice as favorable is assigned a 2; a tenfold preference is given a 10; and so on. The rank scores are then re-relativized as rank/(maximum-minimum), yielding scores on [0,1]. The elicitation can be done via survey or face-to-face by a trained facilitator.

There are other ways to estimate utility (Clemen (2001), Gregory et al. (2012); and Huang et al. (2011) provide an overview). While some are perhaps appealing in their suggestion of more quantitative precision, these also require more expertise to implement (and see below). Here we focus on simple methods that can be used by environmental managers without technical expertise in valuation or elicitation.

Monetizing Values

A different approach to stakeholder valuation is to attempt to *monetize* these values. Tools include estimates based on market values, stated preference methods (willingness to pay or accept), revealed preferences (e.g., contingent valuation), avoided costs, and so on (reviewed in NESP (2016)). This approach to valuation—well established in economics—makes explicit the value of performance provided.

This can be useful in cases where benefit-cost analysis (BCA; Boardman et al. 2006; Hanley and Barbier 2009) is desired or required. But monetizing can be difficult to apply to nonmarket goods and services. Biodiversity is often problematic in this way. And objectives that cannot be monetized easily are often simply ignored, (i.e., defaulting to no value). We return to this issue shortly. In the USA, additional federal guidance has recently been provided to help practitioners incorporate ecosystem services into BCA (OIRA 2024). There is a growing community of practice supporting BCA (e.g., www.benefitcostanalysis.org).

It should be emphasized that utilities elicited in any way are particular to the levels of ecological performance being considered (e.g., 2–20 bird species), the stakeholders whose preferences are being solicited, and the time span that these preferences are intended to represent. That is, these are particular to a specific decision context in space and time, and should not be extended or transferred to another context. Similar caveats apply to monetized values. Formal tools for extrapolating such preferences—so-called benefits transfer models—exist (e.g., Boyle et al. 2010; Johnston and Rosenberger 2010, Johnston et al. 2015) but are beyond the scope of our discussion here.

Preferences for Objectives

Just as different stakeholders might have different preferences for levels of ecological performance, they might also have different preferences for particular objectives. Again, to be clear, *all* objectives are preferred at some level, or they would not be declared as objectives at the beginning of the decision process. But stakeholders might value these differently, especially when it becomes apparent that trade-offs will need to be made. That is, one might value actual bird species diversity more than potential diversity as indexed by biophysical habitat proxies; another might put a higher premium on connectivity than diversity; most birders, of course, prefer rare birds to common species.

Relative preferences for objectives can be elicited in the same way that utility functions are estimated. Again, a ratio method can be used (as above), or the objectives can simply be weighted using a "slider bar" technique that allocates relative preferences within the constraint that the values must sum to 1.0.

Geospatial Models, Revisited

It might be instructive here to revisit common practice in implementing multi-criteria ranking systems in a GIS. In geospatial models, the first set of weights is applied within an objective, by specifying the relative importance of the indicators to the aggregate value of the objective. As MAUT (Eq. 8.3) suggests, a common scheme is linear—which is to say that incremental changes in the indicators convey a linear increase in the objective score. An alternative is to subjectively bin the scores into categories or ranges of values on the indicators. This frees the user to apply nonlinear schemes to the scoring function (as in Fig. 8.8), with the caveat that the rationale and details of these assignments might not be transparent to others. Again, in such models the assignments typically are assigned by the practitioner/programmer, rather than by stakeholders.

Another issue with GIS-based ranking tools is that it is often not obvious what the values represent. Explicitly, rankings based on a weighting function (Eq. 8.3 or user-defined) rate each site relative to the other sites—scores are scaled relative to the scores observed over all sites (i.e., scaled between the minimum and maximum values observed over all sites). But these are not incremental or marginal values compared to other sites; they are free of that context. It also is not clear what the counterfactual might be in many instances: is the value of a site scored relative to its being lost (e.g., developed) or is relative to its current management (i.e., without formal protection)?

These considerations do not negate the value of geospatial models of this sort; but they do invite a full explanation of the process and what the relative scores represent.

Models that accept user-specified weightings for alternative objectives are more closely aligned with structured decision-making, as it is much easier to invite different weightings of the objectives to illustrate the consequences to the rankings. It is recommended that examples of alternative weightings be used to illustrate and communicate these alternatives weightings as part of reporting and communication (see below).

Decisions Under Uncertainty

Thus far, we have considered the decision process as if it were conducted under certainty—that we have confidently estimated indicators and weights within and across objectives. Real applications tend to be fraught with uncertainty on many points: imprecision in the empirical indicators, ambiguities about stakeholder preferences, and so on.

Gregory et al. (2012) offer considerable guidance (especially their Chap. 8) to decision-making under uncertainty. They note that there are two sorts of uncertainty of concern here:

1. *Epistemic* uncertainties are about knowledge or facts. These stem from natural variability in underlying processes, measurement error or bias in indicators, model uncertainties, or disagreements among experts in their judgments about available information. These uncertainties can be reduced, in some cases, with more or better data. But these refinements are necessary only if more data is likely to change the decision. In decision analysis, estimates of the *value of information* are targeted precisely at this issue (Gregory et al. 2012; Bennett et al. 2018).

   When more data cannot resolve epistemic uncertainties, practitioners can use scenarios to bracket the decision space. In this, a set of contrasting scenarios are devised to represent a reasonable range of expectations, and the decision analysis then can assess the range of consequences. For example, we might devise a set of scenarios based on a range of assumptions about development pressure or climate change.
2. *Linguistic* uncertainties are about communication failures. These stem from ambiguous terms, imprecise or value-laden language (e.g., if the same term means different things to different stakeholders), or context dependencies. Linguistic uncertainties, fortunately, can be minimized simply by being very careful and clear about communicating. Shared reference points such as objectives hierarchies, means-ends models, and decision tables are very helpful in this.

   A critical point is that deliberations should be organized to provide insight into the effect of uncertainty on the decision process itself and to guide the process toward decisions that are as robust as possible. Again, Gregory et al. (2012) provide several practical tools for dealing with uncertainty.

8.3.5 The Decision

The ultimate aim in structured decision-making is to populate the decision matrix (Table 8.3) with stakeholder preferences for each of the options. The values tabled in each cell of the matrix are the product of stakeholder utility for that level of performance (i.e., relative to other levels of performance) and their preference for that particular objective relative to other objectives. Summing these values over each column summarizes relative stakeholder preferences for each management alternative being considered.

From a decision-making standpoint, it is tempting to force the decision to the alternative favored in the decision matrix (i.e., assuming there is a clear "winner"). In decision contexts where there is a single "decider," this is a reasonable approach. In this instance, an additional value of the decision matrix is that it documents the reason for the decision and in this way can help communicate this rationale to stakeholders.

But in many instances, there might be multiple stakeholder groups with very different preferences. While it might be tempting to somehow "average" these preferences into a single aggregate decision matrix, it is often more useful to generate multiple decision matrices—one for each stakeholder group. These competing matrices can then be used to explore differences and, ideally, to find a way to reconcile the differences. Even if the final decision goes against the preferences of some stakeholders, the inclusive process and the information in the matrices can help explain the rationale and so minimize feelings of disenfranchisement.

8.3.6 Reporting and Communication

One of the key values of a structured decision-making process is its objectivity and transparency. By being explicit about the objectives and how they are weighted, the decision process is essentially self-documenting. But for many real applications, the process is still a bit messy because there might be several iterations in which different preferences or weighting schemes are explored. Communicating all of this information requires a systematic reporting process to match the decision process. This must begin with a clear statement of the decision context:

- ☑ What is the overall goal of the project? Who are the stakeholders? What are the spatial domain and time frame of the decision process?
- ☑ What are the specific objectives, to reach the goal?
- ☑ What empirical indicators are used to capture the objectives, and which data were used to estimate these (i.e., the objectives hierarchy)?
- ☑ What stakeholder groups provided the weightings for the indicators, and how were these assigned?
- ☑ How were the likely outcomes of alternative management actions or policies developed (i.e., the means-ends models)? What is the level of empirical confidence in the models? Where are the greatest uncertainties?
- ☑ How were objectives weighted relative to each other? Who made these determinations?

The presentation of the decision process should detail which steps were followed to arrive at a final decision. Often, it is helpful to show preliminary or partial results to set the stage for the final assessment:

- ☑ Rankings or maps of candidate sites in terms each of the objectives by itself (e.g., separate maps of sites ranked on diversity, area or patch geometry, connectivity, etc.).
- ☑ Rankings or maps based on combinations of objectives. These might include a set reflecting equal weights across all attributes, or (as appropriate) different rankings

reflecting the preferences of different stakeholder groups (or assumptions about these).

☑ If appropriate, illustrations of the sensitivity of the rankings to uncertainty or imprecision in the weightings. These can be explored by "jittering" the weightings in small increments, to gauge the robustness of rankings. Alternatively, the decision can be analyzed using a range of weights representing presumed lower and upper bounds. Ideally, the results would be robust to minor imprecision or uncertainty.

☑ The final weighting scheme and rankings implied by this, along with the justification for the final decision.

Again, a huge benefit of structured decision-making is the transparent and deliberate nature of the process and that it can be inclusive of stakeholders with different preferences for objectives. Communicating this clearly is part of the process.

In their review of 20 years of multi-criteria decision analysis applications, Esmail and Geneletti (2017) suggested that practitioners were still somewhat lax in the transparency of decisions (e.g., how weightings were generated), though applications have certainly improved over the years.

8.4 Targeting Ecosystem Services

Ecosystem services are goods and amenities that ecosystems provide to humans (Daily et al. 1997). These include commodities (food and fiber), natural protections such as flood regulation by wetlands or the reduction of wave energy by natural shorelines, supportive services such as pollination, and cultural services including recreation and spiritual values associated with open space or sacred sites (MEA 2005).

Recently, ecosystem services have emerged as a focus on conservation planning and land management in general (Chan et al. 2006; Wainger and Mazzotta 2011; Goldstein et al. 2012; NESP 2016; Watson et al. 2019). For example, in the USA many federal agencies that manage land now consider ecosystem services explicitly in their planning process (Ruhl and Salzman 2020; see Olander et al. (2021) for an example from the US Forest Service's Forest Planning process). Many nongovernmental organizations (e.g., many land trusts) also use the idea (if not the language) of ecosystem services to shape conservation plans. A large volume of practical guidance has been developed, by academics and by agency practitioners (e.g., NESP (2016), an online guidebook[2] geared toward federal agencies in the USA).

---

[2]Much of the material in this section is adopted from the NESP (2016) guidebook (I was one of the many coauthors).

Targeting ecosystem services is a natural extension of the multi-criteria decision framework outlined above. In particular, the use of the term "services" explicitly demands that we identify the stakeholders to whom these services should flow (Wainger and Mazzotta 2011). Watson et al. (2019) mapped the supply (production) and demand (beneficiaries) of selected ecosystem services and showed that for some (e.g., pollination), the beneficiaries could be far removed from where the services were produced. Carbon sequestration is a service that benefits everyone on Earth (no matter where the carbon is sequestered), while watershed protection has downstream beneficiaries (though these might be far downstream). Identifying beneficiaries is a major challenge in targeting ecosystem services, because each service can benefit different people in different places.

Identifying the beneficiaries, in turn, invites their full participation as stakeholders in the decision-making process. While this might seem obvious at this point of our discussion, it has not been the primary mode of decision-making in most environmental management settings. More typically, these decisions have been made on the basis of ecological performance (real or presumed) and often by a decision-maker authorized to act on behalf of stakeholders. This is not to say that such decisions have been made poorly or in bad faith. Rather, the recent focus on ecosystem services invites us to a fuller investment in structured decision-making as outlined here.

8.4.1 Benefits, Beneficiaries, and Stakeholders

One consequence of adopting ecosystem services as targets is that it invites the presumed beneficiaries into the decision process earlier. In particular, this means inviting stakeholders to participate in identifying goals and objectives, as well as preferred outcomes. This is by contrast to an after-the-fact communication and solicitation of stakeholder reactions to decisions that have already been proposed.

After the initial identification of alternative objectives and possible outcomes, stakeholders return to the decision process in sharing their preferences for the outcomes anticipated of the management or policy alternatives (i.e., as suggested by means-ends models and as reflected by estimates of utility). At this point, a heterogeneous pool of stakeholders might lead to a very different decision process as compared to a decision made by an authority acting on an ecological mandate or even on behalf of presumed stakeholders. For example, most land management agencies have a rather narrow mandate—to protect species, to protect cultural sites, to facilitate the allocation of multiple-use demands on public lands, and so on. Exploring alternatives to a narrow mandate to be more inclusive of alternative visions can be a positive outcome of stakeholder input (and see examples at the website of the National Ecosystem Services Partnership, www.nespguidebook.com).

8.4.2 Linking Ecology to Social Outcomes

A second consequence of adopting an ecosystem services perspective is that the decision process entails an explicit switch from ecological performance values to estimates of stakeholder preferences for those performance values, that is, embracing not only the multi-attribute nature of the decision but also the utility aspect of multi-attribute utility theory, a fuller version of structured decision-making (Gregory et al. 2012). In this, there is a necessary hand-off of the decision from natural scientists to social scientists.

While this is as it should be—a fully interdisciplinary solution—it can also be problematic because many agencies or decision-makers lack the expertise or capacity to perform both the natural- and social-science pieces. For example, in the USA many natural resource agencies are staffed primarily with natural scientists, and might not have the expertise in valuation methods (whether based on utility or monetization). This can severely limit the full implementation of structured decision-making (Johnston et al. 2015).

Olander et al. (2017) addressed this bottleneck by recommending the use of *benefit-relevant indicators* (BRIs). BRIs are natural-science indicators that are framed in terms that should resonate easily with stakeholders. For example, we might index water quality in technical terms such as dissolved oxygen, *E. coli* concentrations, turbidity, or nutrient loadings (some of which measures might, in fact, be legally mandated). But most stakeholders would have limited appreciation for these technical measures. A BRI more interpretable to stakeholders might be limitations on whether or how much fish could be safely consumed from those waters, or whether the beaches might be closed to swimming. Using BRIs, decision-makers trained in the natural sciences might make better decisions, even without the formal analysis of stakeholder valuation. Of course, ultimately we will need to build capacity for including social-science methods in the decision-making process (Johnston et al. 2015).

Another advantage of benefit-relevant indicators is that these can be used in *cost-effectiveness analysis* of management or policy alternatives. While benefit-cost analysis (BCA) requires a monetized value to stakeholders, cost-effectiveness analysis (CEA) is framed in terms of ecological uplift (change in performance) relative to cost, e.g., how large a reduction in beach closures for a given cost. CEA can be applied with any estimate of ecological performance. Cost-effectiveness stops short of BCA but can lead to much more effective allocation of resources for conservation (Naidoo et al. 2006; Messer and Allen 2018). A common reporting in conservation practice is "bucks and acres," which is simply how much land was protected for the money spent. But this performance might be more usefully communicated with BRIs. Many conservation organizations now report performance in these terms: miles of hiking trails, stream frontage for fishing or recreation, and so on. These measures are much easier to communicate effectively and they resonate more strongly with stakeholders.

8.5 Illustrations: The Means-Ends Process

In this section, we explore two illustrations of environmental management decision contexts: stream restoration and forest management to reduce the risk of catastrophic fire. In each, the emphasis is on framing the decision in terms of alternative management options and the possible ecological consequences of each. In this, the process seeks to expand the consideration of possible co-benefits or trade-offs, relative to the focal objective. While each is a realistic decision context facing managers today, the examples are hypothetical and focused on articulating the decision options iteratively. The illustrations developed here focus on ecological outcomes; the social consequences to stakeholders are not considered here—but these would be the next step.

8.5.1 Example: Stream Restoration

Stream restoration might have multiple goals, ranging from ameliorating very local issues within a stream reach (e.g., bank incision) to larger-scale issues of watershed integrity. In this example, we begin with the local issue of improving stream conditions to improve fishing but expand the example to consider ancillary co-benefits and potential trade-offs. We will not specify an objectives hierarchy at the outset, to emphasize that the process of exploring alternatives often can lead to revised objectives: the process is iterative.

To begin, we want to improve fishing and we might posit that buffering the stream with riparian forest could improve stream condition by shading the stream, thus reducing temperature, increasing dissolved oxygen concentrations, and improving habitat condition for fish that require such conditions (e.g., native trout). This buffering would also increase infiltration, reduce surface runoff into the stream, and reduce sedimentation and other inputs (nitrogen, phosphorus, pollutants) into the stream (Fig. 8.9, top tier).

The act of increasing buffer width along the stream also provides the possible co-benefit of increasing habitat area (i.e., forest) and connectivity (as streams are natural networks) (Fig. 8.9, middle tier). And so the management option to improve water quality might also provide ancillary co-benefits in terms of biodiversity support: riparian bird species along with other wildlife species, pollinators, and so on.

At the same time, expanding the riparian buffer increases vegetation cover and, in particular, transpiring leaf area. In water-limited systems, the increased transpiration from the expanded riparian buffer might actually decrease streamflow or water available for other extractive uses (e.g., irrigation, municipal water supplies) (Fig. 8.9, lower tier).

The net result of this is a decision context that must now weigh the possible benefits of the management option on its intended purpose, as well as the

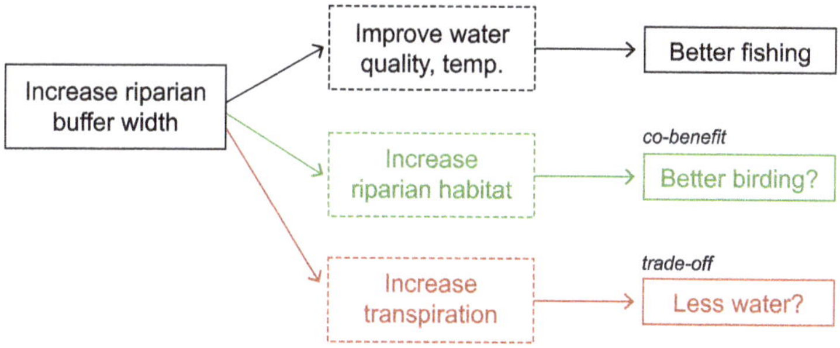

Fig. 8.9 Path model elaborating the possible ecological consequences of buffering a stream with riparian vegetation. Beyond the implications for water quality, this action might also have co-benefits in terms of riparian wildlife habitat while also risking a trade-off with lower water yield (e.g., for irrigation or other human water uses)

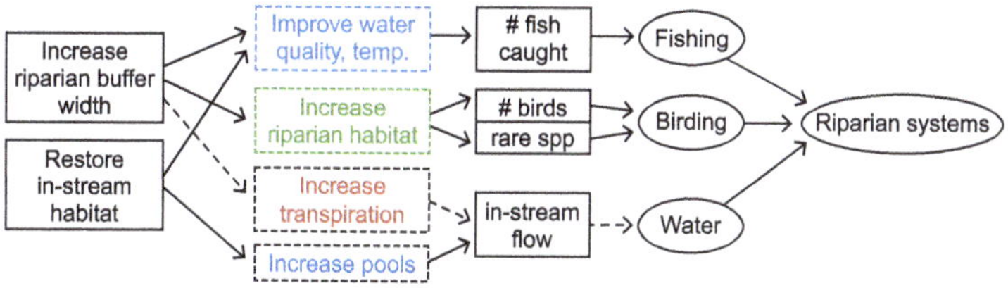

Fig. 8.10 Means-end diagram for the "fishing" example, expanded to consider another management alternative, to directly manipulate in-stream habitat features. Solid lines are positive effects; dashed lines, negative. This alternative also has implications for water yield but less so for riparian wildlife habitat. By contrast, riparian buffers might increase water quality but decrease yield

extenuating circumstances (or unintended consequences) of ancillary co-benefits or trade-offs. The value of articulating the path model is that it invites a full exploration of these other consequences. In this example, the initial objective of improving fishing has expanded to embrace other co-benefits and possible trade-offs. An objectives hierarchy, assembled or expanded during the deliberations, might now include fishing, birding, and water yield explicitly.

Of course, the real point of structured decision-making is to compare across alternative decision options. For example, in this case we might wish to compare the effects of riparian buffers with a more direct action to improve in-stream habitat (e.g., by planting coarse debris or rocks to create deep pools). This expands the means-end diagram to compare the two alternatives explicitly (Fig. 8.10).

In this illustration, the option of in-stream habitat manipulation has its own direct effects on fishing (by creating pools that might serve as refugia during droughts), but it also might have an influence on water yield. It will not influence riparian habitat or transpiration.

To underscore the sequencing of this process, we began on the right side with an *end*, to improve fishing. We then worked on the left side, to pose a *means* that might

Table 8.4 Template for decision matrix corresponding to Fig. 8.10

Option	Status quo	Buffers	In-stream
Fish caught	#	+	+
Total birds	#	+	
Rare birds	#	+?	
In-stream flow	cfs	–	+?

Under status quo are the units of measurement (cfs = cubic feet/second); these entries would all be 0's for the status quo

lead to that end, by elaborating the mechanisms and processes in the middle, the *path*.

We then explored the path to see if there were other potential consequences of the action and swept from left to right to a new *end* (birding). This process iterated, back and forth, until we had filled in the diagram. This process should be highly collaborative and often can be an intense learning experience.

In Fig. 8.10, the convention is that concepts are drawn in ovals and empirical indicators are drawn in rectangles (recall the path models of structural equation models in Chap. 7). Here, the diagram includes the means-end model (empirical) on the left, while the right side of the diagram represents an objectives hierarchy that has emerged from the iterative process of elaborating the means-end diagram. The means-end model and objectives hierarchy are joined by the indicators.

Subsequent steps in the decision process would follow through on the means-end diagram (Fig. 8.10) by first estimating the changes in ecological or hydrological performance expected from each action (Table 8.4). These estimates would be in ecological terms: fish caught, birds observed, and water flow.

In an actual application, Table 8.4 would be supplanted by several tables in the same format. The first would capture ecological (or hydrological) performance, in empirical terms as indicated in the table. These are difficult to compare directly, and so a second table would be generated to capture stakeholder preferences for these anticipated changes. Phrasing preferences in terms of utility would also translate these to dimensionless units (e.g., on [0,1]) that can be compared across objectives. Note that this processing is *within* objectives.

Next, the decision table would be converted into a table of stakeholder preferences *across* objectives. This is now a two-layer scoring so that, e.g., the entry for "Rare birds" in each alternative column reflects (1) stakeholder preference for the incremental change in rare bird observations given that management alternative and (2) stakeholder preference for rare birds as compared to changes anticipated for fish caught, total birds, or in-stream flow.

Again, it might be noted that in geospatial applications of this logic, what is tallied is the expected contribution or value of a site in terms of each objective—and probably not the incremental value of each site relative to the sites already in the reserve system. That is, the focus is on multiple criteria but not so much on the marginal-value aspect of site selection.

8.5.2 Example: Forests and Fire Risk Management

In forests of the western USA (as elsewhere globally), forests are experiencing increases in fire frequency and severity. This reflects decreasing fuel moisture as a result of climate change, as well as (at least in the western USA) increasing fuel loads as a result of decades of fire suppression. Forest management to reduce the risk of catastrophic fire entails removing or rearranging fuels to reduce vertical connectivity (e.g., by thinning the understory to remove ladder fuels) and to reduce horizontal connectivity (i.e., to reduce fuel-bed contiguity). Reducing vertical fuel connectivity is intended to reduce the chance that a surface (ground) fire might "crown" and reach into the overstory, thus dramatically increasing fire *intensity* (how hot the fire burns, flame length) and *severity* (impacts on the forest system, e.g., tree mortality). Reducing horizontal fuel connectivity is intended to reduce the likelihood of fire spread, reducing fire *magnitude* and (indirectly) fire severity.

The benefit of articulating these objectives in a means-end diagram is that it invites a consideration of ancillary co-benefits or trade-offs (Fig. 8.11). For example, altering the vertical structure of forests would be expected to have substantial impacts on wildlife habitat diversity; e.g., bird species diversity responds strongly to the vertical layering in a forest (MacArthur et al. 1961; see Urban (2023), Chap. 4). Changes in horizontal structure of the forest might also affect wildlife by influencing processes such as territoriality (based on identifiable and defensible borders) or activity ranges. Finally, reducing forest biomass could have some impact on transpiration and so might influence in-stream flows and water yield; this would depend on where along hillslope gradients the management was implemented and the extent to which the forest was water-limited previously.

The point of this example is that means-ends diagrams often invite expertise from various disciplines that might not already collaborate. So, while forest fire ecologists might interact regularly in devising and implementing fuels management treatments, these experts might not be as engaged with wildlife ecologists or forest hydrologists...much less the stakeholders affected by decisions related to fire, wildlife, and water. Here the decision-structuring process invites this broader collaboration and engagement.

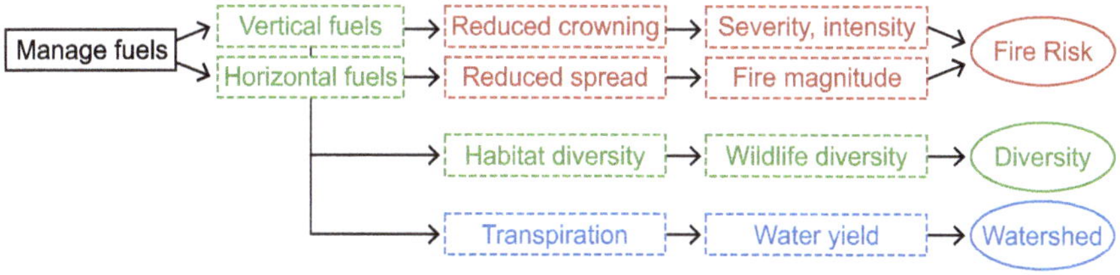

Fig. 8.11 Means-end diagram for fire management in western forests. Here, the focus is on reducing fuels (vertically and horizontally), but the diagram invites an exploration of ancillary impacts on forest wildlife (which respond strongly to forest structure) as well as possible impacts on water yield mediated by changes in transpiration

8.5.3 *Reality Check*

Structured decision-making can be a complicated process, compared to an authorized agent making a decision on behalf of (presumed) stakeholders. This complexity is warranted in a rather small fraction of actual decisions. Hemming et al. (2022) provided a schematic breakdown of cases: They suggested that, out of a hypothetical pool of 10,000 decisions, fully 90% are either inconsequential or so easy ("no-brainers") that a complicated decision process is not worth the effort. In many cases (10%) that do require more effort, the added clarity that comes from "thinking through" the options might suffice. In a few cases (2.5%?), at least a partial decision analysis is warranted; these might entail devising an objectives hierarchy and a means-ends diagram to marshal the process along and to attend complexities due to uncertainty about objectives or alternatives. In a very few instances (<1%?), a full decision analysis might be required; these few cases might be driven as much by stakeholder interest and politics as by the complicated details of the decision alternatives themselves. In every case, a sensitivity to the process of structured decision-making can provide the guidance that will help decide *how* we should decide any given case.

Decisions are often based on models with uneven empirical support. As part of the decision process, weighing the evidence supporting means-ends models is a crucial element of the decision process. Tallis et al. (2017), Olander et al. (2018), and Qiu et al. (2018) provide general guidance relative to means-ends models applied to ecosystem services. Mason et al. (2018) illustrate an approach for a general model for services provided by salt marshes: how the general model can be adapted locally, with an evidence library supporting various details of the general model.

8.6 Further Reading

This chapter has focused on site prioritization as a fundamental task in conservation at the landscape scale. There are other aspects of conservation, of course, and these are addressed in more general texts (e.g., Meffe and Carroll 1994; Pickett et al. 1997; Hunter and Gibbs 2007; Primack 2012) and at least one with the perspective of landscape ecology (Gutzwiller 2002).

Conservation priorities depend on the particular context of any given landscape. In general, for the common case of forests in a human-modified landscape, the priorities are to increase the absolute amount of habitat, to improve its condition or habitat quality, to increase its connectivity, and to reduce the contrast between habitat and the nonforest matrix. Arroyo-Rodríguez et al. (2020) provide a review and useful synthesis of the ecological concepts that underpin planning in human-modified landscapes managed for conservation as well as ecosystem services.

Decision analysis is a well-developed practice and there are several general references (e.g., Hammond et al. 1999; Clemen 2001; Gregory et al. 2012). Likewise, there are general texts on the economic side of decision-making, including valuation techniques (e.g., Champ et al. 2003; Freeman et al. 2014) and benefit-cost analysis (e.g., Boardman et al. 2006; Hanley and Barbier 2009). The US National Research Council (NRC 2005) and Ninan (2014) specifically address the valuation of ecosystem services.

Esmail and Geneletti (2018) reviewed 20 years of environmental applications of MCDA and offer recommendations to improve the practice. Johnston et al. (2015) and Hemming et al. (2022) provide helpful introductions to structured decision-making specifically for conservation practitioners.

8.7 Summary and Prospectus

Site prioritization can emerge from a simple ranking of candidate sites based on species diversity or some other index of relative value. But, often, prioritization invokes multiple objectives, objectives that might be subject to leveraged co-benefits or forced trade-offs. Multi-criteria decision frameworks can help generate and communicate these decision contexts. At this level, decisions are often framed in ecological or biophysical terms, and decisions can be made by invoking a user-specified system of weights across the multiple criteria.

In many cases, heterogeneous stakeholder groups will have different preferences for the alternatives, and these preferences can be incorporated into a structured decision-making process by considering relative preferences for the ecological outcomes suggested by the alternative management or policy options.

Clarifying the decision alternatives in terms of an objectives hierarchy and a means-ends diagram that links options to outcomes can be useful because these can be communicated readily to stakeholders. The level of rigor, objectivity, transparency, and inclusion of stakeholders all contribute to better decisions.

It is becoming increasingly easy to implement decision support tools using geospatial indicators in a GIS. This can greatly facilitate site prioritization. But it also can lead to a static ranking of sites if the process is not updated to reflect a changing decision context—e.g., which sites have been protected, how connectivity might have changed, increasing threats from development, climate change, a growing interest in environmental equity, and so on. Structured decision-making is only useful if it is updated and revisited over time.

Systematic conservation planning is a well-established practice now, but it is still evolving. For example, there are emerging approaches for incorporating connectivity into conservation planning (Daigle et al. 2019; Hanson et al. 2021). And increasingly, practitioners are addressing climate change directly in conservation plans (e.g., Reside et al. 2018; Eaton et al. 2019; reviewed by Urban 2023, Chap. 10).

These applications can get increasingly complicated, and again the value of systematic approach and structured decision-making is its rigor, transparency, and inclusivity. We return to the assessment and communication of (sometimes complicated) change scenarios in the final chapter of this book.

References

Anderson MG, Barnett A, Clark M, Prince J, Olivero Sheldon A, Vickery B (2016) Resilient and connected landscapes for terrestrial conservation. The Nature Conservancy, Eastern Conservation Science/Eastern Regional Office, Boston

Anderson MG, Clark M, Olivero AP, Barnett AR, Hall KR, Cornett MW, Ahlering M, Schindel M, Unnasch B, Schoss C, Cameron DR (2023) A resilient and connected network of sites to sustain biodiversity under a changing climate. PNAS 120(7):e2204434119

Arroyo-Rodríguez V, Fahrig L, Tabarelli M, Watling JI, Tischendort L, Benchimol M, Cazetta E, Faria F, Leal IR, Melo FPL, Morante-Filho HC, Santos BA, Arasa-Gisbert R, Arce-Peña N, Cervantes-López MJ, Cudney-Velenzuela S, Galán-Aveco C, San-José M, Vieira ICG, Slik JWF, Nowa-kowski AJ, Tscharntke T (2020) Designing optimal human-modified landscapes for biodiversity conservation. Ecol Lett 23:1404–1420

Ball IR, Possingham HP, Watts M (2009) Marxan and relatives: software for spatial conservation prioritisation. In: Moilanen A, Wilson KA, Possingham H (eds) Spatial conservation prioritisation: quantitative methods and computational tools. Oxford University Press, Oxford, pp 185–195

Bennett JR, Maxwell SL, Martin AE, Chadès I, Fahrig L, Gilbert B (2018) When to monitor and when to act: value of information theory for multiple management units and limited budgets. J Appl Ecol 55:2102–2113

Boardman AE, Greenberg D, Vining AR, Weimer DL (2006) Cost-benefit analysis: concepts and practice, 3rd edn. Prentice Hall, Upper Saddle River

Boyle KJ, Kuminoff NV, Parmeter CF, Pope JC (2010) The benefit-transfer challenges. Ann Rev Resour Econ 2:161–182

Champ PA, Boyle KJ, Brown TC (2003) A primer on nonmarket valuation: the economics of nonmarket goods and resources. Springer, New York

Chan KMA, Shaw MR, Cameron DR, Underwood EC, Daily GC (2006) Conservation planning for ecosystem services. PLoS Biol 4(11):e379

Clemen RT (with T. Reilly) (2001) Making hard decisions, 2nd ed. Rev. Dux-bury Press, Pacific Grove

Csuti B, Polasky S, Williams PH, Pressey RL, Camm JD, Kershaw M, Keister AR, Downs B, Hamilton R, Huso M, Sahr K (1997) A comparison of reserve selection algorithms using data on terrestrial vertebrates in Oregon. Biol Conserv 80:83–97

Daigle RM, Metaxas A, Balbar AC, McGowan J, Treml EA, Kuempel CD, Possingham HP, Beger M (2019) Operationalizing ecological connectivity in spatial conservation planning with Marxan connect. Methods Ecol Evol:11, 570–579

Daily GC, Alexander S, Erlich PR, Goulder L, Lubchenco J, Matson PA, Mooney HA, Postel S, Schneider SH, Tilman D, Woodwell GM (1997) Ecosystem services: benefits supplied to human societies by natural ecosystems. Issues in Ecology No. 2. Ecological Society of America, Washington, DC

Eaton MJ, Yurek S, Haider Z, Martin J, Johnson FA, Udell BJ, Charkhgard H, Kwon C (2019) Spatial conservation planning under uncertainty: adapting to climate change risks using modern portfolio theory. Ecol Appl 29:e01962

Ferrier S, Pressey RL, Barrett TW (2000) A new predictor of the irreplaceability of areas for achieving a conservation goal, its application to real-world planning and a research agenda for further refinement. Biol Conserv 93:303–325

Freeman AM, Herriges JA, Kling CL (2014) The measurement of environmental and resource values: theory and methods, 3rd edn. RFF Press, Washington, DC

Game ET, Kareiva P, Possingham HP (2013) Six common mistakes in conservation priority setting. Conserv Biol 27:480–485

Goldstein JH, Caldarone G, Duarte TK, Ennaanay D, Hannahs N, Mendoza G, Polasky S, Wolny S, Daily GC (2012) Integrating ecosystem-service tradeoffs into land-use decisions. PNAS 109: 7565–7570

Gregory R, Failing L, Harstone M, Long G, McDaniels T, Ohlson D (2012) Structured decision making: a practical guide to environmental management choices. Wiley Blackwell, Oxford

Groves CG, Jensen DB, Valutis LL, Redford KH, Shaffer ML, Scott JM, Baumgartner JV, Higgins JV, Beck MB, Anderson MG (2002) Planning for biodiversity conservation: putting conservation science into practice. BioScience 52:499–512

Gutzwiller KJ (ed) (2002) Applying landscape ecology in biological conservation. Springer, New York

Hammond JS, Keeney RL, Raiffa H (1999) Smart choices. Harvard Business School Press, Cambridge, MA

Hanley N, Barbier E (2009) Pricing nature: cost-benefit analysis and environmental policy. Edward Elgar Publishing, Cheltenham

Hansen JO, Vincent H, Schuster R, Fahrig L, Brennan A, Martin AE, Hughes JS, Pither R, Bennett JR (2022) A comparison of approaches for including connectivity in systematic conservation planning. J Appl Ecol 59:2507–2519

Hanski I, Ovaskainen O (2000) The metapopulation capacity of a fragmented landscape. Nature 404:755–758

Hanson JO, Schuster R, Morrell N, Strimas-Mackey M, Watts ME, Arcese P, Bennett J, Possingham HP (2021) Prioritizr: systematic conservation prioritization in R. R package version 7.0.1. Available at https://CRAN.R-project.org/package=prioritizr

Hemming V, Camaclang AE, Adams MS et al (2022) An introduction to decision science for conservation. Conserv Biol 36:1–16

Huang IB, Keisler J, Linkov I (2011) Multi-criteria decision analysis in environmental sciences: ten years of applications and trends. Sci Total Environ 409:3578–3594

Hunter ML, Gibbs JP (2007) Fundamentals of conservation biology, 3rd edn. Blackwell, Malden

Johnston RJ, Rosenberger RS (2010) Methods, trends and controversies in contemporary benefit transfer. J Econ Surv 24:479–510

Johnston RJ, Rolfe J, Rosenberger RS, Brouwer R (2015) Benefit transfer of environmental and resource values: a guide for researchers and practitioners. Springer, Cham

Keeney RL, Gregory RS (2005) Selecting attributes to measure the achievement of objectives. Oper Res 53:1–11

Keitt T, Urban D, Milne B (1997) Managing fragmented landscapes: a macroscopic approach. Conserv Ecol 1:4

Kukkala AS, Moilanen A (2013) Core concepts of spatial prioritisation in systematic conservation planning. Biol Rev 88:443–464

Margules CR, Pressey RL (2000) Systematic conservation planning. Nature 405:243–253

Mason S, Olander L, Warnell K (2018) Ecosystem services conceptual model application: NOAA and NERRS salt marsh habitat restoration. National Ecosystem Services Partnership Conceptual Model Series No. 3. Nicholas Institute for Environmental Policy Solutions, Duke University, Durham. https://nicholas-institute.duke.edu/conceptual-model-series

Meffe GK, Carroll CR (1994) Principles of conservation biology. Sinauer, Sunderland

Messer KD, Allen WL (2018) The science of strategic conservation: protecting more with less. Cambridge University Press, Cambridge

Millennium Ecosystem Assessment (MEA) (2005) Ecosystems and human wellbeing. Island Press, Washington, DC

Moilanen A, Wilson KA, Possingham H (2009) Spatial conservation prioritization: quantitative methods and computational tools. Oxford University Press, Oxford

Moore JL, Folkmann M, Balmford A, Brooks T, Burgess N, Rahbek C, Williams PH, Krarup J (2003) Heuristic and optimal solutions for set-covering problems in conservation biology. Ecography 26:595–601

Naidoo R, Balmford A, Ferraro PJ, Polasky S, Ricketts TH, Rouget M (2006) Integrating economic costs into conservation planning. Trends Ecol Evol 21:681–687

National Ecosystem Services Partnership (NESP) (2016) Federal resource management and ecosystem services guidebook, 2nd edn. National Ecosystem Services Partnership, Duke University, Durham. https://nespguidebook.com

National Research Council (NRC) (2005) Valuing ecosystem services: toward better environmental decision-making. National Academies Press, Washington, DC

Ninan KN (ed) (2014) Valuing ecosystem services: methodological issues and case studies. Edward Elgar Publishing, Northampton

Office of Information and Regulatory Affairs (OIRA) (2024) Guidance for assessing changes in environmental and ecosystem services in benefit-cost analysis. US Office of Management and Budget, Washington, DC. Available at www.white-house.gov/wp-content/uploads/2024/02/ESGuidance.pdf. February 2024

Olander LP, Johnston RJ, Tallis H, Kagan J, Maguire LA, Polasky S, Urban D, Boyd J, Wainger L, Palmer M (2017) Benefit-relevant indicators: ecosystem services measures that link ecological and social outcomes. Ecol Indic 85:1262–1272

Olander L, Mason S, Warnell K, Tallis H (2018) Building ecosystem services conceptual models. National Ecosystem Services Partnership Conceptual Model Series No. 1. Nicholas Institute for Environmental Policy Solutions, Duke University, Durham. https://nicholasinstitute.duke.edu/conceptual-model-series

Olander L, Warnell K, Warziniack T, Ghali Z, Miller C, Neelan C (2021) Exploring the use of ecosystem services conceptual models to account for the benefits of public lands: an example from National Forest Planning in the United States. Forests 12:267

Önal H, Briers RA (2006) Optimal selection of a connected reserve network. Oper Res 54:379–388

Pickett STA, Ostfeld RS, Shachak M, Likens GE (eds) (1997) The ecological basis of conservation. Chapman and Hall, New York

Poiani KA, Richter BD, Anderson MG, Richter HE (2000) Biodiversity conservation at multiple scales: functional sites, landscapes, and networks. BioScience 50:133–146

Possingham H, Ball I, Andelman S (2002) Mathematical methods for identifying representative reserve networks. In: Ferson S, Burgman M (eds) Quantitative methods for conservation biology. Springer, New York, pp 291–306

Pressey RL, Humphreys CJ, Margules CR, Vane-Wright RI, Williams PH (1993) Beyond opportunism: key principles for systemic reserve selection. Trends Ecol Evol 8:124–128

Prim RC (1957) Shortest connection networks and some generalizations. Bell Syst Tech J 36:1389–1401

Primack RB (2012) A primer of conservation biology, 5th edn. Sinauer, Sunderland

Qiu J, Game ET, Tallis H, Olander LP, Glew L, Kagan JS, Kalies EL, Mich-anowicz D, Phelan J, Polasky S, Reed J, Sills EO, Urban D, Weaver SK (2018) Evidence-based causal chains for linking health, development, and conservation actions. BioScience 68:182–193

Tallis H, Kreis K, Olander L, Ringler C (2017) Bridge collaborative practitioner's guide: principles and guidance for cross-sector action planning and evidence evaluation. The Nature Conservancy, Washington, DC

The Nature Conservancy (1996) Conservation by design: a framework for mission success. The Nature Conservancy, Washington, DC

Urban DL (2002) Prioritizing reserves for acquisition. In: Gergel SE, Turner MG (eds) Learning landscape ecology: a practical guide to concepts and techniques. Springer, New York, pp 293–305

Urban DL (2023) Agents and implications of landscape pattern: working models for landscape ecology. Springer, Cham

Urban DL, Keitt TH (2001) Landscape connectivity: a graph-theoretic perspective. Ecology 82: 1205–1218

US Forest Service National Forest System Land Management Planning, Final Rule, 36 C.F.R. 219 2012. Available at http://www.fs.usda.gov/Internet/FSE_DOCU-MENTS/stelprdb5362536.pdf

Wainger L, Mazzotta M (2011) Realizing the potential of ecosystem services: a framework for relating ecological changes to economic benefits. Environ Manag 48:710–733

Watson KB, Galford GL, Sonter LJ, Koh I, Ricketts TH (2019) Effects of human demand on conservation planning for biodiversity and ecosystem services. Conserv Biol 33:942–952

Zurrell D, König C, Malchow A-K, Kapitza S, Bocedi G, Travis J, Fan-dos G (2022) Spatially explicit models for decision-making in animal conservation and restoration. Ecography 2022: e05787

Chapter 9
Landscape Change

Abstract Perhaps the most fundamental observations of landscapes are changes in land cover or condition over time. The initial task, given such observations, is to determine whether the observed changes are real—an issue of extracting an ecological signal from the background noise. In this chapter we begin with trend detection and then proceed to capture trends in simple models that can be extrapolated into the future. Simple models are often too simple to be satisfyingly realistic, but they provide a useful point of departure for models that are too realistic to be simple. Considering a variety of simple and less simple models of landscape change leads to a more general consideration of the modeling process (and our role in this). While appreciating that most practitioners will not aspire to be modelers, most landscape-scale applications involve models of some kind. This chapter aims to instill an appreciation of the modeling process and workflow and the role of models in landscape ecology and management.

9.1 Introduction

Perhaps the most basic observation of landscapes, or natural systems in general, is that they change over time. This is despite a natural inclination to think of landscapes as being "slow" and nearly static over timescales corresponding to human lifespans (or the careers of landscape ecologists?) (Sprugel 1991). As they say, the only constant is change.

Observing change over time invites several reactions. The first is confirmatory: is the change we think we are seeing actually real? *Change detection* is the first hurdle in exploring landscape change. This is largely a matter of detecting a significant signal, given the background noise in any ecological data series. We begin this chapter with a consideration of change detection as a signal-noise problem. This is the essence of a monitoring program. Given an observed change, the question arises whether the change is substantial enough to warrant a management response, or even a reassessment of our expectations of the system: the "react" stage of adaptive management. We finish this book with ecological assessment, in Chap. 10.

D. L. Urban, *Landscape Ecology: A Task-Oriented Perspective*,
https://doi.org/10.1007/978-3-031-72251-6_9

A natural response to trends observed through monitoring is to extrapolate these trends into the future, a *forecast*. A forecast is a model, and models of landscape change are the main focus of this chapter.

On landscapes, a common signal of interest is the change in a focal land cover type (e.g., forests, or developed land covers). While such interests are easily motivated, they are complicated by the reality that land cover change is a zero-sum game: if something is increasing, then something else *must* be decreasing at the same time. So it makes sense to consider land cover change in terms of simultaneous changes in all types over times. An empirical summary of such joint changes is perhaps the simplest model of landscape change. Our first entry into models of landscape change will be with such models: models that are simple and sometimes perhaps too simple to be realistic (not that unrealistic models cannot be useful!).

A perhaps inevitable response to simple models is to try to improve them, to address perceived inadequacies. This leads to what we will term extended models of landscape change, which often are models that are too realistic to be simple.

The evolution of models from "too simple to be realistic" to "too realistic to be simple" invites a more general discussion of modeling and (perhaps unavoidably) of modelers. We close this chapter with a pragmatic discussion of modelers and modeling, with the aim of cultivating an appreciation of the modeling process—especially for landscape ecologists who are not personally very interested in modeling. Even those who are not particularly invested in models will encounter them regularly in their work, and so a cautious appreciation of models and modeling is a useful professional skill to develop.

9.2 Models in Landscape Ecology

Landscape ecology as a discipline has always been heavily invested in models of various kinds (Risser et al. 1984; Urban 2023). It will be useful to begin this chapter with some foundational issues in modeling, before delving into actual practice. We will return to some general issues after considering a few examples.

9.2.1 Why Use Models?

Ecologists use very many kinds of models (Zurrell et al. 2022), and so no simple list of reasons for modeling can suffice. Models can serve:

- As a logical system for organizing ideas. We all do this, however informally or unconsciously. One might argue that models of this sort are how we translate *observations* into *information*: observations only take on meaning in the context of some model.

- As a framework for comparisons. We might use a model as a way to compare different ecosystems, or similar ecosystems in different regions. In this, the model forces us to collect measurements to conform to its structure and parameterization, in effect providing the experimental design for the comparison.
- To interpolate or extrapolate in space. We do this routinely when we map the potential distribution of a focal species or resource using some sort of species distribution model (Chap. 2). This can be done with nonspatial, implicitly, or explicitly spatial models (Peters et al. 2004). An implicitly spatial model is one that is nonspatial in its formulation but provides spatially structured output because it is driven with spatially structured data as input. Spatial models explicitly incorporate spatial processes or constraints.
- To explore realistic or hypothetical scenarios. We do this to evaluate alternative management strategies, or to explore the implications of natural or anthropogenic changes to systems (e.g., land use change). This is especially important for landscapes, which are not often amenable to experimental approaches for logistical (large size, difficulty of control) or ethical reasons (e.g., working with rare or endangered species).
- To explore attribution. For example, could *this mechanism* be responsible for the observed state of the system? Attribution is explored via controlled model experiments using a "all else being equal" design.
- To make specific predictions (forecasts). While this might seem the most straightforward and common application of models, it tends to be the least frequent application in practice. This is largely because the high levels of uncertainty in most models (especially landscape models) tend to steer modelers away from specific predictions. (This is why climate modelers present their model runs as *scenarios* rather than *predictions*.) Nonetheless, it is precisely these forecasts—associated with quantitative estimates of uncertainties—that increasingly are demanded by managers and policy-makers (Clark et al. 2001).

9.2.2 Stages of Model Development

Regardless of the purpose, models tend to be built in a sequence of logical stages, beginning with the initial conceptualization and proceeding through successively more technical steps of actually encoding the model to its ultimate application for purposes as itemized above. The workflow of model-building might be outlined loosely as follows (Fig. 9.1).

Conceptualization This is the stage where we realize that we need a model and decide what the model should do. That is, what question should the model address, and what would constitute a useful answer to the question? What processes will be included in the model (and so, what will be left out)? This process often leads to a "box and arrow" (Forrester) diagram of the model, which depicts its components and how these interact (recall prototype structural equation models in Chap. 7 and

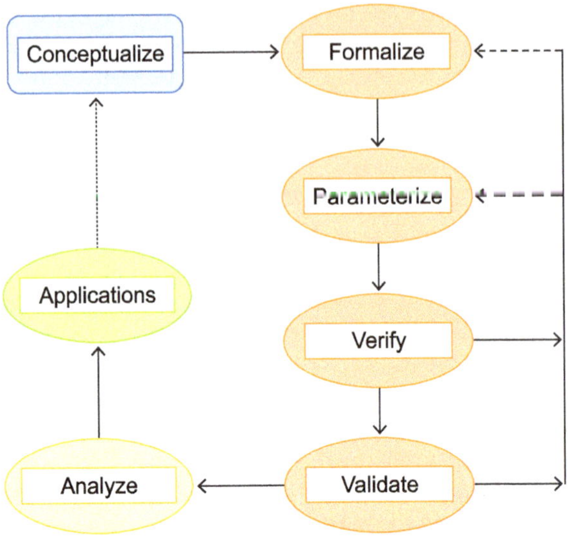

Fig. 9.1 Workflow for model development, implementation, and application. Model failures in verification or validation lead to revisions in formalization and/or parameter estimates (or perhaps in the conceptual model); applications include model analysis, but can also act as further validation tests, while perhaps also inviting extensions to the original conceptual model

means-ends models in Chap. 8). To the extent that this conceptual model represents our best understanding of the system of interest, this also might constitute the "plan" stage of the adaptive management process.

It is at this stage that the aphorism "All models are wrong but some are useful" (Box 1976) comes into play. A model designed for any given application will unavoidably be inappropriate (or worse) for other applications. It is crucial to be explicit about what the model is intended to do, as these decisions will also affect what it will *not* do.

Formalization How will the key processes be represented in the model? What are the state variables? What equations will represent these (are they linear or nonlinear)? What other relationships or interactions will need to be specified?

This is where the code gets written and the equations get translated from the language of statistics or mathematics into a computer language (C++, Java, R, Python?). In principle, choices made at this stage do not affect the behavior of the model, though they might influence its ease of use.

Parameterization This is the stage in which actual numbers get assigned to the equations that formalize the model. That is, while the formalism might be expressed in terms of named coefficients:

$$P(y) = b_0 + b_1x_1 + b_2x_2 + \ldots + \varepsilon \tag{9.1}$$

here we need to specify what *values* the *b*'s take on. In practice, the form of the equation and its coefficients are often fit simultaneously, in that the data will not fit the equation if the equation is inappropriate. Still, it is a good policy to implement the model in general terms (i.e., code it in terms of named coefficients and parameters) and then to provide the numerical estimates of the parameters separately. This separation also will make it easier to interpret the performance of the model (see below).

The previous steps result in a working model, one that can be used. The next few steps actually use the model:

Verification Model verification is the first of two model tests. Strictly, verification demonstrates the model does what it was intended to do, using the data used to build the model. Goodness-of-fit statistics in regressions are a familiar example of model verification (the regression explanatory power, R^2, is one summary of this).

Note that this is *not* a strictly independent test of the model, as the same data used to fit the model are also used to test it. Thus, verification is a necessary but not sufficient test of the model.

Validation Model validation is a test using data that were *not* used to build the model and in this sense represents an independent test of the model. The further removed from the model construction the test is, the more demanding. For example, we might build a forest simulation model for the Southern Appalachians in the southeastern USA and test it with data from elsewhere in the Appalachians. A successful test would build some confidence in the model. But a successful test of the model using data from New England or the Pacific Northwest, or Sweden, would be a more demanding and thus more satisfying test of the model. We considered some approaches for model validation when we explored species distribution models (Chap. 2).

Analysis Often, it is appropriate to do some systematic analysis of the model, to gain insight into its behavior: *Why does it do what it does?* One of the compelling reasons for developing models is that the model itself often can offer general insights as a consequence of its (simple) formalization. We turn to this in more depth in Sect. 9.5.

Applications Once a model is validated, it is ready for the applications it was designed to support. These include many types of formal analyses (above) but also include extrapolations in space or time, the exploration of alternative scenarios, and so on. Again, we return to some examples in Sect. 9.5.

Clearly, model validation is an ongoing process: as the model is pushed further in applications, the successful applications (which are also validations) help to define its *domain of applicability*, the realm of applications within which it provides trustworthy behavior. We will pick up this theme of model evaluation after working through some examples.

Even before we explore some examples of landscape models, the stages of development (above) suggest guidelines for reporting and communication because the reporting will recapitulate the stages of model development. We return to reporting in Sect. 9.5.3.

On Models and Modelers (Reassurance)

At this point, some reassurance is perhaps warranted. Most landscape ecologists and managers will not ever become full-time modelers. But many will work with models on occasion (as we have suggested in previous chapters!). And essentially all of us encounter models as consumers of information: models developed by others and used to inform policy or practice. Examples include regressions of many kinds, species distribution models (Chap. 2), path diagrams or causal chains used to develop or motivate management alternatives (Chaps. 7 and 8), and information conveyed in public discussions of issues such as land use planning and climate change.

Our dive into models is intended as a tutorial on how models work and some guidance for their application. An appreciation for these concepts should help end users be more informed and confident consumers of models. This is the real aim of this chapter.

In the next section, we consider data on landscape change such as that collected during landscape monitoring. These sorts of data invite their own interpretation, but they also serve as a foundation for developing more formal models of landscape change.

9.3 Monitoring and Trend Detection

Monitoring data are generated by repeating a sampling program, either from inventory or a more targeted purpose (Chap. 1). The result is a set of the same variables as observed at the same places over time—a stack of data sets.

Monitoring is essentially a task of signal detection: is there a trend that can be identified and extracted from the background noise? We should all be so lucky as to work with monitoring data like the iconic CO_2 time series from Mauna Loa (Keeling et al. 1976; data now curated at https://gml.noaa.gov/ccgg/trends). In this data set, the trend is so noise-free that the trend is evident on casual inspection; even the increasing amplitude of seasonal variation is readily apparent. But most monitoring data are not so idyllic!

On landscapes, we often work with trends in land cover, land use, or land condition (Homer et al. 2004; Fletcher and Fortin 2018). None of these is as simple and straightforward as carbon dioxide. Land cover change illustrates the issue nicely.

9.3.1 Monitoring Land Cover Change

A common observation of landscapes is that their composition and configuration changes over time. Perhaps this reflects that reality that landscape ecologists actually live in landscapes, and so cannot avoid noticing change. One such change, in most parts of the world, is a trend toward more development and urbanization. It is hard to deny this trend, but the trend itself provides a compelling (if sobering) entry into the empirical reality of landscape change.

It is instructive to unpack the trend of "increasing development" into its component parts. First, the perceived trend is a *concept*, and to document the trend requires *data*. Data on land use and land cover are not the same: land cover is superficial (what is actually seen on the ground?), while land use is about practices on the ground (is the farmland plowed?). For landscapes, these are typically subsumed into a classified map of land use/land cover (LULC), based on satellite imagery. To document a trend, we would require such classified imagery over time.

Several realities complicate this seemingly simple demand. First, satellite imagery has not been readily available everywhere over a long time period (much as current availability might suggest otherwise), and so data on historical LULC are sometimes hard to find. Second, LULC classification tools have themselves evolved over time, so that early maps might have been generated using different tools than later maps. This is good, in terms of the classification models getting better over time; but it means that the older classifications cannot be compared directly to newer data products.

Trends in land cover in the Triangle region of the North Carolina Piedmont (USA) illustrate many of the issues in monitoring landscape change. This region has experienced explosive development since the 1970s, a time span neatly covered by the Landsat satellite mission. Sexton et al. (2013) used a multi-year time series of Landsat images to calibrate a LULC classification to the 2001 National Land Cover Dataset (NLCD) map product (Homer et al. 2004). Sexton et al. then extrapolated the 2001 classification backward (to 1985) and forward (to 2005) in time.

The classification was complicated by the availability of cloud-free images, minor geographic errors of registration (which blur otherwise clear distinctions on the ground), interannual variability in image brightness, and seasonal variability in spectral signatures (which, for example, separate agricultural land uses from similarly "bright" developed land covers). In sum, the extension of the classification in time retained an overall classification success that was equal to or better than the NLCD itself. . .and yet these small complications accrue to complicate the detection of a widely appreciated trend (Fig. 9.2).

Medium-density development shows a strong trend over time. A trend such as this naturally invites a more formal set of questions about its statistical robustness: Is it significant? (Here, $P < 0.0001$.) How much of the variance does it explain? (Adjusted R^2 is 0.86.) Is the trend linear, or is it increasing or decreasing over time? (Here, the data will not support a nonlinear model.)

It might be emphasized here that other land cover types do not show statistically significant trends over the same time period—even though such trends (e.g., an increase in development, a loss of forest cover types) are widely assumed to be occurring. In the case of high-density development, this might be because this class is an end-member and additional development might well be occurring in sites that are already classified as high-density. Low-density development is actually a mix of land covers (lawns, buildings, roads, shade trees, etc.), and slight variations in the spatial registration over time might result in a given low-density pixel being

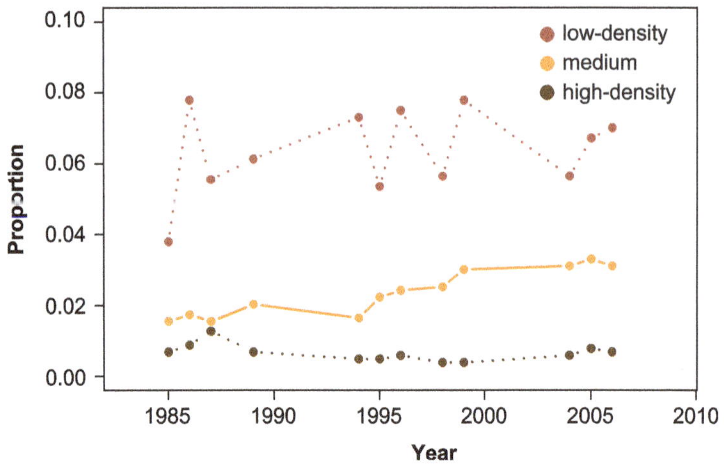

Fig. 9.2 Trend in three density classes of developed land cover in the Triangle region of the NC Piedmont (Sexton et al. 2013). The trends are confounded by various sources of uncertainty and noise; the dashed lines are not significant as linear trends ($P > 0.10$), while the solid line is ($P < 0.0001$)

classified as many other cover classes in any given year; that is, this is a very noisy class with an intrinsically high error rate (Sexton et al. 2013). All of which is to say, actual land cover change data can be substantially more complicated than our expectations.

A second consequence of an observed trend in one land cover type is the follow-on question of associated changes in other land covers. If medium-density development is increasing, then something else must be decreasing. And so it makes sense to consider LULC change as a zero-sum game and to model changes of all types simultaneously.

9.4 Models of Landscape Change

In this section, we first develop a simple model of landscape change, framed in terms of the relative proportions of various land cover types. Such models are often dismissed as being too simple to be realistic, and yet their very simplicity is often what makes them useful and interpretable. In the following section, we extend this simple model to address more realistic issues on landscape change. . .with the result being models that are too realistic to be simple and can be difficult to interpret or communicate to others.

9.4.1 *Simple Models*

An intuitive model of land cover change can be generated by capturing the rates of transitions among land cover classes or states over time. These might be land cover and land use classes, or successional stages, or any other discrete types.

Model Development

The minimum requisite data comprise tallies of the number of samples occupied by each cover type, for two time periods. A cross-tabulation of these tallies generates a *change* or *transition matrix*, a tally of all incidences in which a sample location changed from one type to another during that time period (lack of change is tallied as well, as the diagonal of the change matrix).

This tally matrix **T** is of simple counts and so reflects sample sizes. The sample might be random points within the study area or, in this age of geospatial remotely sensed data, might be an exhaustive tally over the study area (Table 9.1).

To adjust for sample size, the elements of the tally matrix can be divided by the row totals, which converts the tallies into proportions. These proportions can also be interpreted as likelihoods or rates, the chances that a location in state i at the first measurement would be in state j at the next measurement. These rates are in time steps of the measurement data, reflecting whatever time has elapsed between measurements. While the transition rates can be evaluated at this time step, it is often convenient to convert them to annual or other regular intervals (e.g., decadal). This assumes that the transitions during that interval were more or less regular (or can be treated as if they were). The result of this normalization is a new matrix of transition *probabilities* or rates, a *transition matrix* **P**.

The transition rates imply a graphical model of the system, in which each transition describes the likelihood or rate of observed transitions (Fig. 9.3). In principle, any pairwise transition might occur; but in reality only some of these are observed and it is the pattern of these transitions that imply the dynamics of the system. For example, in a succession model where the states are seral stages, flow tends to be forward, in steps from one type to the next but without stages changing "backward" except for disturbances or natural mortality events. Likewise, transitions cannot "skip" stages. In such a system, there are nonzero rates just above the diagonal and in the first column, while the rest of the transitions do not occur. Similarly, in models of land use or land cover change, some transitions are much more common than others, and some transitions—once they occur—are essentially permanent.

Table 9.1 An example of a tally matrix of observed transitions among discrete states, from each state (rows) to each other state (columns)

From/to state	1	2	3
1	T_{11}	T_{12}	T_{23}
2	T_{21}	T_{22}	T_{23}
3	T_{31}	T_{32}	T_{33}

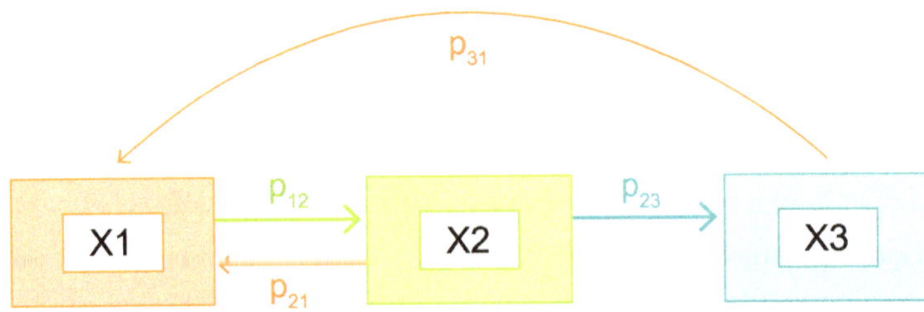

Fig. 9.3 A flow diagram (or Forrester diagram) of a system of three states and observed transitions among them. This example suggests a successional dynamic with overall flow to the right, with some recycling (via morality or disturbance) backward

The transition matrix is actually a complete and accurate summary—an empirical model—of the observed transitions observed over time. To use the model, we will require some additional information. A tally of the proportions of the samples that were in each cover type at each time completes the data requirements.

One simple model of landscape change invokes the transition matrix as applied to the observed tallies of the proportions of the landscape in each cover type:

$$x_{t+1} = x_t P \tag{9.2}$$

where **x** is a vector of the cover types (in proportions) and **P** is the transition probability matrix. This is a first-order Markov chain, or Markov model (Usher 1992; Urban and Wallin 2002).

A Markov model can be projected forward in time, to predict the state of the system after an interval of *k* time steps:

$$x_{t+k} = x_t P^k \tag{9.3}$$

In the limit, this system likely will equilibrate so that further iterations of the transition matrix yield the same vector of states. This equilibrium distribution of states can be calculated from the transition matrix, but in practice it is often approximated by projections of the model, that is, by simulation (and see below).

Assumptions of a Markov Model

As a simple model, a Markov makes simplifying assumptions. In this case, there are three main assumptions that matter:

1. The state of the system in the next time step (time $t + 1$) depends only on the current state (time t). That is, earlier states do not influence the transitions; there is no system memory.
2. The transitions depend only on the state of the system at the observed location. That is, for a given sample, we only need to know about the state of that sample—and, in particular, *not* about the state of its neighbors. There are no neighborhood or spatial effects.

3. The observed transitions themselves are stationary, they do not change over time. That is, it is legitimate to extrapolate them forward in time.

In practice, for a landscape-scale model, it would be difficult to meet these assumptions; indeed, violations of these assumptions are precisely what we find interesting about landscapes! Such violations do not invalidate the model, but they do mean that we need to be careful and transparent about how we interpret model behavior.

One of the reasons we like simple models, of course, is that they *are* simple. When they fail, they fail simply and transparently. In particular, demonstrating such failures is one of the compelling applications of simple models: we can learn a lot by observing how models fail (and see below).

Model Evaluation

Model evaluation is a fuzzy concept because what we evaluate depends on the model, and what is interesting to evaluate depends on why the model was developed. In general, however, there are a few things of interest.

A first step in evaluation is to ask "Did the model do what we expected or designed it to do?". In the case of a Markov model, we expect it to reproduce the observed pattern of transitions. This is a model *verification*, a test of the model with the data used to parameterize it. This is not an independent test, and a successful verification is necessary but not sufficient to build our confidence in the model. For a simple Markov model, the verification *must* work: the model is purely empirical, and if it does not reproduce the observed transitions, it can only be due to "operator error" (miscalculations or coding mistakes) along with (perhaps) some minor rounding error.

Beyond verification, validation is a test of the model using data that were *not* used in developing it. This is an independent test, and if the model validates successfully, it helps bolster our confidence to use it in further applications. (We return to model validation later in this chapter.)

One thing we would like to know of a model such as the Markov example is "What are the longer-term implications of the transitions observed over the short term?" For example, if we build a model based on a decade or so of observed transitions, what do these imply over several more decades? In the case of the Markov model, the answer is provided by projecting the model (analytically or by simulation) to see if and when it reaches a steady state. In particular, the steady-state distribution of cover types is the main *result* of the model.

There are other aspects of model evaluation. But we will defer these until after we have looked at some example applications.

Examples

In this section we consider, superficially, a few examples of simple (Markov) models of landscape change. The aim here is to illustrate the range of insights that can be garnered from simple models.

Forest Harvest Rates in the PNW Urban and Wallin (2002) developed a simple Markov model to explore forest change in the Pacific Northwest of the USA. This

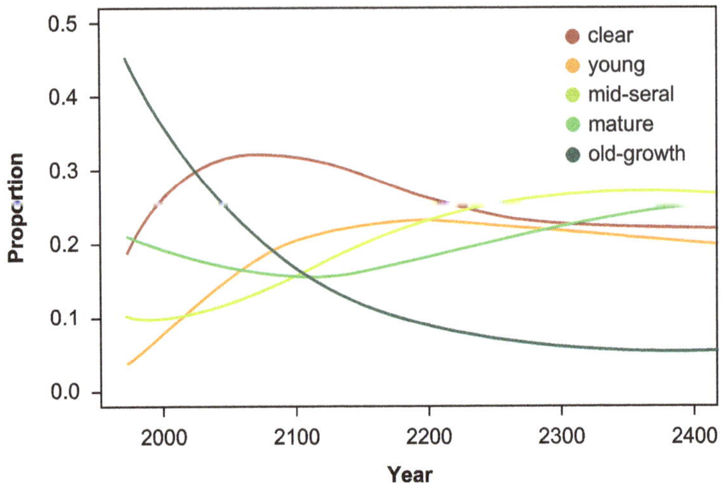

Fig. 9.4 Forest dynamics under transition rates estimated from a small sample of points in the Western Cascades, Oregon, USA (Urban and Wallin 2002). The projection clearly suggests a substantial reduction in old growth over time

was during a time when the amount and fate of old-growth forest was a contentious issue (Ruggerio et al. 1991; FEMAT 1993; USDA/USDI 1994; Marcot and Thomas 1997). One of the confounding issues was that it was not clear how much old growth existed, and the main question was whether forest harvest rates in place at the time were sustainable longer term. The first issue was resolved in part by novel (at the time) analyses of remotely sensed imagery, which allowed an objective estimate of the amount of old growth relative to other age classes of forests (Cohen et al. 1995). The exercise[1] was designed to explore the longer-term consequences of forest transitions observed during the late 1970s into the early 1990s.

The model was constructed from a small sample ($n = 200$) of points extracted from satellite images classified to forest age classes. Forward transitions between age classes then represent succession, while reversions to earlier ages represent disturbances (including timber harvest) or natural mortality.

Extrapolation of the model over time showed another clear implication: the harvest rates over this time period would lead to a dramatic loss of old-growth forests over timescales relevant to regional forest planning (Fig. 9.4).

Two results of this exercise are pertinent here. First, the model verified correctly but failed to validate using data beyond the period used to parameterize the model (1991 versus 1972–1984). This failure clearly demonstrates that the assumption of stationarity was violated; indeed, harvest rates increased over this time period (Fig. 9.5).

[1] It should be emphasized here that this application was designed as a teaching exercise. The sample size was quite small (the original exercise was done by hand), and results are thus a bit noisy and not very precise—though the essential results are probably robust.

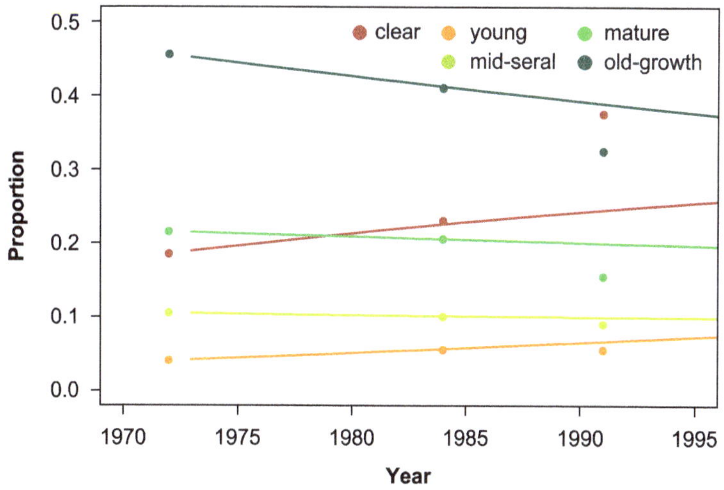

Fig. 9.5 Validation of a simple Markov model of forest clearing for the Pacific Northwest of the USA (Urban and Wallin 2002). Here, the "misses" for age classes in 1991 compared to the years used in model parameterization (1972–1984) show that the transition rates changed over time: harvest rates increased

These simple results are worth noting because similar results (using more robust data sets) were instrumental in changing forest management policy in this region. The models were simple and their simplicity was a benefit in this application.

Succession in Managed Versus Natural Forests Hall et al. (1991) classified satellite imagery into forest cover types for two regions in the Upper Midwest of the USA: a wilderness area and an adjacent area subject to forest management. They generated Markov transition matrices for both study areas over a 10-year interval. Their study is a nice illustration of the interpretative value of simple models.

They compared transition matrices for the two regions and found that the ecological transition rates (i.e., the upper off-diagonal that captures succession) were similar for the two regions, while the lower off-diagonals (disturbances) were different; these differences were obvious on inspection. They found that roughly half of the observations (pixels in the imagery) changed state over the 10 years, while both regions were near the steady state implied by their transition matrices. Again, much of this insight is available from inspection of the transition matrices.

This study also illustrates some of the analytic insights available from the Markov model. This began with the expected steady-state distribution of cover types (i.e., as analytic result rather than via simulation). Their analyses also included estimates of the time a sample would be expected to remain in its current state (*holding time*) and the time it would take for a sample to leave that state and cycle through other states back to the initial state (*cycle time*). Only simple models offer this tractability.

Watt's Unit Pattern Watt (1947) provided a timeless and profound conceptual model for plant ecology in his notion of the *unit pattern*. In this, he posited that the observed distribution of abundances of plant community types (successional stages) reflected the duration of those types over time. This "pattern-process paradigm" has been a foundation of community ecology ever since, and it provides the logical basis for "space-for-time substitution" as a means of studying plant community dynamics such as succession by observing the distribution of seral states at any given time.

Urban (2023, Chapter 2) used field survey data from Watt's earlier studies to generate a simple model of succession for British beechwoods. One notable point in terms of model construction was using the duration of successional stages to estimate transition times between stages. Although not formalized during Watt's career, the approach is relatively generic: If the duration of a stage is T in time (e.g., years), then the transition rate of that type to the next in succession can be roughly estimated as $t = 1/T$. This simple approximation generated a Markov model of beechwoods dynamics.

Two takeaway messages come from this model. First, the steady-state distribution of successional stages *is* Watt's unit pattern: the "community in harmony with itself" (Watt 1947:19). This is a space-for-time substitution, translating the abundances of types observed on the ground in a synoptic survey into the temporal dynamics of succession.

Second, Watt presented his conceptual model of the steady state, in part, to underscore how difficult it might be to observe this in nature. Especially, chance events such as disturbances would propagate anomalies into the distribution of seral stages, which anomalies might persist for many decades or even centuries.

Simulations of a simple model based on Watt's data illustrate both points (Fig. 9.6). First, the steady-state distribution of seral stages depends only on the transition rates inferred from the durations of those stages; the model converges on the same distribution no matter what the starting conditions. Second, an arbitrary perturbation of the system (mimicking a large disturbance) injects a transient wave of readjustment that might last for a century or more. Thus, Watt's simple model introduces a conceptual foundation for space-for-time substitution while also warning us that it might be really difficult to apply this model to real systems.

Succession in the NC Piedmont A similar application in succession modeling can be developed for the Piedmont of North Carolina, USA. This example focuses on an iconic depiction of forest succession (Fig. 9.7), from Odum (1953) and Johnston and Odum (1956). The succession diagram, in turn, was based largely on detailed descriptions of plant communities of various successional ages by Oosting (1942). Many of these field surveys were on sites now maintained as part of the Duke Forest.

This approach uses a *chronosequence* (a collection of sites of varying ages) to make inferences about succession, a common method of space-for-time substitution.

As with Watt's beechwoods, a simple Markov model can be generated based on reported duration and transition times for the community types. Oosting's (1942) studies emphasized early-succession types, especially old fields from abandoned agriculture (these are floristically quite interesting, if short-lived). The transition

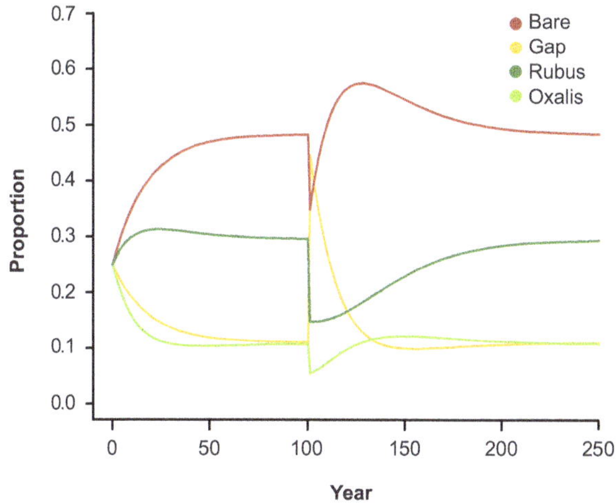

Fig. 9.6 Projection of a simple Markov model of succession in British beechwoods (after Watt (1947)). The forests are dominated by beech (*Fagus*) and the stages are defined in terms of the understory; the sequence is gap (created by the death of a canopy-dominant tree), bare (very dense shade), *Oxalis*, and then *Rubus*. The steady state was perturbed by an arbitrary disturbance in year 100, with the resulting transient persisting for another 100+ years. (Reproduced from Urban (2023), permission conveyed via Copyright Clearance Center, Inc.)

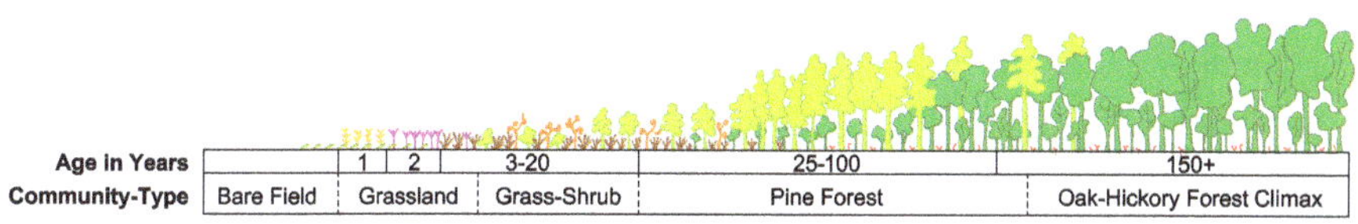

Fig. 9.7 Schematic of secondary succession in the Piedmont of North Carolina, based largely on observations of the Duke Forest (redrawn from Odum (1953), Johnston and Odum (1956); After Oosting (1942)). Oosting's work summarized plant communities in terms of their successional ages, including the duration of those stages. (Permission conveyed via Copyright Clearance Center, Inc.)

rates are only approximate, as the authors reported ranges of values for the durations of seral stages.

The Duke Forest model generates a successional sequence consistent with that depicted in Fig. 9.7 (Fig. 9.8). This should not be surprising: the model *should* do this, by definition. The simulation shows a rapid turnover of old fields and shrub types and a later transition from pine to oak-hickory forests. The oak-hickory forest begins to assert its dominance after about 100 years, as the hardwood understory emerges through the senescent pines.

This example is introduced here because it is an iconic model of forest succession. . .and a model we will revisit in the next chapter.

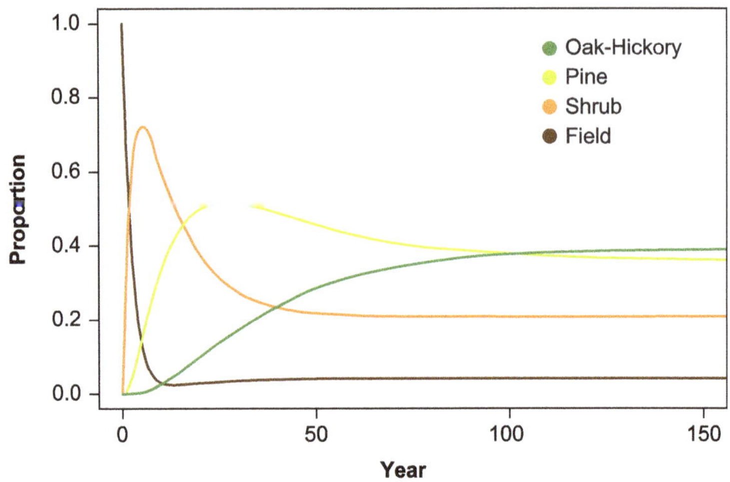

Fig. 9.8 Secondary succession in forests of the North Carolina Piedmont, as projected using a first-order Markov model estimated from the transition times shown in Fig. 9.7

9.4.2 Extended Models of Landscape Change

The models illustrated in the previous section are appealing in their simplicity, but this simplicity can sometimes detract from their broader acceptance. For example, the dynamics from such models can be a bit *too* clean and smooth, and it is easy to show that some of the basic assumptions are often not met by real systems. There are model extensions or modifications that can address these limitations.

More realistic temporal dynamics can be implemented by using a semi-Markovian approach that allows time lags before transitions and distributes the transitions over time (e.g., Acevedo et al. 1995, 2001). Nonstationary transitions can be addressed by computing a sequence of different transition matrices over time (i.e., updating the rates) or by modeling the transition rates to be explicitly functions of time. If history matters, a higher-order Markov model might be used, in which the transition likelihoods depend not only on the current state but also on previous states. Foster (1992) has shown just how pervasive and important such historical legacies can be. Finally, spatial influences can be incorporated, as contagion effects (e.g., patchy disturbances or development patterns) or as gradient effects (e.g., processes that depend on topographic position).

The trade-off in such model extensions is that the models are more realistic but no longer as readily tractable. Here we consider some examples of extended models, to illustrate the types of models that result from such improvements. We then turn to the issue of how to evaluate models that are too realistic to be simple.

Agent-Based Models

We can develop a popular approach to land use change modeling by returning to the species distribution models (SDMs) we explored in Chap. 2. If we envision development as the expansion of "human habitat," then an SDM such as a logistic regression can be fitted to describe human habitat affinities. The data required are two sets of samples: (1) sites that have been observed to be developed over a given time interval (known positives, or presences in SDM terms) and (2) a second set of samples that represent either (2a) locations that were observed to *not* be developed (known negatives or absences, in SDM terms) or (2b) a set of samples representing locations that *might have* been developed (pseudo-absences, in an SDM). Typically the pseudo-absences would be a set of random points over the study area but excluding sites that could not possibly be developed for logistical or political reasons. (Recall Chap. 2, Sect. 2.2.3, for the logic behind these data choices.)

Set up this way, the model is fitted exactly as with other SDMs. In application, a logistic regression is a popular choice, with the predictor variables selected to capture elements that might distinguish human preferences (e.g., distances to work or shopping centers or other amenities) as well as logistical variables (such as proximity to existing infrastructure, topography). In social-science applications, these models are often termed *hedonic models*, in that they model human preferences (e.g., Geohegan et al. 1997; Irwin and Geoghegan 2001; Veldkamp and Lambin 2001).

We now can extend this starting model. To extend a hedonic model for a landscape, the model can be embedded in another model. In the extended model, virtual "developers" or homebuyers, termed *agents*, are presented with a sample of available sites. Each site's suitability is evaluated using the hedonic model. The virtual agent then chooses an appropriate site (e.g., with the highest suitability or hedonic value) and that location is developed. This process is repeated for a very large number of agents. The development process must be constrained to a target allocation (e.g., based on observed development rates or how much development is expected in the future); the model suggests *where* that development might occur.

An agent-based model can be extended further by including different types of agents: individual homebuyers (perhaps with various preferences represented by different hedonic models), commercial developers, and so on. This constitutes a *multi-agent system* (Parker et al. 2003).

Agent-based models and multi-agent systems have been a popular choice in simulating land use and land cover change (Verburg et al. 2006; NRC 2014). The challenge, of course, is in estimating the hedonic models. Empirical estimates are typically based on surveys administered to local or regional populations. These can be very data-hungry models.

Hybrid Models

Modeling land cover change is intriguing but challenging, in ways that capture most of the compelling issues in modeling landscape change. While multi-agent systems might nicely capture the *kinds* of sites that are developed, and the expected amount of development, they are not designed to simulate the *spatial pattern* of development. To capture the spatial signature of land use change, other models are used.

One common way to capture the contagious nature of development is to use a *cellular automaton*. An automaton is a cell-based model in which transition rules depend on the state of the cell (pixel) and the state of the cell's neighbors (e.g., its four or eight neighbors). In land use change modeling, an automaton might be used to force development to be a contagious process that spreads through neighborhoods.

An automaton still needs to be informed by the suitability of sites for development, and so these models tend to be hybrids that use multiple components. Pickard et al. (2017) reviewed several such models, comparing their performance when implemented for the same study area and input data sets. They considered four models, which nicely illustrate the variety of approaches to this common task. Summaries here are taken largely from Pickard et al. (2017).

GEOMOD This model (Hall et al. 1995) includes a cellular automaton and a site suitability component and requires a user-specified target allocation (amount of development). Optionally, development can be stratified geographically (e.g., by county or other internal region), and also optionally, development can be constrained to occur only at the edges of developing regions.

SLEUTH SLEUTH (Slope, Land cover, Exclusion, Urbanization, Hillshade) (Clark et al. 1997; Jantz et al. 2010; Chaudhuri and Clarke 2013) includes a cellular automaton and a set of transition rules that control spontaneous development, new centers of spreading development, growth at developed edges, and development influenced by roads. Transitions are constrained by elevation and topography (which are primary constraints on road-building).

LCM The Land Change Model (Lein 2003) works from three drivers: (1) a target allocation (amount of change), (2) transition potential based on site suitability (e.g., via an SDM tool), and (3) change predictions based on an internally calibrated Markov model of "from" and "to" (developed) land cover types.

FUTURES The FUTure Urban-Regional Environment Simulator (Meentemeyer et al. 2013) consists of three linked modules. A *demand* module estimates a target amount of development based on a model that relates historical development amounts to changes in population. A *site potential* module summarizes site suitability (e.g., via an SDM tool). A *patch-growth algorithm* handles the spatial configuration, by using an empirical library of observed spatial patterns (patch sizes and shapes) of developed land covers. During the simulation, the demand function is refreshed at each iteration to generate land demand for the next time interval.

Pickard et al. (2017) noted that these models were quite heterogeneous in terms of their ability to capture (verify) various aspects of land cover change. In principle, a model of land cover change should be able to predict the *amount* of development, the *location* of new development, the *pattern* of what other land covers tend to be developed (i.e., what development comes from), and the *configuration* of developed land covers. No existing model can do all of those things perfectly.

In another review, Sohl et al. (2016) compared a different set of land use change models, in very large-scale applications over the USA. Similar to Pickard et al.'s results, they found inconsistent results over the modeled scenarios and no clear solution to reconciling these.

For our purposes here, these reviews provide two key insights. First, they underscore the crucial need, at the outset of any modeling application, to be very clear about the intent of the application. A model's failure to reproduce any given aspect of a real system is not a failure if predicting that aspect was not an objective in the first place.

Second, the sheer complexity of these models invites us to ask "What might one do with a model this complicated?". Even quite sophisticated models of development often cannot reliably predict actual development (if they could, land use change modelers would all be real estate investors and wealthy beyond imagination!). So what can we learn from models that are too realistic to be simple? We turn to this question in the next section.

9.5 Model Applications

A model that includes processes implemented in realistic detail is unlikely to be interpretable at a glance; nor is it likely to yield a simple solution that describes model results (outputs) in terms of inputs.[2] And yet there is a great deal we can learn from such models, by analyzing them to discover why they behave the way they do. In this section we consider some general forms of model analysis that can provide such insights.

One way we learn from models is to ask the models themselves why they act the way they do. Formal analyses include sensitivity and uncertainty analysis. In other cases, we pose model experiments—"What if?" scenarios—to explore model behavior and expectations, which are typically variations on an "all else being equal..." theme.

9.5.1 Model Sensitivity and Uncertainty

Model sensitivity and uncertainty both refer to how much model output varies in response to slight variations in model input. In the case of sensitivity, the emphasis is just that: an input parameter is *sensitive* if slight variations in that parameter elicit large variations in model output. A parameter is *uncertain* if it is sensitive to variations within our ability to estimate that parameter. Both aspects are readily estimated through model analysis.

[2] At a seminar early in my career, I presented a fairly complicated model, a spatially explicit individual-based bird metapopulation model. In the discussion that followed, someone in the audience asked "Why would anyone want a model like that?" It's a fair question.

Sensitivity Analysis

In a simple model such as a regression, parameter sensitivity is straightforward: A regression coefficient by definition expresses the incremental change in the dependent variable we expect, given an incremental change in the independent variable. These can be compared across predictor variables only if the parameters are standardized (e.g., to z-scores) so that possibly different measurement units can be reconciled. Given standardized coefficients, parameters with the largest absolute values are the most sensitive.

But in many models, the effects of an input on an output are not as direct. For example, in a path model (Chap. 7) with indirect effects, the influence of a given variable might show up in multiple places in output, or its influence might be expressed (or even negated) through various intermediary pathways. In simulation models, it can be unclear how the influence of any given input might propagate through the system.

In sensitivity analysis, these complications are resolved through a systematic analysis of the parameters. In the analysis, a large number of simulations are conducted with the model. In each instance, the value of each input parameter is varied by an arbitrary amount, say ±10% of its nominal (mean) value. These perturbations of the parameter values can be stochastic (random within the ±10%) or systematic (varied incrementally over that range of values); the choice might depend on how many parameters the model includes. The outputs from all of these simulations are then collected, and the outputs are regressed on the inputs. The result of this analysis—a multiple regression problem—is a summary of the partial explanatory power of each input parameter, within the constrained ranges of variation. Again, by definition, a parameter is sensitive if minor changes in its value elicit large changes in output.

Sensitivity analysis is a powerful way of summarizing why a model does what it does: A model responds to its sensitive inputs. This is then also a powerful way to discover which parts of a model are most important and, by extension, which parts might warrant further scrutiny or follow-up study.

Uncertainty Analysis

Model uncertainty is related to sensitivity. A sensitive parameter elicits substantial variation in model output. An *uncertain* parameter is one that cannot be estimated with much precision. This becomes an issue with models if an uncertain parameter is also sensitive: in that case, we cannot really distinguish between valid model output and output due to errors of estimation.

Uncertainty analysis proceeds essentially like sensitivity analysis. The difference is that instead of perturbing parameters by a uniform range (e.g., 10% of their means), the parameters are perturbed by a range of their respective errors of estimation (e.g., ±1 standard error). The rest of the analysis proceeds as with sensitivity analysis. The result is a set of model outputs and a set of inputs that represent estimation errors. Parameters that explain a large proportion of variation in the output are sensitive within their errors of estimation; they cause high uncertainty in model output.

Uncertainty analysis is revealing and useful because it tells us which part(s) of the model could be improved by better empirical estimates of the associated parameters. A parameter that is uncertain but not sensitive is not really a critical concern. By contrast, a parameter that is both uncertain and sensitive warrants further attention. In this way, model analysis can marshal future research to improve the model most efficiently.

9.5.2 Model-Based Scenarios

Often, a motivation for model construction is to devise a tool that will allow us to ask "What if?" questions. These might look like model predictions, but they are more restricted in form. These are forecasting scenarios, and the model predictions are couched in a set of assumptions embodied in the model itself but also specified explicitly in the modeled scenario (Clark et al. 2001).

Model-based scenarios come in various forms. But three common applications include the assessment of management alternatives, retrospective analysis to try to attribute possible causation, and explicit forecasts of future scenarios. Many such modeling exercises, if based on geospatial inputs, can be mapped explicitly to highlight locations that are particularly compelling or interesting for any given application.

Alternative Management Options

A common application of models is to use them to ask "What if" questions about alternative management scenarios. These might be management interventions or policies that drive such interventions. Often, the comparison includes a "do nothing" alternative to represent the status quo.

We have already considered these applications in Chap. 7, using structural equation models (Sect. 7.4.1 and Fig. 7.10). We also highlighted alternative management scenarios using means-ends diagrams (Sect. 8.3.2) as a part of structured decision-making in Chap. 8 (recall examples in Sect. 8.5).

Shoemaker et al. (2019) used the FUTURES model (Sect. 9.4.2) to explore policy scenarios to influence land use change in terms of the provision of ecosystem services under urbanization. They focused on watershed protection (pollution abatement), protecting sensitive habitats, and carbon sequestration. No scenario simultaneously improved all three services, suggesting necessary trade-offs in land use planning (recall Chap. 8 for trade-offs in decision-making).

Miller and Urban used a detailed forest simulator (Miller and Urban 1999a) to explore fire management alternatives, contrasting prescribed burns and mechanical fuel reductions (Miller and Urban 2000). They were able to use the model to assess a range of prescribed fire intensities, finding that it would require burns hotter than normal practice to achieve the same fuel reductions as mechanical thinning. These are model-based experiments, which are a logistically feasible way to "preview" on-the-ground experiments that are more logistically difficult and expensive.

Historical Effects and Attribution

Models are often used to evaluate possible explanations of observed outcomes, with the causes typically being in the past and unobserved. This is not really hindcasting (which would mean running a model backward in time), but rather detective work aimed at *attribution*: what might have caused the situation we observe? A substantial amount of work in landscape ecology is about model-based attribution.

One appealing aspect of structural equation models (Chap. 7) is that they allow us to explicitly describe alternative paths by which influences can propagate through a system, either as direct or indirect effects. Fitting the model from data yields estimates of the relative importance (and sometimes plausibility) of the various paths, in terms of relative explanatory power and significance levels. The means-ends models used in structured decision-making would provide similar inferences, if actually fitted to data (as compared to being used as conceptual or heuristic guides).

We use many types of models in attribution applications. These include, rather prominently, a long history of studies that aim to infer the relative importance of various aspects of landscape pattern in shaping ecological responses. The long-running debate over the relative importance of habitat loss (i.e., habitat area itself) versus the spatial configuration of that habitat (especially fragmentation itself, or connectivity) is a telling example of the challenges of model-based attribution when the competing explanations are correlated (e.g., Fahrig 2003, 2017; see also Urban 2023, Chapter 5).

This same issue of area versus configuration, and confounded influences, is repeated in analyses of urban streams as affected by watershed pattern: Is it the total area of impervious surfaces that matter, or how these are arranged and connected? This is a complicated case—the *urban stream syndrome* (Walsh et al. 2005)—but one that is yielding to analytic approaches borrowed from terrestrial applications (recall Sect. 7.4.2 and see Urban 2023, Chapter 9).

Future Scenarios

Model forecasts into the future are typically cast as *scenarios* rather than precise predictions. A scenario is a *forecast* bounded by a stated set of assumptions and with full disclosure of uncertainties (Clark et al. 2001).

Projecting species distribution models into the future is a familiar example of a model-based forecast. In this case, it might not be obvious to the reader just how many assumptions are being made: that habitat suitability as modeled is the main constraint on species distribution, that dispersal is not limiting, that interactions with other species are not important, and so on. Again, open communication about the assumptions underlying the forecast is critical.

Forecasts are especially challenging for systems characterized by substantial natural variability or processes beyond our control. Land use change models are a good illustration: While we might understand the main processes that drive land use change, we cannot really know about future events that could easily change the trajectory of change. For example, a new (unexpected) development project might spawn contagious development around it; real estate markets influenced by larger-scale macroeconomic trends might alter the demand for land; and so on. Such events induce *path dependencies* into processes such as land use change: observed changes depend on particular events that can change trajectories, and these can be difficult to anticipate (Brown et al. 2005).

What forecasts like this provide is an appreciation of the *kinds* of outcomes we might expect. For example, land use change models might offer insights into the patterns of development expected from policies favoring infill in urban centers rather than suburban sprawl. Climate-change models are a familiar example of future scenarios: they predict changes in storm frequency and intensity, but they do not predict individual storms.

Mapping Model Outputs

A compelling feature of many models used in landscape ecology is that they are driven by geospatial data and so their output can be mapped in a geographic information system (GIS). Through mapping, the user can better understand not only *why* the model behaves as it does (through model analyses, above) but also *where* it does what it does. This can be enormously informative.

We have already considered one example of mapping model predictions, when we mapped the output from a species distribution model (Chap. 2, Sect. 2.4.4, Fig. 2.9). In particular, mapping misclassifications can show, at a glance, cases that might be ecologically interesting. Such cases might include species occurrences in locations predicted to be "nonhabitat" (false negatives), but which are occupied because of a dispersal from nearby source habitats (Pulliam 1988). Similarly, unoccupied sites predicted to be "habitat" (false positives) might be so because they are isolated. Both misclassifications, though strictly model failures, are expected from metapopulation theory and accessible by mapping model predictions.

Model sensitivities and uncertainties also can be mapped, if these analyses are conducted in terms of geospatial variables. We considered one example in Chap. 1, in which model sensitivity analysis was used to help guide the design of a climate-change monitoring program (Chap. 1, Sect. 1.3.4 and Fig. 1.10; Urban 2000).

In regression-based models, model parameters (i.e., fitted coefficients) can be mapped into geographic space. This can be done using geographically weighted regressions (Brunsdon et al. 1998; Fotheringham et al. 2002; Dale and Fortin 2014) or varying coefficient functions in models (Osborne et al. 2007); these are considered as extensions to species distribution models in Supplement 2S.1.3. If the coefficient in a regression model varies substantially over the study area, this implies that relative importance of the variable's contribution to habitat suitability also changes spatially. For example, we might expect a regression coefficient representing temperature to vary from north to south over a geographic range of a species. Mapping model parameters can capture this graphically.

Finally, some model outputs are readily interpretable only if they are mapped. Models of habitat connectivity are perhaps an obvious example. Some such maps are rather straightforward to interpret (e.g., the locations of managed corridors), while others might require some analysis (e.g., using network theory: Urban and Keitt 2001, McRae et al. 2012, Dale and Fortin 2014, Carrea Ayram et al. 2016, Lechner et al. 2017; and see Urban 2023, Chapter 6). For example, Ashander et al. (2022)

used network analysis to help design a monitoring program for efficient and early detection of invasive species spread.

Models and Model Applications

Perhaps most models, once developed, are used for multiple purposes: model experiments, attribution, and forecasts. Extrapolating a species distribution model over a large study area and then projecting it under a future climate scenario is one example.

We have already touched on several applications of a forest simulator developed by Urban et al. (2000): comparing fire management options (above, Miller and Urban 2000) and using sensitivity analysis to help guide a climate-change monitoring program (Urban 2000, Chapter 1). The model has also been used to explore climate-change scenarios, focusing on couplings between climate, forest process, and fire (Miller and Urban 1999b). These applications used a model that was developed initially to help synthesize existing data and marshal future research (Stephenson and Parsons 1993).

Of course, the converse of "one model, many applications" is also true: many applications rely on multiple models integrated together. Projecting a species distribution model into a future climate couples the SDM to various emissions scenarios, which scenarios are themselves simulated with one or more climate models. Forecasting land use and land cover change usually embeds another model that classifies land cover from satellite imagery. In all such applications, sensitivities and uncertainties also integrate over the models (termed *error propagation*), and reporting and communication must be open and transparent about all of the sources of uncertainty in the application.

In sum, models have many uses. A key to applications is to be very clear in communicating how the application was developed and how it should be interpreted. We turn to this in the following section. We then return to a more general consideration of model applications and how we might learn how to judge which applications might be appropriate for any given model.

9.5.3 Reporting and Communication

Although models are—by definition—simplifications of reality, models can be complicated and unwieldy in terms of the details that might need to be communicated to fully explain any given application, especially to a nontechnical audience.

Often, an audience does not need to know all of the minute details that went into model specification and parameterization (although these details do need to be available for those who want to learn more). More important are higher-level motivations about why the model was constructed the way that it was and how its applications are supported. This naturally leads to a multi-level presentation of the model: first the conceptual model in broad outlines, then the overall structure and logic, and then the details of implementation and parameterization. For many consumers of models, the first two layers are more critical than the details.

For many modeling applications, this leads to a checklist of what to report, a scheme that might be nested, perhaps with some details relegated to appendices or supplements. Information reported here would naturally follow the steps of model-building (Sect. 9.2). Reporting should include:

☑ What are the specific objectives of the modeling exercise? How did these objectives guide decisions about what processes or constraints to include and what to exclude from the model? This high-level explanation is crucial because it communicates not only what the model will do but also what it will not do.
☑ How was the model implemented? In particular, how was it bounded or restricted in spatial and temporal scale, to focus on the application at hand? What design criteria guided these decisions?
☑ How was the model coded? What platform? What programming language? These decisions should not influence the model's behavior, but they might have an impact on how user-friendly the model is or how easy it is to interface the model's input and output with other platforms (especially a GIS).
☑ What data were used to parameterize the model, and how was this process conducted?
☑ How was the model verified? Was it validated? If so, how independent were the validation data from the original parameterization and verification of the model?
☑ What analyses were conducted with the model? Sensitivity? Uncertainty? Was there a general solution to the model available via parametric analysis or simulation? Not all of these analyses would apply to all models, but collectively they confer some appreciation of how the model works, why it does what it does.
☑ Finally, how was the model applied and how were the applications structured to be as informative as possible? For example, if the model was used to explore alternative scenarios, the details of these scenarios need to be reported: what was varied or manipulated, what was held constant, and what else was assumed in the scenario?
☑ In discussing model results, what assumptions or parameterizations in the model should temper any interpretation of the results? That is, how should the reader interpret the results? The modeler will always have deeper insights into model behavior than a consumer of the model, and the modeler needs to help the reader understand the application as much as possible.

9.6 On Models and Modelers

The stages of model-building are important to recognize, because they admit varying levels of model success and failure. In particular, a model might fail (i.e., might not reproduce behaviors observed in real systems) because it is inappropriate conceptually, because its formalization is inadequate, or because it is poorly parameterized.

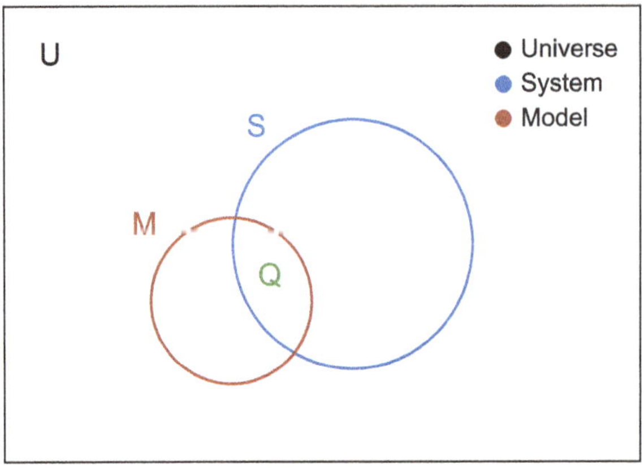

Fig. 9.9 Venn diagram representation of model "goodness-of-fit". (After Mankin et al. (1975)). The universe of all possible behaviors of the system is *U*, and *S* is those behaviors that actually have been observed of the system. The set of model behaviors, *M*, intersects *S* as *Q*, which is those observed behaviors that the model has reproduced

In the first case, a conceptual failing suggests a fundamental misunderstanding of the system. An inadequate formalization suggests that, while the conceptual model might be correct, the form of the relationships is not quite close enough (e.g., a key relationship is implemented as linear when in fact it is nonlinear). Finally, an inadequate parameterization might merely indicate that the data used to fit the curves were insufficient. It is important to recognize these levels of model failure, because a conceptual failure is a failure at a very basic (and personal!) level, while an inadequate parameterization can be rectified by simply collecting more data: model failures are not all equally damning.

9.6.1 Model "Goodness"

This discussion of model failure invites a more general discussion of how to evaluate the relative "goodness" of a model. It is tempting to demand that a model reproduce the behaviors of a real system rather exactly, and this expectation corresponds to a conventional definition of model "goodness-of-fit" as applied to regression analysis. But all models are, by definition, simplifications of reality and so all models ultimately will fail this test.

Mankin et al. (1975) provided an insightful rubric for evaluating models. They used Venn diagrams to illustrate the intersection between model behavior and the observed behavior of the focal system (Fig. 9.9). In this, the encompassing space *U* represents the set of all behaviors that *might* be observed of the real system (some which have not yet been observed). The model demonstrates a set of behaviors *M*.

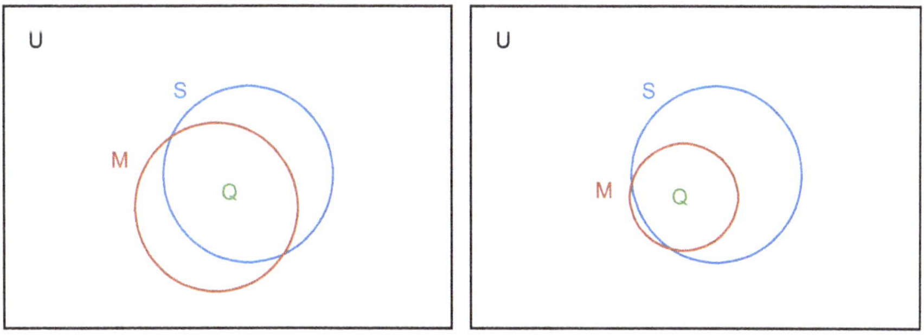

Fig. 9.10 Illustrations of a model (left) with high adequacy but perhaps low reliability and (right) high reliability but lower adequacy

The real system has been observed to demonstrate a set of behaviors *S*, that is, we have *data* representing *S*. The intersection of sets *M* and *S* is *Q*, the subset of behaviors observed of the real system that the model can reproduce. Mankin et al. noted, to begin, that a model can be *useful* if *Q* exists at all. This subset *Q*, in fact, defines the model's *domain of applicability*. To anticipate a bit, even if this domain is very small, the model will be useful if we never stray from this domain in applying the model. That is, even a "bad" model can be useful if we apply it carefully. Thus, our task in model evaluation is to define this domain as a guide to confident applications. And we might fine-tune the adage from *All models are wrong but some are useful* (Box 1976) to *All models are wrong but any model might be useful—if used carefully*.

Mankin et al. defined two more nuanced measures of model "goodness." A model's *adequacy* increases as *Q* increases relative to *S*: the model can do more of the things that the system is known to do. A model's *reliability* increases as *Q* increases relative to *M*: more of what the model does is shared by the system (equivalently, the model tends to *not* do things that the system has not been observed to do). These distinctions are useful because they underscore the crucial insight that a model can be quite *useful* even if it is often wrong. For example, a model with high adequacy might reproduce system behavior but also exhibit many behaviors that have not been observed of the system (Fig. 9.10, left); alternatively, a model might have a very limited range of behaviors, all of these corresponding to behaviors observed of the real system (Fig. 9.10, right). In the former case, the model might often be "wrong" (make predictions not in *S*), but, importantly, we will never learn anything *new* if we never push our understanding of the system toward behaviors that have not yet been observed. In this sense, the latter model will be right as long as the applications stay within its domain, but the model also will always be boring in the sense that its predictions will always be familiar.

In other disciplines, the trade-offs between model reliability and adequacy are sometimes easier to see. For example, physicists routinely use models that make predictions that are outside the set of observed behaviors: discoveries of new planets or new subatomic particles are often anticipated by models long before we have the technology to actually detect (measure) the new entity. In the strictest sense, these discoveries are made by models that are wrong—they make predictions beyond the boundaries of the known system—but we learn new and exciting things from this. On the other hand, models in pharmacology and medicine must be constrained so that their behaviors are within the accepted boundaries of the system: exceptions to this might well be a matter of life or death.

9.6.2 Model Evolution

The Venn diagrams for model evaluation also underscore an important practical issue in model evaluation: this is an ongoing process, in which the accumulated weight of evidence defines the domains illustrated in the diagrams. Thus, a model test (a comparison of model prediction or output to measured data) establishes a single point in this space (i.e., *somewhere* in Fig. 9.9), and the domains of S, M, and especially Q are defined only by accumulating a large number of points as test cases. Thus, our understanding of a model, the definition of its domain of applicability, and our trust in it evolve over time as the model is used.

Models also evolve in another sense, a sense that perhaps has more to do with modelers than with models. Models tend to start simply, as modeling is, in its very essence, a process of abstraction: the art of capturing the essential features of a system as parsimoniously as possible. In testing, simple models of ecological systems often are found wanting—they are inadequate. And so the natural tendency is to add more complexity to the model, to make it more realistic (i.e., more adequate). This tendency captures a fundamental dilemma in modeling, the trade-off between simplicity and realism. As we have already noted, simple models tend to be too simple to be realistic, while realistic models tend to be too realistic to be simple. This trade-off can be expressed in terms of *sins of omission* and *sins of commission* (Fig. 9.11). Sins of omission are model failures caused by oversimplification of the system (i.e., leaving out important details). Sins of commission are model failures caused by inclusion of details that are poorly understood or inadequately represented (e.g., where the process is real but its parametric uncertainty is very high). Importantly, model uncertainty increases in the aggregate as the level of

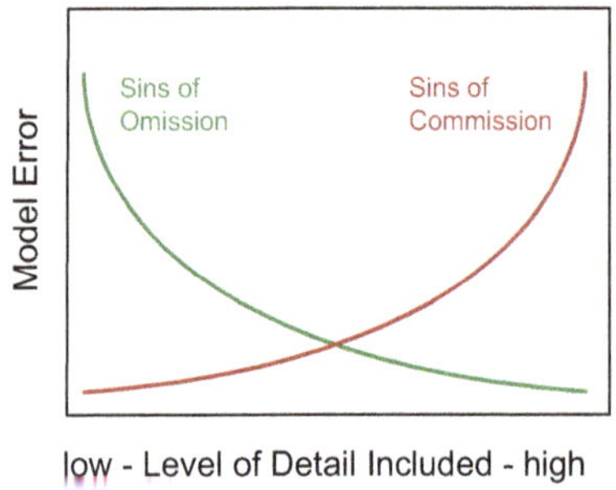

Fig. 9.11 Trade-offs in model complexity, in terms of model inclusions and exclusions (after Gardner and Urban (2003))

detail increases, simply because of the increasing number of (perhaps poorly estimated) processes included in the model. The trade-off curve implies an optimal level of model complexity (i.e., the intersection of the two curves), but, in fact, this balance can only be approached by actually implementing and evaluating a range of models for the same system.

Gardner and Urban (2003) discussed this tendency for models to evolve toward greater complexity.[3] This natural evolution, unfortunately, also tends to produce models that are increasingly unwieldy, and more uncertain, even as the model's apparent adequacy increases. Missing from this evolutionary sequence is the step at which modelers reassess the model, to simplify it if possible. Again, models are simplifications of reality and the greatest utility of models comes from the insights that can be garnered from this simplification. Urban (2005) reviewed some approaches for simplifying complicated models.

9.7 Further Reading

Modeling has ebbed and flowed as a topic in ecology. There is a journal dedicated to this topic (*Ecological Modelling*), but much of the foundational work is rather old (but still useful), or new and quite specific to particular application domains (e.g., land use change modeling, species distribution modeling).

Haefner (1996) offered a general but useful treatment of modeling for biologists. His book includes high-level guidance on model development, analysis, and evaluation as well as pragmatic advice on implementation (e.g., how to choose an appropriate curve to model a nonlinear relationship).

Mladenoff and Baker (1999) edited a collection of approaches to modeling forest systems at the landscape scale. Dale (2003) is another guide for ecologists and natural resource managers. Canham et al. (2003) is a compendium on the use of models in ecosystem science. This book covers a wide range of material, including some also covered in this chapter (e.g., trade-offs between realism and simplicity). In that volume, Urban (2003) offers a commentary on the two-way partnership between modelers and consumers of those models.

Zurrell et al. (2022) provide a wide-ranging review of ecological models and applications. While their focus is on spatial applications in animal conservation and restoration, their review and typology of modeling applications—and their practical recommendations—are useful more generally.

[3]Bob Gardner and I developed this idea in conversations with Bob O'Neill. I still think of Fig. 9.11 as "O'Neill's conjecture" because while we think this is true, it is really hard to demonstrate these trends empirically for any actual model.

9.8 Summary and Prospectus

While most landscape ecologists might not self-identify as modelers, the reality is that much of landscape ecology and management relies—implicitly or explicitly—on models of various types. An appreciation of how models work, and how they can be applied usefully, is important to anyone working in this arena. Models can be developed for many purposes, from conceptual guidance to very specific predictions, but all models share a common ontology and a set of best practices for implementation, evaluation, and reporting.

Many models in landscape ecology originate in a data set that captures some aspect of landscape behavior or change over time. Models help us synthesize and extrapolate these observations. This might be by extrapolating, with a species distribution model, a set of species observations from a limited census, to a map of suitable habitat over the entire study area. It might be capturing a trend in monitoring data and extrapolating that trend into the future, using models ranging from simple regressions to complicated multi-agent simulators. Application domains tend to evolve toward a shared set of modeling approaches and standards, a community of practice.

An appreciation of the modeling process helps inform practitioners who encounter models developed and applied by others. In the best of cases, any modeling application is presented in sufficient depth and nuance that the consumer can understand why the model was constructed as it was, why it does what it does, and what its behaviors or predictions imply (how these should be interpreted, what uncertainties should temper any interpretation). This communication is a two-way conversation: modelers need to be open and helpful, and end users of models need to be informed consumers. In this sense, all of us are modelers.

References

Acevedo M, Urban DL, Ablan M (1995) Transition and gap models of forest dynamics. Ecol Appl 5:1040–1055

Acevedo MF, Ablan M, Urban DL, Pamarti S (2001) Estimating parameters of forest patch transition models from gap models. Environ Model Softw 16:649–658

Ashander J, Kroetz K, Epanchin-Niell R, Phlps NBD, Haight RG, Dee LE (2022) Guiding large-scale management of invasive species using network metrics. Nat Sustain 5:762–769

Box GEP (1976) Science and statistics. J Am Stat Assoc 71:791–799

Brown DG, Page S, Riolo R, Zellner M, Rand W (2005) Path dependence and the validation of agent-based spatial models of land use. Int J Geogr Inf Sci 19:153–174

Brunsdon C, Fotheringham S, Charlton M (1998) Geographically weighted regression. J R Stat Soc D 47:431–443

Canham CD, Cole JJ, Lauenroth WK (eds) (2003) Models in ecosystem science. Princeton University Press, Princeton

Carrea Ayram CA, Mendoza ME, Etter A, Salicrup DRP (2016) Habitat connectivity in biodiversity conservation: a review of recent studies and applications. Prog Phys Geogr 40:7–37

Chaudhuri G, Clarke KC (2013) The SLEUTH land use change model: a review. Int J Environ Resour Res 1:88–104
Clark KC, Hoppen S, Gaydos L (1997) A self-modifying cellular automation model of historical urbanization in the San Francisco Bay area. Environ Plann B Plann Des 24:247–261
Clark JS et al (2001) Ecological forecasts: an emerging imperative. Science 293:657–661
Cohen WB, Spies TA, Fiorella M (1995) Estimating the age and structure of forests in a multi-ownership landscape of western Oregon, USA. Int J Remote Sens 16:721–746
Dale VH (ed) (2003) Ecological modeling for resource management. Springer, New York
Dale M, Fortin M-J (2014) Spatial analysis: a guide for ecologists, 2nd edn. Cambridge Univ. Press, Cambridge
Fahrig L (2003) Effects of habitat fragmentation on biodiversity. Annu Rev Ecol Evol Syst 34:487–515
Fahrig L (2017) Ecological responses to habitat fragmentation *per se*. Annu Rev Ecol Evol Syst 48: 1–23
FEMAT (1993) Forest ecosystem management: an ecological, economic, and social assessment. Report of Forest Ecosystem Management Assessment Team, July 1993. FEMAT, Portland, Oregon
Fletcher R, Fortin M-J (2018) Spatial ecology and conservation modeling: applications with R. Springer Nature Switzerland AG, Cham
Foster DR (1992) Land-use history (1730–1990) and vegetation dynamics in central New England, USA. J Ecol 80:753–772
Fotheringham AS, Brunsdon C, Charlton M (2002) Geographically weighted regression—the analysis of spatially varying relationships. Wiley, Chichester
Gardner RH, Urban DL (2003) Model testing and validation: past lessons and present challenges. In: Canham CD, Cole JJ, Lauenroth WK (eds) Models in ecosystem science. Princeton University Press, Princeton, pp 184–203
Geohegan J, Wainger LA, Bockstael NE (1997) Spatial landscape indices in a hedonic framework: an ecological economics analysis using GIS. Ecol Econ 23:251–264
Haefner JW (1996) Modeling biological systems: principles and applications. Chapman and Hall, New York
Hall FG, Botkin DB, Strebel DE, Woods KD, Goetz SJ (1991) Large-scale patterns of forest succession as determined by remote sensing. Ecology 72:628–640
Hall C, Tian H, Qi Y, Pontius R Jr, Cornell J (1995) Modelling spatial and temporal patterns of tropical land-use change. J Biogeogr 22:753–757
Homer C, Huang C, Yang L, Wylie B, Coan M (2004) Development of a 2001 national land-cover database for the United States. Photogramm Eng Remote Sens 70:829–840
Irwin EG, Geoghegan J (2001) Theory, data, methods: developing spatially explicit economic models of land use change. Agric Ecosyst Environ 85:7–23
Jantz CA, Goetz SJ, Donator D, Clagett P (2010) Designing and implementing a regional urban modeling system using the SLEUTH cellular urban model. Comput Environ Urban Syst 34:1–16
Johnston DW, Odum EP (1956) Breeding bird populations in relation to plant succession on the Piedmont of Georgia. Ecology 37:50–62
Keeling CD, Bacastow RB, Bainbridge AE, Ekdahl CA Jr, Guenther PR, Waterman LS, Chin JFS (1976) Atmospheric carbon dioxide variations at Mauna Loa Observatory, Hawaii. Tellus 28: 538–551
Lechner AM, Sprod D, Carter O, Lefroy EC (2017) Characterizing landscape connectivity for conservation planning using a dispersal guild approach. Landsc Ecol 32:99–113
Lein JK (2003) Sensing sprawl: towards the monitoring of urban expansion using Dempster-Shafer theory. Geocarto Int 18:61–70
Mankin JB, O'Neill RV, Shugart HH, Rust BW (1975) The importance of validation in ecosystems analysis. In: Innis GS (ed) New directions in the analysis of ecological systems, part 1. Simulation Councils proceedings series, vol 5. Simulation Councils, La Jolla, California, pp 63–71

Marcot BG, Thomas JW (1997) Of spotted owls, old growth, and new policies: a history since the Interagency Scientific Committee report. USDA Forest Service General Technical Report PNW-GTR-408. PNW Forest & Range Expt. Station, Portland

McRae BH, Hall SA, Beier P, Theobald DM (2012) Where to restore ecological connectivity? Detecting barriers and quantifying restoration benefits. PLoS One 7:e52604

Meentemeyer RK, Tang W, Dorning MA, Vogler JB, Cunniffe NJ, Shoemaker DA (2013) FUTURES: multilevel simulations of emerging urban-rural landscape structure using a stochastic patch growing algorithm. Ann Assoc Am Geogr 103:785–807

Miller C, Urban DL (1999a) A model of surface fire, climate, and forest pattern in the Sierra Nevada, California. Ecol Model 114:113–135

Miller C, Urban DL (1999b) Forest pattern, fire, and climatic change in the Sierra Nevada. Ecosystems 2:76–87

Miller C, Urban DL (2000) Modeling the effects of fire management alternatives in Sierra Nevada mixed-conifer forests. Ecol Appl 10:85–94

Mladenoff D, Baker WL (eds) (1999) Spatial modeling of forest landscape change: approaches and applications. Cambridge University Press, Cambridge

National Research Council (NRC) (2014) Advancing land change modeling: opportunities and research requirements. National Academies Press, Washington, DC

Odum EP (1953) Fundamentals of ecology. W.B. Saunders Co., Philadelphia

Oosting HJ (1942) An ecological analysis of the plant communities of Piedmont, North Carolina. Am Midl Nat 28:1–126

Osborne PE, Foody GM, Suárex-Seoane S (2007) Non-stationarity and local approaches to modelling the distribution of wildlife. Divers Distrib 13:313–323

Parker DC, Manson SM, Janssen MA, Hoffman MJ, Deadman P (2003) Multi-agent systems for the simulation of land-use and land-cover change: a review. Ann Assoc Am Geogr 93:314–337

Peters DP, Herrick JE, Urban DL, Gardner RH, Breshears DD (2004) Strategies for ecological extrapolation. Oikos 106:627–636

Pickard B, Gray J, Meentemeyer R (2017) Comparing quantity, allocation, and configuration accuracy of multiple land change models. Land 6(3):52

Pulliam HR (1988) Sources, sinks, and population regulation. Am Nat 132:652–661

Risser PG, Karr JR, Forman RTT (1984) Landscape ecology: directions and approaches. Illinois Natural History Survey Special Publ. 2, Champaign

Ruggerio LF, Aubry KB, Cary AB, Huff MH; (tech. coords.) (1991) Wildlife and vegetation communities of unmanaged Douglas-fir forests. USDA Forest Service General Technical Report PNW-GTR-285. PNW Forest & Range Expt. Station, Portland

Sexton JO, Urban DL, Donohue MJ, Song C (2013) Long-term land cover dynamics by multi-temporal classification across the Landsat-5 record. Remote Sens Environ 128:246–258

Shoemaker DA, BenDor TK, Meentemeyer RK (2019) Anticipating trade-offs between urban patterns and ecosystem service production: scenario analyses of sprawl alternatives for a rapidly urbanizing region. Comput Environ Urban Syst 74:114–125

Sohl TL, Wimberly MC, Radeloff VC, Theobald DM, Sleeter BM (2016) Divergent projections of future land use in the United States arising from different models and scenarios. Ecol Model 337:281–297

Sprugel DG (1991) Disturbance, equilibrium, and environmental variability: what is 'natural' vegetation in a changing environment? Biol Conserv 58:1–18

Stephenson NL, Parsons DJ (1993) A research program for predicting the effects of climatic change on the Sierra Nevada. In: Veirs SD Jr, Stohlgren TJ, Schonewal-Cox C (eds) Proceedings of the fourth conference on research in California's National Parks. U.S. Department of the Interior Park Service Transactions and Proceedings Series 9, pp 93–109

Urban DL (2000) Using model analysis to design monitoring programs for landscape management and impact assessment. Ecol Appl 10:1820–1832

Urban DL (2003) A community-wide investment in modeling. In: Canham CD, Cole JJ, Lauenroth WK (eds) Models in ecosystem science. Princeton University Press, Princeton, pp 466–470

Urban DL (2005) Modeling ecological processes across scales. Ecology 86:1996–2006

Urban DL (2023) Agents and implications of landscape pattern: working models for landscape ecology. Springer Nature AG, Cham

Urban DL, Keitt TH (2001) Landscape connectivity: a graph-theoretic perspective. Ecology 82: 1205–1218

Urban DL, Wallin DO (2002) Introduction to Markov models. In: Gergel SE, Turner MG (eds) Learning landscape ecology: a practical guide to concepts and techniques. Springer-Verlag, New York, pp 35–48

Urban DL, Miller C, Stephenson NL, Halpin PN (2000) Forest pattern in Sierran landscapes: the physical template. Landsc Ecol 15:603–620

USDA Forest Service and USDI Bureau of Land Management (1994) Record of decision for amendments for Forest Service and Bureau of Land Management planning documents within the range of the northern spotted owl. USDA Forest Service and USDI Bureau of Land Management, Portland

Usher MB (1992) Statistical models of succession. In: Glenn-Lewin DC, Peet RK, Veblen TT (eds) Plant succession: theory and prediction. Chapman and Hall, London, pp 215–248

Veldkamp A, Lambin EF (2001) Predicting land use change. Agric Ecosyst Environ 85:1–6

Verburg PH, Kok K, Pontius RG, Veldkamp A (2006) Modeling land-use and land-cover change. In: Lambin EF, Geist H (eds) Land-use and land-cover change. Global change—the IGBP series. Springer, Berlin, Heidelberg, pp 117–135

Walsh CJ, Roy AH, Feminella JW, Cottingham PD, Groffman PF, Morgan RP II. (2005) The urban stream syndrome: current knowledge and the search for a cure. J N Am Benthol Soc 24:706–723

Watt AS (1947) Pattern and process in the plant community. J Ecol 35:1–22

Zurrell D, König C, Malchow A-K, Kapitza S, Bocedi G, Travis J, Fandos G (2022) Spatially explicit models for decision-making in animal conservation and restoration. Ecography 2022: e05787

Chapter 10
Ecological Assessment

Abstract Landscapes change over time, due to natural processes (succession), disturbances (natural or human-induced), or deliberate interventions by managers. Assessing observed changes and evaluating them relative to an expectation—ecological assessment—is a crucial step in the adaptive management process. In this chapter we begin by posing a framework in which to evaluate and visualize change in ecological data sets, such as might be collected as part of a monitoring program. Ordination (Chap. 4) provides this framework. The assessment is articulated in terms of change vectors, movement over time in an ordination that summarizes ecological condition. For discrete events, including management interventions conducted as experiments, this framing leads to impact evaluation, which entails various forms of before/after, control/intervention, or before/after/control/impact inferential designs. Ecological assessment is the final task in the adaptive management cycle. As such, this is the step that defines adaptive management and makes ecosystem management a science-based enterprise.

10.1 Introduction

In the previous chapter we explored landscape change, with an emphasis on capturing trends in land cover and projecting these to show their longer-term implications. The changes we observed were those typical of a monitoring program, again with the emphasis on land cover. But monitoring programs can track ecological changes more subtle and multidimensional than land cover. For example, we might track species composition on a set of monitoring sites (sample quadrats) over time and ask questions about compositional changes due to succession, natural or human-caused disturbances, or climate change. Or we might have implemented forestry, restoration, or other interventions in the form of management experiments. In these cases, we will want to know whether the treatments are having the intended effects. In each of these cases, the focus is on *interpreting* the observed changes.

In this chapter we delve into the assessment of ecological change. The changes might be due to any number of drivers, but the approach will be the same: we will want to detect the changes, describe them, and interpret the changes in terms of

D. L. Urban, *Landscape Ecology: A Task-Oriented Perspective*,
https://doi.org/10.1007/978-3-031-72251-6_10

formal or informal expectations. In terms of the overall workflow of this book, this task represents the final stage ("react") of the logical flow of adaptive management, or the coming together of a braided stream of various applications (Preface, Fig. 2) that all invite this sort of interpretation.

The "react" stage of the adaptive management cycle might imply that this stage is a simple assessment of the underlying model, and the answer will be either "yes" or "no"—the monitoring data are either consistent with the model or not. But ecological change is rarely so straightforward. Often, ecological assessment begins with change detection (as in Chap. 9), to determine whether the observed changes can be distinguished from the background noise of natural variability. Beyond this, an underlying model of the system often suggests a direction and perhaps magnitude of change, and so the detection is gauged relative to these expectations. These are rarely simple "yes/no" answers but rather matters of degree. Further, ecological systems often present surprises over time, and so new questions or hypotheses might arise from the data, beyond those that drove the collection of the data. In short, this can get messy.

In this chapter, we approach ecological assessment as a multistage process. The first task is to construct an interpretative framework in which change can be observed and summarized effectively. Ordination (Chap. 4) provides this framework. The next stage is to explore the observed changes and to refine hypotheses about the nature of observed changes—those anticipated or expected, as well as any that emerge from the data but without prior expectations. The final stage assesses these observed changes inferentially, with an aim to either accept or reject the hypotheses. In the simplest sense, these inferential tests would be forms of "before/after, control/intervention" (BACI) designs as used in management experiments. But, again, natural systems often do not behave so nicely, and we will need to consider other approaches.

10.2 Ordination as a Framework

Ecological data are multivariate, redundant (correlated), and noisy. In Chap. 4, we embraced ordination techniques as a way to summarize such data, extracting the main trends into a low-dimensional reference space while suppressing noise. In particular, we looked at nonmetric multidimensional scaling (NMS) as a tool that maps ecological dissimilarities as directly as possible into ordination space. In this ordination space, samples that are close together are ecologically similar while samples that are far apart are ecologically dissimilar. Other ordinations can also provide such a reference space, depending on the data, but NMS does this by design.

The basis for using NMS as a reference space for ecological assessment is that observed changes in this space—movement of individual samples over time—can be interpreted readily as ecological change. In this section we develop a conceptual appreciation for how ecological change manifests in ordination space and how such changes can be interpreted.

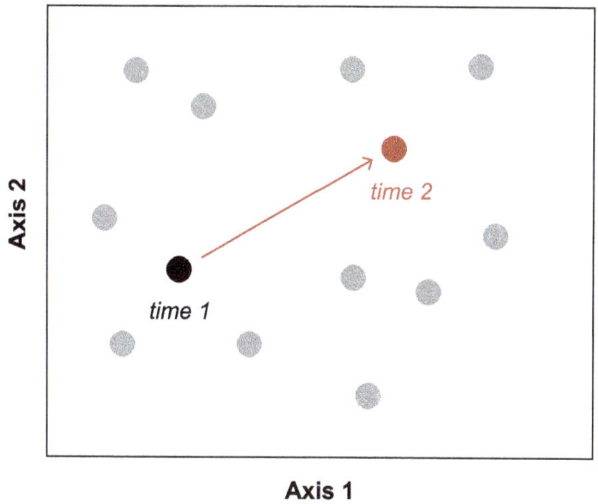

Fig. 10.1 A change vector in an NMS ordination space. The vectors connect the same sample as measured over two measurement intervals and so represent the amount of ecological change over that time

10.2.1 Change Vectors

Samples can be ordinated into NMS space over time in either of two ways. In many cases (and the simplest case), the data already include observations collected over time. That is, the data matrix is samples *x* variables (e.g., species) *x* time. In this case, the pooled samples are ordinated and the reference space represents all ecological conditions as observed over time. In the second case, there is a reference set of samples that defines the reference space, and new samples are added to this space as they are collected. Of course, in this latter case, the ordination could also be generated anew using the new samples; the choice might depend on how wedded the investigator is to the original reference space (and how much any new samples might diverge from this space). To begin, we will consider ordinations based on species composition (with NMS, this means based on compositional dissimilarity; recall Chap. 4, Sect. 4.3.2). But the approach can be generalized to an ordination on any variables.

Given an ordination of samples over time, any sample can be connected to itself at a subsequent measurement time, by drawing an arrow from its position at one time to the next (Fig. 10.1). This arrow is a *change vector*. A series of vectors over multiple measurement intervals would comprise a *change trajectory*. This trajectory might be linear or curved, and the first task in ecological assessment is to interpret the change vectors and trajectories ecologically.

Change vectors are not new (Goff and Zedler 1972), and they have been used in various applications including restoration (e.g., Zedler and Calloway 1999) and succession (e.g., Bergeron and Dubuc 1989; Smith and Urban 1988). Here we will review the basics as these apply to ecological assessments. This interpretation is relevant to monitoring programs (Chap. 1) and efforts to model landscape change (Chap. 9).

Vector Length
If sample separation in NMS space represents ecological dissimilarity, then movement in this space represents ecological change. The *length* of a change vector represents an amount of ecological change (as dissimilarity) over the observed measurement interval. A longer vector represents more change; a short vector represents less change. As these are defined for a given measurement interval, these are change over time: *rates* of change. Longer vectors are more rapid change; shorter vectors are slower.

Vector Direction
Change vectors have length but they also have direction. Because the axes of the ordination have ecological interpretations, then direction of the change vector also has an ecological interpretation or *meaning*. For example, if one of the axes represents a moisture gradient, then movement along this dimension implies a change in moisture status of the sample—or, more precisely, changes in species compositional response to moisture.

In restoration ecology, the reference ordination space often includes samples representing reference or target conditions (i.e., healthy or natural conditions) as well as degraded sites that are candidates for restoration. In these cases, the reference sites are typically separated from the degraded sites on one or both ordination axes. And, so, a change vector in this space has not only meaning but also *value*: we think some of the sites are better than others, and so a change can be better or worse.

Change Trajectories
Connecting a series of change vectors together over time generates a *change trajectory*. Both the lengths and directions of vectors within this trajectory retain their meanings, but the trajectory itself might have a shape (linear, curved), and this shape might also have meaning.

In the particular case of restoration, we want the change trajectory to move from degraded to target conditions, and we hope that this movement is direct (effective) and rapid (efficient) (Fig. 10.2). This figure represents a conceptual model of how to restore southwestern pine forests that are overly dense due to decades of fire suppression (Allen et al. 2002), and it is helpful here to connect this to our previous work. Here, the axes are conceptual constructs ("Structure," "Process"), while the attributes in smaller text on each axis would be empirical indicators for each axis. We might construct the axes in a variety of ways: by ordination, factor analysis, or weighted averaging (Chap. 4). The upper right domain of the figure ("Current conditions") represents degraded conditions, while the lower left domain is the desired (reference) condition. The aim of restoration is to "move" samples from the upper right to the lower left. In this case, the expectation is that this might occur in stages over time (shaded ellipses).

In some systems, there might be some debate about how to get to the target condition most efficiently. We might, for example, try to restore the structure of the system with the expectation that process will follow; alternatively, we might try to restore process and hope that this will restore system structure. These alternatives imply different trajectories of change: nonlinear trajectories moving initially either

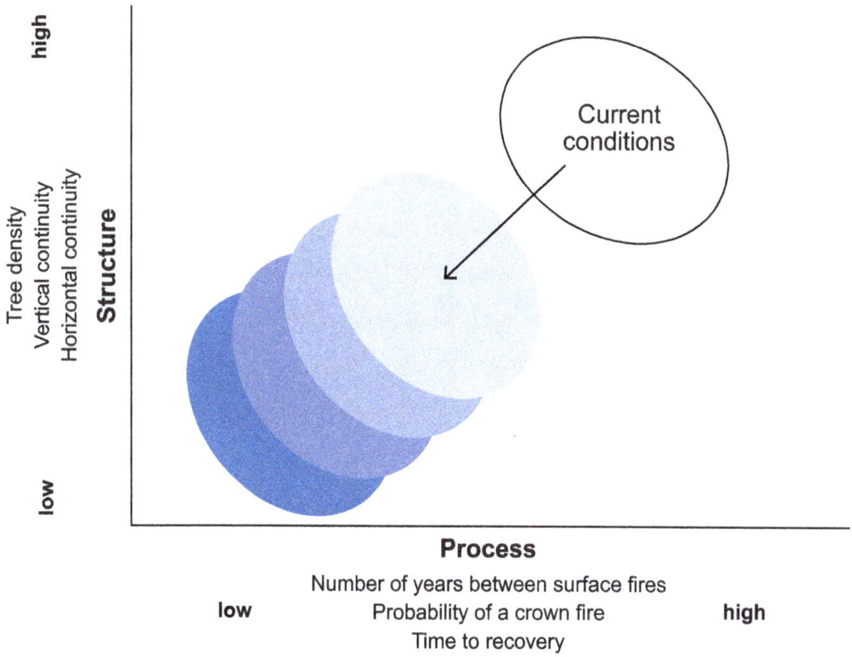

Fig. 10.2 A change trajectory within an ordination that represents degraded sites to be restored, along with target or reference conditions. An effective restoration treatment would move the degraded sites toward target conditions directly and rapidly. (Redrawn with permission from Allen et al. (2002); permission conveyed via Copyright Clearance Center, Inc.)

down (structure first) or to the left (process first) and then curving toward target conditions. These are testable hypotheses within this reference space.

10.3 Workflow: Evaluating Change

The task of ecological assessment is a stepwise process of posing expectations and then evaluating these with appropriate statistical tests. While it might seem that the expectations would be self-evident, this might not be the case in multivariate systems with substantial noise or natural variability. To be sure, it would be desirable for the hypotheses to be stated up front, based on an underlying model. But we will retain the right to discover additional or revised hypotheses along the way, by exploring the reference space.

The workflow (Fig. 10.3) thus begins with a conceptual model, builds the reference space (an ordination space), and then iteratively explores sample movement (change vectors) in this space to pose testable hypotheses. Such tests—significant or not—might then suggest revised or new hypotheses to explore.

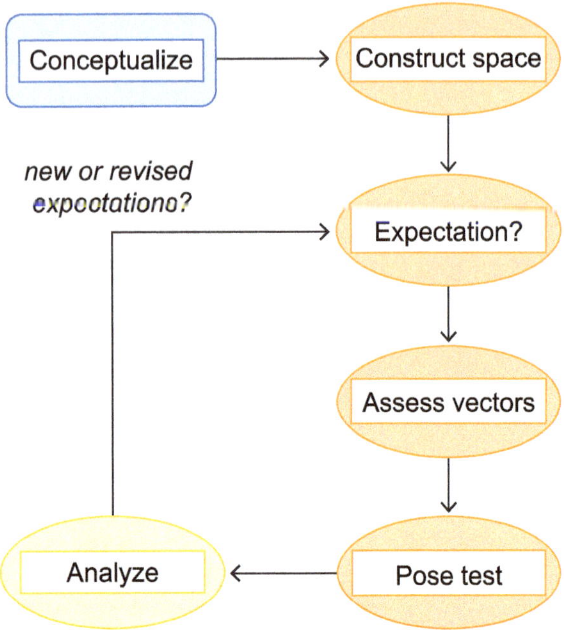

Fig. 10.3 Workflow for evaluating ecological change using ordination as a visualization tool

To be clear, in this approach, the ordination space serves the purpose of condensing the multivariate richness of the system into a manageable, low-dimensional construct. This makes it easier to visualize change and to pose questions or hypotheses about the changes observed in this space. The actual evaluation—the statistical tests—will be done using the richness (and noise) of the full data set.

10.3.1 Interpreting Change and Setting Expectations

It is impossible to declare at this point what changes might be interesting in the constructed reference space: the interesting questions will depend on the system and management context. But some possible questions can be identified in terms of patterns that might be observed within such a space. There are several cases.

Vector Lengths in Different Domains of the Space
Vector lengths imply rates of ecological change over time, and there might be instances where it is reasonable to expect samples in some parts of the space to change at different rates compared to other regions.

For example, in a space where the axes represent environmental gradients of some sort, we might expect drier sites to change more slowly than more mesic sites because of the more favorable growing conditions (and higher growth rates) on the better sites. The same logic might apply to temperature gradients, or other conditions.

In these cases, the tests will focus on vector lengths sorted somehow into regions within the reference space. A box-and-whisker plot is a simple way to display and explore these. Because the ecological distances underlying change vectors are not independent, a formal test of group differences would entail a permutation version of MANOVA (Anderson 2001), a Mantel test, or equivalent (recall Chap. 5 and Supplement 5S.4 and Chap. 6). Oksanen et al. (2021) provide options for permutation tests under a variety of inferential designs.

Vector Lengths over Time

Change vectors over time might speed up or slow down, and this can represent interesting ecology. For example, in succession we often expect changes to be comparatively rapid initially, slowing down later as the system approaches a natural endpoint or steady state (e.g., Odum 1969).

In such cases, the test of interest is a comparison of vector lengths over time. Again, box-and-whisker plots are an easy way to explore these; a more formal test would entail correlating the vector lengths with time as a variable (e.g., "year") or contrasting vector lengths for different time intervals (as discrete factors). While the differences over time (vector lengths) might meet assumptions of normality required of a parametric correlation, the underlying distances are not independent, and a more cautious test would be a randomization version of the parametric test.

Vector Directions

The direction of a change vector often conveys ecological information and often with a subjective value attached to the direction (e.g., restoration toward the desired condition). In such cases, we might naturally ask whether observed changes are heading in the right direction.

Testing the directions of change vectors is really feasible only in the 2D case, and is complicated by the circular nature of directions (ranging from 0° to 360°). To compare directions, the observed changes first must be converted to angles, relative to the ordination axes (i.e., with the vertical axis representing northing and the horizontal axis easting). To solve the circularity issue, these angles can then be simplified to absolute departures (e.g., deviations from north). In cases where there is a clear expectation for the direction of the vectors, the angles can be expressed as departures from this direction (e.g., in Fig. 10.2, the desired restoration trajectory is "southwest").

In some applications, there might be groups whose directions are expected to differ. For example, in a restoration project done as an experiment with control and treatment samples, we would expect the treatment plots to show a clear and consistent movement toward the desired future condition, while the control plots might move less (shorter vectors) and in a less organized (perhaps random) way (Fig. 10.4).

Membership in a Group or Domain

Finally, we might like to know whether a sample can be assigned with confidence to a group or a domain within the reference space. For example, in restoration, we might want to know whether a sample under restoration has entered the domain of the target conditions: Has the restoration been successful?

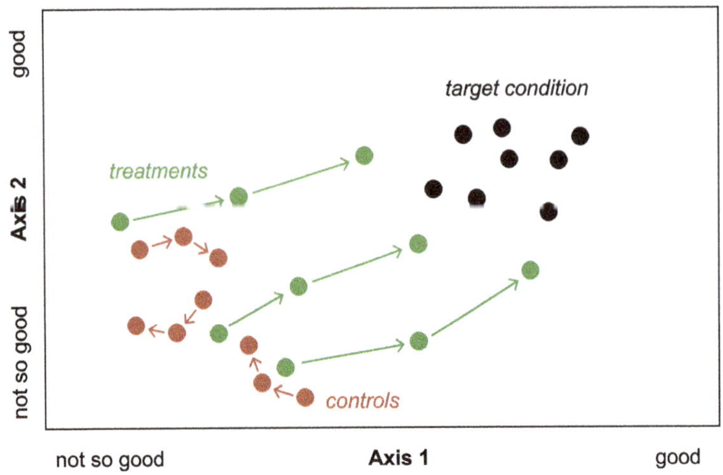

Fig. 10.4 Schematic of a restoration project as framed in ordination space. The axes might be based on species composition (e.g., indicator species) or biophysical factors. The target condition is identified as a domain of samples (black dots) representing the desired or reference condition. Paired treatment/control samples are shown as nearly adjacent in the space, suggesting their ecological similarity (a matching exercise simplified in this space). Under treatment, we expect the treatment samples to move toward the target domain, while the controls might show less and perhaps random change. A treatment might be considered successful when it moves to fall within the domain of the target condition. Here, the reference samples (black dots) are not shown moving, but we would expect them to change as well, perhaps the way the control sites move

These tests entail assessing the locations of selected samples relative to a grouped sample mean or centroid. For example, "Is a restoration treatment sample within the domain defined by the target or reference condition?". The question is whether that sample is significantly different from the reference group centroid, given the variability within that group. In the simplest case, this is a one-sample t-test, but in practice it is more often a multivariate test based on within-group distances or dissimilarities.

If the underlying distances that define the ordination are biophysical, they might be calculated as Mahalanobis distances (Chap. 3, Sect. 3.2.4). Mahalanobis distances approximately follow a Chi-squared distribution (with degrees of freedom equal to the number of raw variables), and so these distances can be converted into probabilities based on the collection of reference samples. Any treatment sample then can be assigned a probability that it falls within a user-defined confidence ellipse around the reference samples. If the underlying variables are species composition, the distances might have an unknown distribution, but the percentiles can be estimated empirically to assign probabilities similarly.

In each of these cases, the aim is to pose an expectation about change vectors observed in ordination space and to state this as a hypothesis that can be assessed readily using familiar statistical tests. The tests typically would be conducted using the full-dimensional data (e.g., based on all species or environmental factors), not the distances observed in ordination space. The ordination serves the sole purpose of simplifying and organizing things so that changes can be observed and communicated more readily.

10.3.2 Presentation and Reporting

While the details of any particular application will vary, all applications of this kind share a similar burden of presentation and reporting. The necessary details include an overview of the application, the details of constructing the reference space, how the hypotheses were generated ecologically and how these were translated into tests within the reference space, and the outcomes of these tests. Many of the components of this presentation have been detailed previously (e.g., how to present an ordination), while other details will be specific to an application.

To describe the construction of the reference space, these details are critical:

☑ A description of the data: what the samples are (samples, variables), sample sizes, any data editing, or transformations of the original data.
☑ What ordination was used to construct the reference space. If NMS, the choice of distance measures, how the number of axes was decided, how the axes were defined (e.g., via species scores or correlations with the original variables), variance captured on each axis and cumulatively (and recall reporting guidelines in Chap. 4).
☑ As changes in ordination space will be interpreted as ecological change, a Shepard diagram (plotting ecological distance versus ordination distance; see Chap. 4) will illustrate the strength of this relationship.

To summarize the analysis of change:

☑ A description of how change vectors were calculated (what time step?) and any other post-processing that might affect interpretations.
☑ A narrative explanation of observed change vectors or trajectories.
☑ A clear statement of expectations to be tested, ecologically and in terms of change vector lengths, directions, trajectories, or group membership.
☑ Details on the statistical test of the hypotheses, framed in terms of the data used to test these (i.e., full data set as compared to the ordination reference space), including an explanation of how the results of such tests are to be interpreted ecologically. These might be communicated in terms of the ordination reference space, if this simplifies matters.
☑ Takeaways from the tests, in narrative ecological terms.

10.4 Illustrations

We will delve into this general approach using two examples. The first begins as a heuristic for ecological restoration and then looks at a real example. The second case is on forest succession. Both examples are applications where the general approach described here is in common use—although not always consistently so. We then turn to the range of analytic approaches that might be used to assess various versions of before/after studies of ecological change.

10.4.1 Ecological Restoration

A restoration project conducted as a management experiment nicely illustrates many of the aspects of ecological assessment (reviewed by Urban 2006). To begin, the ordination framework provides a simple means of choosing pairs of treatment/control samples. Because samples that are close together in ordination space are ecologically similar, choosing nearly adjacent samples as treatment/control pairs is a form of *pre-matching*; the adjacent pairs will be as similar as possible on all aspects except the treatment and should help satisfy the "all else being equal" clause that often confounds management experiments.

Once the treatments are done, we have clear expectations about how the system should respond: the treatment plots should move toward the target domain, while the controls should show slower and less organized movement in the reference space (we will revisit this expectation later). The restoration project is successful if/when the treatment plots fall within the reference domain (Fig. 10.4).

Matthews and Spyreas (2010) used a framework like this to assess a wetland restoration project in Illinois (USA). They constructed an NMS ordination framework based on Bray-Curtis dissimilarities on plant species. They proposed four possible restoration trajectories within this space, depending on whether the restored sites converged or diverged within this space and whether they progressed toward the target conditions or diverged away from the targets. Their first case (Fig. 10.5a) is the one illustrated in Fig. 10.4. The second case (b) would occur if sites were restored with a set of similar plant species which were subsequently filtered by different environmental conditions on each site, leading to divergence. In the third case (c), the sites converge toward the same endpoint but *not* the target conditions; and the fourth case (d) is essentially unordered change. Within the reference space, they identified domains of target conditions as well as degraded conditions.

They found that restored sites initially progressed toward the target conditions, but over time they diverged and moved more toward the degraded sites. This was a result of invasion by nonnative species, which essentially displaced the restoration sites from their initial trajectory (Fig. 10.6).

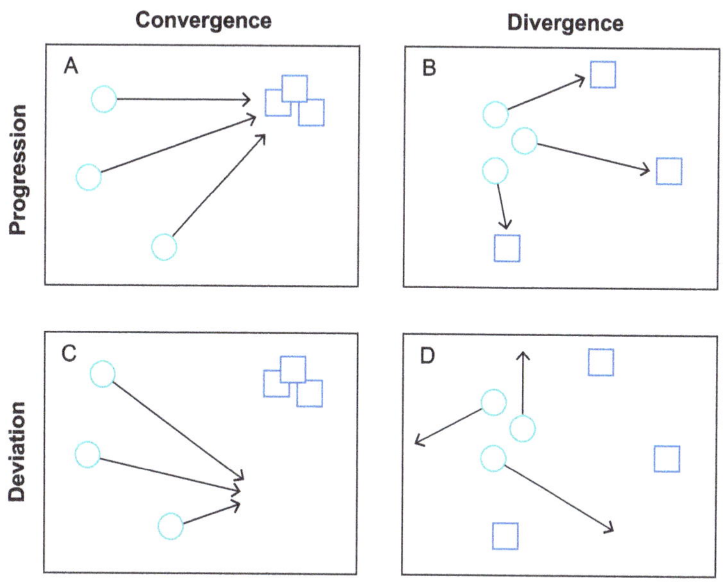

Fig. 10.5 Possible trajectories for a restoration project, depending on whether the restored sites converge on the same condition or diverge toward different conditions (columns) and whether the sites progress toward the desired target or deviate toward some other endpoints (rows). (Figures 10.5 and 10.6 redrawn with permission from Mathews and Spyreas (2010); permission conveyed via Copyright Clearance Center, Inc.)

Fig. 10.6 Change trajectories, in NMS space, observed in the wetland restoration project assessed by Matthews and Spyreas (2010). The initial trajectories toward target conditions (open squares) diverged toward degraded conditions (solid squares) due to nonnative invasives

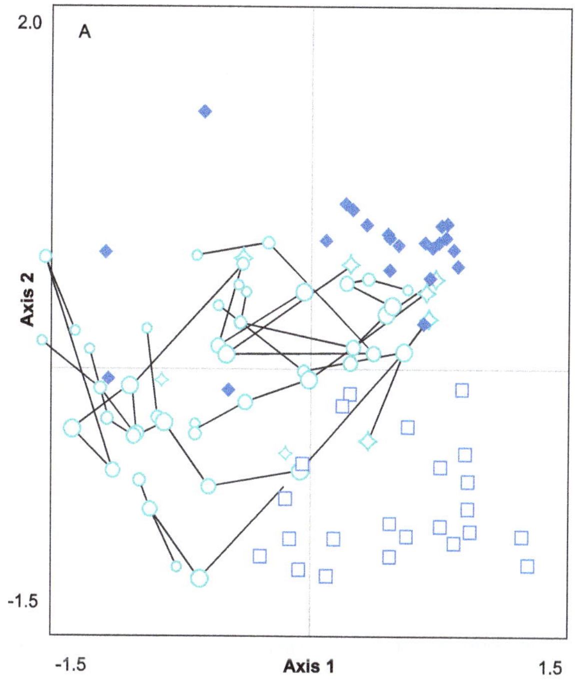

In this example, the ordination framework helped frame expectations about the system and to interpret the observed trajectories—which did not conform to any simple expectation.

10.4.2 Forest Succession

In the previous chapter, we devised a simple Markov model of forest succession in the North Carolina Piedmont (USA), by space-for-time substitution, based on field sites described by Oosting (1942) and interpreted by Odum (1953) and Johnston and Odum (1956) (see Chap. 9, Sect. 9.4.1 and Figs. 9.7 and 9.8). Here we revisit this case study but now based on monitoring data collected in the Duke Forest over a period of roughly 80 years—enough time to actually observe succession. The data and methods are described in more depth by Payne[1] and Peet (2023).

The data set comprised a set of 51 plots initially established in the 1930s. These have been resurveyed regularly over time, at roughly 5-year intervals. There are a total of 48 measurement years spanning 1933–2013, for a total sample size of 625 plot-year samples. Some of the plots have been lost over time (mostly to development) and so the number of remeasurements varies among the plots.

In this data set, there were a total of 66 tree species observed; the analyses shown here included 39 species that occurred on >5% of the samples. Compositional data were relativized by column (species) maxima and row (sample) totals and converted into extended Bray-Curtis dissimilarities (recall Chap. 3).

An NMS ordination was constructed in two dimensions following a step-down procedure that examined solutions from six to one dimensions (Goslee and Urban 2007). In this, a single plot (and its remeasurements) was omitted, as it was compositionally so unusual that it represented its own ordination axis. The two-axis solution captured 73% of the variance in compositional dissimilarity, with 49% of this on the first and 24% on the second axis. The Shepard diagram (not shown) is linear but with scatter about the 1:1 line, suggested that this is a reasonable reference space to use for this illustration. Review Chap. 4 for more on the procedure of NMS.

The samples were classified into forest types by partitioning around medoids (PAM; Kaufman and Rousseeuw (1990)). This analysis was repeated for two to ten groups, and a six-group solution was selected based on among-group contrasts evaluated with Mantel tests. See Chap. 5 for more detail on community classification and PAM.

[1] Chris Payne kindly provided the cleaned and curated data shown here. Some of the results shown here are slightly different from similar analyses in Payne and Peet (2023) because of minor differences in data editing, transformations, and graphical decisions. Chris Payne and Bob Peet kindly reviewed my presentation here.

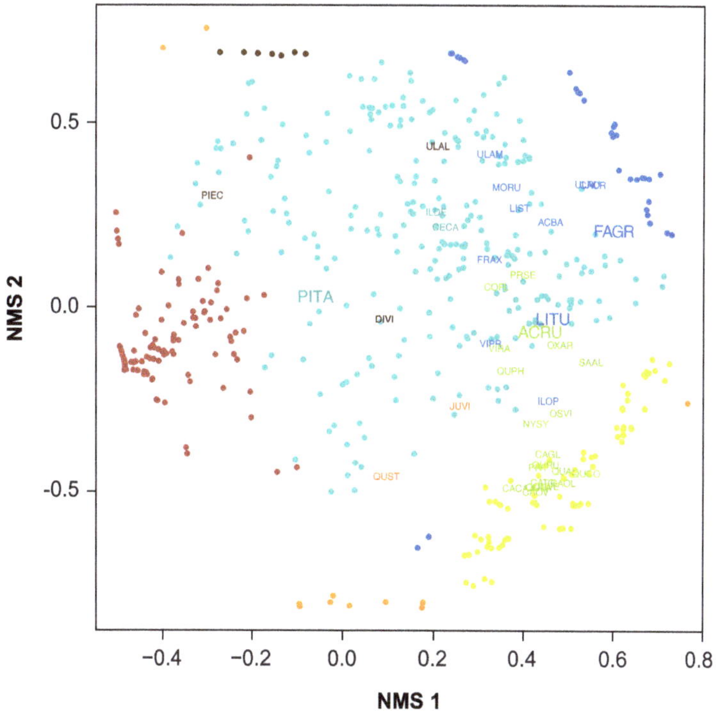

Fig. 10.7 NMS ordination of Duke Forest samples. Samples are color-coded according to classified community types. Species codes are located at their weighted-average positions on the axes (see text)

Species were located into the NMS ordination at their weighted-average positions on each axis (Oksanen et al. 2021), and samples were color-coded by community type. The NMS ordination (Fig. 10.7) reveals a successional gradient on the first axis, with pines (mostly loblolly, *P. taeda* [PIta], in red, but also shortleaf (*P. echinata*, PIec), in black, on the left side of this axis. Mature hardwoods (various oaks and hickories, *Quercus* and *Carya*), in green, occupy the right side. A mixed forest with many other hardwoods occupies the center of the ordination (in cyan). The second axis suggests a soil moisture gradient, with more mesic sites toward the top and more xeric sites near the bottom of this axis. In this figure, the ordination has been cropped slightly on the left side, to ignore an outlier plot and make the image (slightly!) more legible.

In the figure, the six community types include three of less interest here: Shortleaf pine (PIec, in black) changes over time but remains its own type. A xeric hardwood type (in violet) is rather uncommon. A mesic hardwood type (in blue) includes species such as beech (*Fagus grandifolia*, FAgr) and associates typical of bottomland forests. Our focus here is on the loblolly pine/mixed hardwood (cyan, in the figure) and oak-hickory forest types (in green). Note that the mixed forest type also

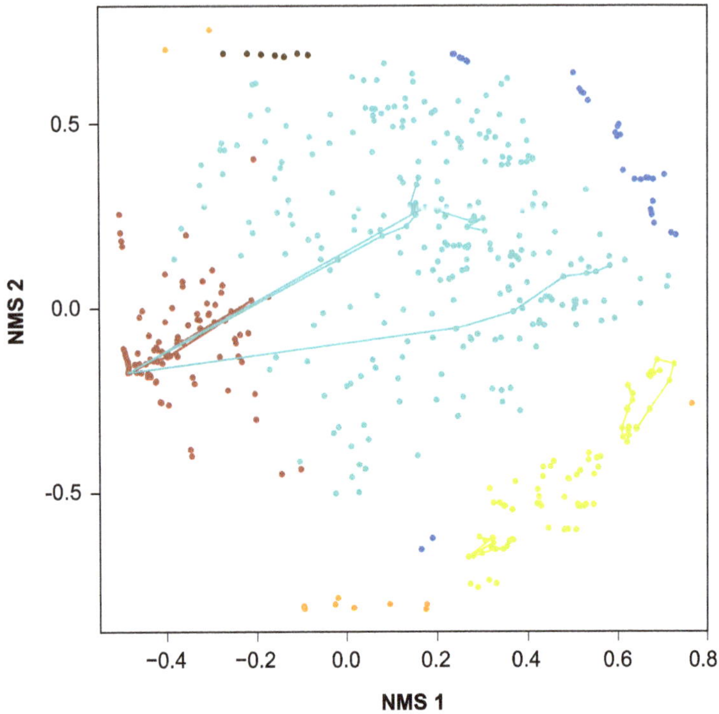

Fig. 10.8 Change vectors for a few selected sample plots in the NMS ordination of data from the Duke Forest. Vectors all flow from left to right and are colored according to the community type in which they end, so a change in color implies a change in forest type. These examples represent loblolly pine and oak-hickory types (see text)

includes mature loblolly pine (PIta; its maximum abundance, in fact, occurs in this type) but also includes many codominant hardwood species.

According to the successional sequence implied by space-for-time substitution (Chap. 9, Figs. 9.7 and 9.8), successional change vectors should move from the left side of the ordination and converge toward the far right side, that is, pines succeeding to forests similar to the presettlement oak-hickory forest. What is observed is a convergence of pine stands toward the middle of the ordination, while the oak-hickory stands show very little movement (as we might expect for a mature forest) (Fig. 10.8).

What is happening in these forests is that a few species have increased dramatically in abundance over time: red maple (*Acer rubrum*, ACru), sweet gum (*Liquidambar styraciflua*, LIst), tulip tree (*Liriodendron tulipifera*, LItu), and beech (*Fagus grandifolia*, FAgr). These species are located near the center of the ordination, and most early-successional samples are converging toward what is a new forest type that was not represented early in the time series. (To give full credit, Oosting (1942) noted the increasing abundance of these species in the understory of pine stands, so perhaps he would not be surprised at this result!) These species show affinities to more bottomland or mesic conditions, and the resulting trend is *mesophication* (Nowicki and Abrams 2008).

Further exploration of this example allows us to assess the rates of change among community types and for younger as compared to older forests. While vector lengths do vary slightly among types ($P < 0.001$), they do not vary much, nor in the directions expected. These tests were conducted on the original compositional dissimilarities, i.e., 39-dimensional as compared to the 2-dimensional ordination. Still, the 2D vector lengths are good proxies for the 39D vector lengths: in a regression, the 2D vectors predict the full-dimensional versions with a slope of 0.955 (ideally, it would be 1.0), adjusted $r^2 = 0.73$, $P < 0.0001$.

This time series includes more than a simple successional sequence; it also includes some major disturbances. We can isolate before/after samples capturing Hurricanes Hazel (in 1954) and Fran (in 1996). Both were very intense storms that passed through the Duke Forest. Hazel struck some younger stands (samples that would have been old fields just prior to the 1930s would have been ~30–50 years old in 1954). Selected change vectors for some of these plots (not shown) move right to left, suggesting that the disturbance "converted" them back to pine stands (other vector directions are observed as well). By contrast, stands were older when Fran struck, and change vectors for some of these stands move mostly left to right—suggesting that disturbance has facilitated conversion to hardwoods (i.e., by removing the overstory and releasing hardwoods in the understory). Payne and Peet (2023) explore the effects of both Hazel and Fran in this system, and Xi et al. (2008a,b, 2012, 2019) have analyzed the effects of Fran on Duke Forest in much more depth.

It can be a bit confusing to look at an ordination space in which all samples are pooled over time. One way to sort this out is to color-code the samples by time, in this case, summarized by decade (Fig. 10.9). This coding makes it clear that younger pine stands occurred only in the early decades and are now gone (i.e., the left side of the ordination space has been vacated), with older samples collecting in the mixed forest type in the center of the ordination. By contrast, the oak-hickory domain of the ordination shows a mix of samples of all ages.

The takeaway message from this example is its value as a reference space, a tool for organizing hypotheses so that they can be readily interpreted and presented. Again, Payne and Peet (2023) discuss this system in much more depth. In particular, they emphasize the combined influences of environmental variables (topography, soil texture, and chemistry), fire suppression, browsing by white-tailed deer, increases in invasive species, declines in other species due to pathogens (Dutch elm disease, dogwood anthracnose), and disturbances (hurricanes, windstorms, ice storms).... This is a complicated story about forest change!

And, finally, it is perhaps worth underscoring here the crucial importance of long-term monitoring data for ecology. In this case, an iconic model of succession based on early data has been evaluated and revised based on later measurements from the same system.

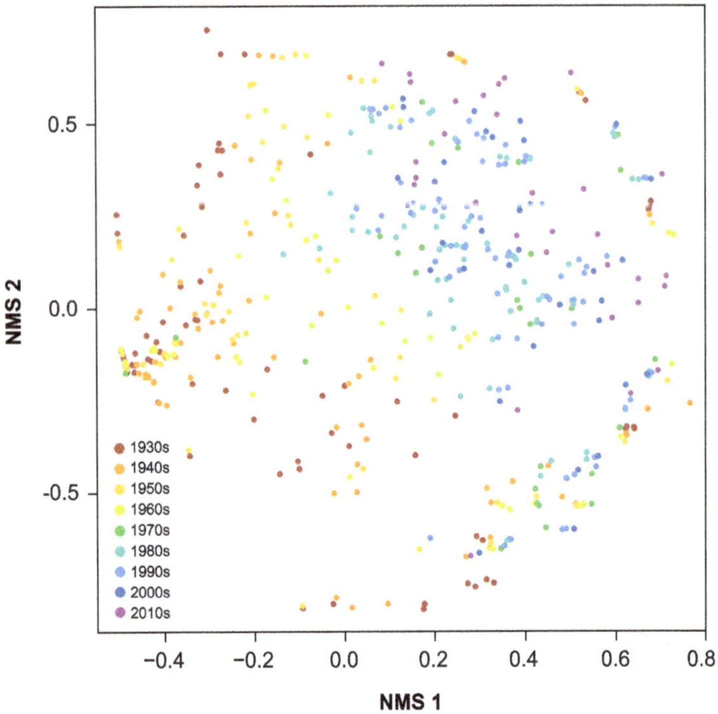

Fig. 10.9 The NMS ordination of Duke Forest samples, with samples color-coded by year of measurement (by decade). This rendition highlights the convergence of successional samples toward the mixed hardwood type in the center, while oak-hickory samples (right side) show a mix of ages

10.5 Ecological Impact Assessment

Monitoring data often reveal trends in univariate or multivariate data sets. While ecologists have long used various statistical tools for evaluating such trends (e.g., Philippi et al. 1988), we have been less successful in adopting tools that are sensitive to the range of possible outcomes. The tools of impact assessment are only recently being applied in environmental applications. Baylis et al. (2016) consider some of the reasons for this slow adoption. Sutherland et al. (2004) emphasized that impact assessment is crucial to evidence-based management. Evidence-based management depends on the critical evaluation of monitoring data, especially the results of management interventions. This, in turn, is the crucial last step in the plan-act-monitor-react adaptive management cycle: to react to what we have observed through monitoring. In this last section, we focus on the *react* stage of adaptive management.

We will generalize this discussion somewhat by defining our terms rather broadly. Monitoring data will comprise measurements, over time, of any focal species, species composition, or measures of environmental condition (e.g., habitat suitability). The impacts of interest might be the results of a deliberate management intervention constructed as an experiment, or the effects of implementing a new policy instrument, the impacts of natural events (fires, floods, human-caused disasters such as oil spills), or climatic events. In every case, we are looking for evidence of the effect of a discrete intervention. This is a more narrow focus than exploring continuous change such as succession or ecosystem response to chronic stressors (e.g., pollution) or climate change. But system responses to discrete events are a huge part of environmental management.

While we all probably recognize the litmus-test standard of a randomized controlled trial (RCT) in evaluating impacts, this approach is typically rather difficult to implement on landscapes because of the sheer logistics. Instead, ecologists rely on *quasi-experimental* approaches. The key to this is to identify a plausible *counterfactual*: a "before" condition that unambiguously isolates the effect of the treatment. Counterfactuals can be constructed in a variety of ways, and these correspond to a variety of inferential designs. In perhaps the simplest case, we have observations before and after the event of interest (e.g., a disturbance), and the counterfactual is that the "after" measurements should match the "before" measurements: a *before/after* (BA) design. Similarly, we might have measurements for a management experiment along with measurements for unmanaged sites: a control/intervention or *control/impact* (CI) design. The counterfactual is that the control sites actually do control other factors so that we can meet the "all other things being equal" assumption. In both cases, the statistical test is that the paired sets of measurements are different (the null hypothesis being that they are not). In such cases, pre-matching of controls to experimental samples can help this inference—an approach that we have adopted based on measures of ecological similarity (Sect. 10.4.1). Schleicher et al. (2020) discuss matching as used in many quasi-experimental designs for conservation science.

But there are a variety of alternative approaches (reviewed by Larsen et al. 2019), many of which can provide more inferential leverage. Perhaps the most intuitive of these is the *before/after, control/intervention* (or impact) (BACI) design. Chevalier et al. (2019) helped clarify interpretations of BACI designs for ecologists. Building on this, Wauchope et al. (2021) have provided a helpful overview of ecological impact evaluation that admits various outcomes of natural or experimental interventions. It is easiest to develop these for the simplest case of a single response variable, the abundance of a focal species. We can expand this to multivariate cases later.

To begin, we should appreciate that there might be a range of responses to a discrete event and its aftermath: the event might elicit (1) an *immediate* change in species abundance; (2) a change in the *average* abundance of the species for "after" as compared to "before" conditions, with or without similar differences in the "control/intervention" contrast; or (3) a change in the *trend* in species abundance before and after, regardless of the direction of those trends (Fig. 10.10). Assessing

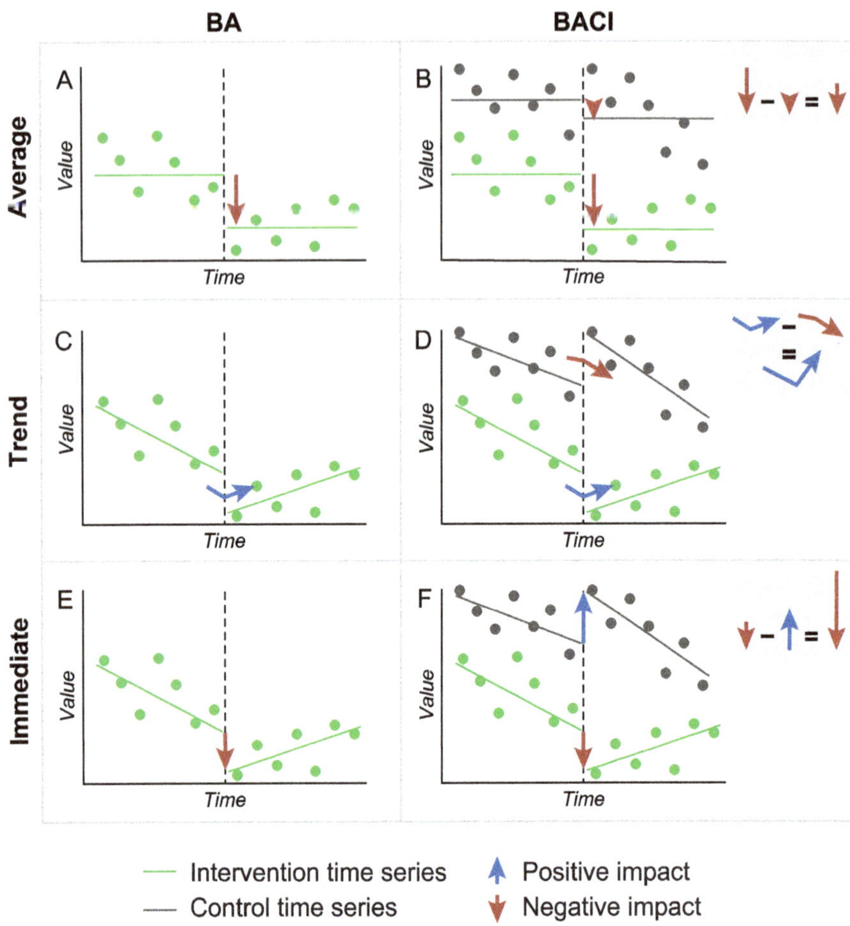

Fig. 10.10 Schematic illustrating the range of outcomes that might be observed from a discrete event (management experiment, natural disturbance) and its impact on the abundance of a focal species (after Wauchope et al. 2021, licensed via Creative Commons). See text for fuller explanation of the cases

these outcomes depends on whether there are control/intervention samples as compared to before/after measures (i.e., BA as compared to BACI designs).

The cases illustrated in Fig. 10.10 do not highlight the "parallel trends" assumption that can facilitate these analyses. If met, this would show parallel trend lines for cases (d) and (f), which would simplify the estimation. It is not clear how often this assumption might be met, and so the assumption that the trends are *not* parallel is shown in Fig. 10.10.

The illustrated cases highlight some possible pitfalls in simple interpretations of change. For example, in Fig. 10.10c, the "after" mean is lower than the "before" mean, and a simple BA design would miss the improved trend in the "after" series. Cases can get more complicated in a BACI design where there are trends in the data both before and after displacements by the intervention (cases (d) and (f) in Fig. 10.10); again, it is easiest if these trends are parallel.

Wauchope et al. (2021) detail the estimation of the inferences in this range of cases, in terms of a regression problem. In the examples below, ecosystem response is denoted by the dependent variable y (here, the abundance of a focal species), and the other terms are as defined below.

BA or CI Design In the simple case, the measurements are paired, either before/after or control/intervention. In this case, the explanatory variable is binary and the main effect captures any observed differences. As a regression, this model takes the form:

$$y = BA + \varepsilon \tag{10.1}$$

where BA is coded 0 for "before" and 1 for "after." Samples are pooled in the analysis so that the effect is averaged within each group. A significant coefficient for the BA term means that the "after" effect is real. The same model applies to a "control/intervention" (CI) model, with the same interpretation.

In practice, the power to detect a difference depends on the internal variability of the samples and the extent to which other confounding sources of variability can be controlled. This is where matching the samples can help. If this is done before-the-fact (e.g., for a management experiment), it is *pre-matching*; if done by aggregating an after-the-fact control (e.g., to interpret an event that has already happened), this is a *synthetic control* approach. In this, a synthetic control is constructed (typically, by weighted averaging) to "look like" the "before" conditions, with the assumption that deviations from this synthetic control can be attributed only to the treatment.

BACI Design The standard for experiments, the BACI design, collects measurements on control and intervention samples, before and after the intervention. This implies a slightly richer model:

$$y = BA + CI + (BA \cdot CI) + \varepsilon \tag{10.2}$$

where the new CI term is also coded 0/1 for control versus intervention. The *average* effects are captured in either the BA or CI terms. Here, the intervention effect as an immediate effect is captured in the coefficient for the *interaction* term $(BA \cdot CI)$. A positive coefficient implies that the intervention samples have higher post-treatment values than the controls, relative to the before/after contrasts. This interpretation holds whether before/after differences are positive or negative.

BACI Time Series In many cases, the pre- and post-event measurements are not single measures but are repeated, a time series. These are termed BACI-PS, for *paired series* versions of the BACI design. This is the case that admits a wider range of responses. The regression model is:

$$y = BA + CI + \text{Time} + (BA \cdot CI) + (BA \cdot \text{Time}) + (CI \cdot \text{Time}) + (BA \cdot CI \cdot \text{Time}) + \varepsilon \quad (10.3)$$

where *Time* is a continuous variable (e.g., "Year"). In this, the interaction between *BA* and *CI* again captures the immediate change in the response variable. The interaction between *CI* and *Time* tests the assumption that trends were similar for control and intervention sites before the intervention (the *parallel trends* assumption), which conditions the interpretation of post-impact trends. The three-way interaction between *BA*, *CI*, and *Time* captures any change in the trends before and after for the control as compared to the intervention samples.

This can get complicated. Equation 10.3 has seven coefficients, each with its own interpretation. The situation can get more complicated if there are time lags in system response, or if the responses are nonlinear. Wauchope et al. (2021) offer some guidance for such cases. Thialt et al. (2017, 2019) have described extensions to the BACI-PS approach, termed *progressive-change* BACI-PS, which allow step changes, lags, as well as linear, asymptotic, or sigmoidal responses post-intervention.

In the examples here, the response variable is univariate, a focal species. But the monitoring data might include observations of several species. While we have spent considerable time in this chapter (and this book!) on multivariate responses, Wauchope et al. (2021) recommend using a single-species approach and aggregating or synthesizing results over all species after the fact. While this is due partly to analytic complexities (the approaches outlined above have not been implemented for multivariate responses), as a practical matter, it might be that all species do not respond in the same way. An after-the-fact summary can capture this for interpretation and presentation.

Clearly, these analyses can get a bit nuanced and data-hungry. The most important implication of this is to underscore the crucial need to plan for the analyses and to collect data that will have the leverage to capture these effects.

10.6 Further Reading

In this chapter we adopted ordination as a framework for assessments. We devoted Chap. 4 to this topic, but McCune and Grace (2002) and Legendre and Legendre (2012) are authoritative references on these tools and provide more in-depth coverage. Restoration ecology is an enormous field, but early efforts to organize the discipline around ecological experiments remain a useful introduction (e.g., Allen et al. 1997, 2002; Falk et al. 2006; Suding 2011). The Society for Ecological Restoration (http://ser.org) provides resources on restoration. Succession is an even larger topic; Payne and Peet (2023) point to many of the key concepts and provide a convenient entry into this huge literature from the perspective adopted in this chapter.

Impact analysis is a crucial topic in many disciplines, some not so closely aligned with our purposes here. Sutherland et al. (2004) provide an accessible introduction from the perspective of evidence-based conservation, and reviews aimed at ecologists in general or specific to conservation practice are provided by Baylis et al. (2016), Chevalier et al. (2019), Larsen et al. (2019), and Wauchope et al. (2021). These papers also offer some review of similar applications by other names as used in other disciplines (the jargon can be confusing).

10.7 Summary and Prospectus

Ecological assessment entails the evaluation of changes observed over time, either continuous changes or those in response to discrete events such as management interventions or disturbances. We have adopted ordination as a framework for assessing such changes, because ordination offers powerful summary and ready communication in graphical form. We begin assessments in exploratory mode, because while we might have specific expectations going into the analysis, ecological data often surprise us. By exploring change vectors or change trajectories within ordination space, we can often frame specific hypotheses so they can be evaluated analytically. New hypotheses often arise from exploratory analyses. We considered two popular application areas: ecological restoration and forest succession.

Assessments are often aimed at ecological responses to discrete events. In such cases, the analysis is some form of impact assessment using a BACI design or, more recently, BACI-PS methods for time-series data. Guidance for ecologists using such tools is emerging, which should support more robust approaches to evidence-based environmental management.

The classic rubric for evidence-based management is the adaptive management cycle "plan-act-monitor-react." We have followed this path throughout this book (with some scenic detours into other areas). The first key to this is the "plan"—which we have invoked at least conceptually in most chapters (and explicitly in Chap. 7) and especially in the form of means-ends models that underpin structured decision-making (Chap. 8). We spent several chapters (especially Chaps. 3, 4, 5, and 6) in developing a facility for ecological data analysis and in interpreting ecological change (Chap. 9). These prepared us to embrace the "react" stage of the cycle in this chapter.

To be sure, there are other tasks in landscape ecology and management beyond those covered in this book. And the tools of the trade will continue to evolve. But the material in this book should provide a solid foundation on which to build.

References

Allen EB, Covington WW, Falk DA (1997) Developing the conceptual basis of restoration ecology. Restor Ecol 5:275–276

Allen CD, Savage M, Falk DA, Suckling KF, Swetnam TW, Schulke T, Morgan P, Hoffman M, Klingel JT (2002) Ecological restoration of southwestern Ponderosa pine ecosystems: a broad perspective. Ecol Appl 12:1418–1433

Anderson MJ (2001) A new method for non-parametric multivariate analysis of variance. Austral J Ecol 26:32–46

Baylis K, Honey-Roses J, Borner H, Corbera E, Ezzine-de-Blas D, Ferraro JJ, Lapeyre R, Martin Persson U, Pfaff A, Wunder S (2016) Mainstreaming impact evaluation in nature conservation. Conserv Lett 9:58–64

Bergeron Y, Dubuc M (1989) Succession in the southern part of the Canadian boreal forest. Vegetatio 79:51–83

Chevalier M, Russell JC, Knapp J (2019) New measures for evaluation of environmental perturbations using before-after-control-impact analyses. Ecol Appl 29:e01838

Falk D, Palmer M, Zedler J (eds) (2006) Foundations of restoration ecology. Island Press, Washington DC

Goff FG, Zedler PH (1972) Derivation of species succession vectors. Am Midl Nat 87:397–412

Goslee SC, Urban DL (2007) The *ecodist* package for dissimilarity-based analyses of ecological data. J Stat Softw 22:7

Johnston DW, Odum EP (1956) Breeding bird populations in relation to plant succession on the Piedmont of Georgia. Ecology 37:50–62

Kaufman L, Rousseeuw PJ (1990) Finding groups in data: an introduction to cluster analysis. John Wiley and Sons, New York

Larsen AE, Meng K, Kendall BE (2019) Causal analysis in control-impact ecological studies with observational data. Methods Ecol Evol 10:924–934

Legendre P, Legendre L (2012) Numerical ecology (3rd English edition). Elsevier, Amsterdam

Matthews JW, Spyreas G (2010) Convergence and divergence in plant community trajectories as a framework for monitoring wetland restoration progress. J Appl Ecol 47:1128–1136

McCune B, Grace JB (2002) Analysis of ecological communities. MjM Software Design, Gleneden Beach, Oregon

Nowicki GJ, Abrams MD (2008) The demise of fire and "mesophication" of forests in the eastern United States. Bioscience 58:123–138

Odum EP (1953) Fundamentals of ecology. W.B. Saunders Co., Philadelphia

Odum EP (1969) The strategy of ecosystem development. Science 164:262–270

Oksanen, J., Blanchet, F. G., Kindt, R., Legendre, P., O'Hara, R. B., Simpson, G. L., Solymus, P., Henry, M., Stevens, H., & Wagner, H. (2021) *vegan*: community analysis package. R version 1.17–11. http://CRAN.R-project/org/program=vegan

Oosting HJ (1942) An ecological analysis of the plant communities of Piedmont, North Carolina. Am Midl Nat 28:1–126

Payne CJ, Peet RK (2023) Revisiting the model system for forest succession: eighty years of resampling Piedmont forests reveals need for an improved suite of indicators for successional change. Ecol Indic 154:110679

Philippi TE, Dixon PM, Taylor BE (1988) Detecting trends in species composition. Ecol Appl 8: 300–308

Schleicher J, Eklund J, Barnes MD, Geldmann J, Oldekop JA, Jones JPG (2020) Statistical matching for conservation science. Conserv Biol 34:538–549

Smith TM, Urban DL (1988) Scale and resolution of forest structural pattern. Vegetatio 74:143–150

Suding KN (2011) Toward an era of restoration in ecology: successes, failures, and opportunities ahead. Annu Rev Ecol Evol Syst 42:465–487

Sutherland WJ, Pullin AS, Dolman PM, Knight TM (2004) The need for evidence-based conservation. Trends Ecol Evol 19:305–308

Thialt L, Kernaléguen L, Osenberg CW, Claudet J (2017) Progressive-change BACIPS: a flexible approach for environmental impact assessment. Methods Ecol Evol 8:288–296

Thialt L, Kernaléguen L, Osenberg CW, Lison de Loma T, Chancerelle Y, Siu G, Claudet J (2019) Ecological evaluation of a marine protected area network: a progressive change BACIPS approach. Ecosphere 10(2):e02576

Urban DL (2006) A model-based framework for restoration. In: Falk D, Palmer M, Zedler J (eds) Foundations of restoration ecology. Island Press, Washington DC, pp 238–256

Wauchope HS, Amano T, Geldmann J, Johnston A, Simmons C, Sutherland WJ, Jones JPG (2021) Evaluating impact using time-series data. Trends Ecol Evol 36:196–205

Xi W, Peet RK, DeCoster JK, Urban DL (2008a) Tree damage risk factors associated with large, infrequent wind disturbances of Carolina forests. Forestry 81:317–334

Xi W, Peet RK, Urban DL (2008b) Changes in forest structure, species diversity, and spatial pattern following hurricane disturbance in a Piedmont North Carolina forest, USA. J Plant Ecol 1:43–57

Xi W, Peet RK, Urban DL (2012) The impacts of a large, infrequent hurricane on understory sapling dynamics and diversity in North Carolina Piedmont forests, USA. Tree For Sci Biotechnol 6: 51–59

Xi W, Peet RK, Lee MT, Urban DL (2019) Hurricane disturbances, tree diversity, and succession in North Carolina Piedmont forests, USA. J For Res 30:219–231

Zedler JB, Callaway JC (1999) Tracking wetland restoration: do mitigation sites follow desired trajectories? Restor Ecol 7:69–73

Index

D. L. Urban, *Landscape Ecology: A Task-Oriented Perspective*,
https://doi.org/10.1007/978-3-031-72251-6

GPSR Compliance
The European Union's (EU) General Product Safety Regulation (GPSR) is a set of rules that requires consumer products to be safe and our obligations to ensure this.

If you have any concerns about our products, you can contact us on

ProductSafety@springernature.com

In case Publisher is established outside the EU, the EU authorized representative is:

Springer Nature Customer Service Center GmbH
Europaplatz 3
69115 Heidelberg, Germany

www.ingramcontent.com/pod-product-compliance
Lightning Source LLC
Chambersburg PA
CBHW070202111225
36514CB00025B/32
* 9 7 8 3 0 3 1 7 2 2 5 0 9 *